S0-BRH-117

COMPREHENSIVE BIOCHEMISTRY

ELSEVIER SCIENTIFIC PUBLISHING COMPANY

335 Jan van Galenstraat, P.O. Box 211, Amsterdam

AMERICAN ELSEVIER PUBLISHING COMPANY, INC.

52, Vanderbilt Avenue, New York, N.Y. 10017

Library of Congress Catalog Card Number 62–10359

ISBN 0–444–41145–3

With 57 plates, 42 figures and 3 tables

PRINTED IN THE NETHERLANDS

COMPREHENSIVE BIOCHEMISTRY

COMPREHENSIVE BIOCHEMISTRY

SECTION I (VOLUMES 1–4)

PHYSICO-CHEMICAL AND ORGANIC ASPECTS OF BIOCHEMISTRY

SECTION II (VOLUMES 5–11)

CHEMISTRY OF BIOLOGICAL COMPOUNDS

SECTION III (VOLUMES 12–16)

BIOCHEMICAL REACTION MECHANISMS

SECTION IV (VOLUMES 17–21)

METABOLISM

SECTION V (VOLUMES 22–29)

CHEMICAL BIOLOGY

SECTION VI (VOLUMES 30–33)

A HISTORY OF BIOCHEMISTRY

COMPREHENSIVE BIOCHEMISTRY

EDITED BY

MARCEL FLORKIN

Professor of Biochemistry, University of Liège (Belgium)

AND

ELMER H. STOTZ

Professor of Biochemistry, University of Rochester, School of Medicine and Dentistry, Rochester, N.Y. (U.S.A.)

VOLUME 31

A HISTORY OF BIOCHEMISTRY

Part III. History of the Identification of the Sources of Free Energy in Organisms

by

MARCEL FLORKIN

ELSEVIER SCIENTIFIC PUBLISHING COMPANY

AMSTERDAM · OXFORD · NEW YORK

1975

GENERAL PREFACE

The Editors are keenly aware that the literature of Biochemistry is already very large, in fact so widespread that it is increasingly difficult to assemble the most pertinent material in a given area. Beyond the ordinary textbook the subject matter of the rapidly expanding knowledge of biochemistry is spread among innumerable journals, monographs, and series of reviews. The Editors believe that there is a real place for an advanced treatise in biochemistry which assembles the principal areas of the subject in a single set of books.

It would be ideal if an individual or a small group of biochemists could produce such an advanced treatise, and within the time to keep reasonably abreast of rapid advances, but this is at least difficult if not impossible. Instead, the Editors with the advice of the Advisory Board, have assembled what they consider the best possible sequence of chapters written by competent authors; they must take the responsibility for inevitable gaps of subject matter and duplication which may result from this procedure.

Most evident to the modern biochemist, apart from he body of knowledge of the chemistry and metabolism of biological substances, is the extent to which he must draw from recent concepts of physical and organic chemistry, and in turn project into the vast field of biology. Thus in the organization of Comprehensive Biochemistry, sections II, III and IV, Chemistry of Biological Compounds, Biochemical Reaction Mechanisms, and Metabolism may be considered classical biochemistry, while the first and fifth sections provide selected material on the origins and projections of the subject.

It is hoped that sub-division of the sections into bound volumes will not only be convenient, but will find favour among students concerned with specialized areas, and will permit easier future revisions of the individual volumes. Toward the latter end particularly, the Editors will welcome all comments in their effort to produce a useful and efficient source of biochemical knowledge.

M. Florkin
E. H. Stotz

Liège/Rochester

PREFACE TO SECTION VI

(Volumes 30–33)

In the many chapters of previous sections of *Comprehensive Biochemistry* covering organic and physicochemical concepts (Section I), chemistry of the major constituents of living material (Section II), enzymology (Section III), metabolism (Section IV), and the molecular basis of biological concepts (Section V), authors have been necessarily restricted to the more recent developments of their topics. Any historical aspects were confined to recognition of events required for interpretation of the present status of their subjects. These latest developments are only insertions in a science which has had a prolonged history of development.

Section VI is intended to retrace the long process of evolution of the science of Biochemistry, framed in a conceptual background and in a manner not recorded in recent treatises. Part I of this section deals with Proto-biochemistry or with the discourses imagined concerning matter-of-life and forces-of-life before molecular aspects of life could be investigated. Part II concerns the transition between Proto-biochemistry and Biochemistry and retraces its main landmarks. In Part III the history of the identification of the sources of free energy in organisms is depicted, and Part IV is devoted to backgrounds in the unravelling of biosynthetic pathways. While these latter parts are concerned with the molecular level of integration, Part V is more specifically directed toward the history of molecular interpretations of physiological and biological concepts, and of the origins of the concept of life as the expression of a molecular order.

The *History* narrated in Section VI thus leads to the thresholds of the individual histories in the recent developments recorded by the authors of Sections I–V of *Comprehensive Biochemistry*.

Liège/Rochester

M. Florkin
E. H. Stotz

CONTENTS

VOLUME 31

A HISTORY OF BIOCHEMISTRY

Part III. History of the Identification of the Sources of Free Energy in Organisms

Introduction

Section I
Anaerobic Phosphorylation
Chapter 17. The Discovery of "Cell-free Fermentation"

Chapter 18. Discovery of the "Coenzyme of Alcoholic Fermentation" and of the Esterification of Phosphate

Chapter 19. The Search, Before 1930, for Metabolic Intermediates Between Glucose and Ethanol, in Alcoholic Fermentation

Chapter 20. Muscle Glycolysis Before 1926

Chapter 21. The Unified Pathway of Glycolysis, and the Process of Phosphate Esterification

Chapter 22. The Scission of Hexosediphosphate into Triosephosphates

Chapter 23. The Unravelling of the Nature of Harden and Young's "Coenzyme" and the Discovery of "Energy-rich Phosphate Bonds"

Chapter 24. The Complete Glycolysis Pathway

Chapter 25. Glycolysis as a Source of Free Energy

Section II
Aerobic Phosphorylation

Chapter 26. Early Theories of the "Biological Oxidations" of Intracellular Respiration

Chapter 27. Iron Catalysis in Biological Oxidations and the Identification of Warburg's Atmungsferment as an Iron-Porphyrin Compound

Chapter 28. The Wieland–Thunberg Theory, p. 213

Chapter 29. Keilin's Rediscovery of Histohematin (Cytochrome)

Chapter 30. The Theory of Szent-Györgyi, p. 237

Chapter 31. The Conversion of Citrate into Succinate, p. 249

Chapter 32. The Original "Citric Acid Cycle", p. 255

Chapter 33. The Oxidative Decarboxylation of Pyruvate and the Biosynthesis of Citrate

Chapter 34. From Fatty Acids to Acetyl-CoA

Chapter 35. From Amino Acids to the Tricarboxylic Acid Cycle

Chapter 36. The Respiratory Chain

Chapter 37. Oxidative Phosphorylation

Section III
Autotrophic Phosphorylations

Chapter 38. Photosynthetic Phosphorylation

Chapter 39. Sources of Free Energy in Chemoautotrophs, p. 431

Section VI

A HISTORY OF BIOCHEMISTRY

List of Plates

(Unless otherwise stated, the portraits belong to the author's personal collection.)

Plate 64. Moritz Traube. By courtesy of Prof. Henri Fredericq, Liège.

Plate 65. Eduard Buchner, from *Le IVe Congrès de Chimie biologique*, Paris, Masson, 1933.

Plate 66. Frederick Gowland Hopkins. By courtesy of Ramsay and Muspratt, Ltd., Cambridge, England.

Plate 67. Louis Pasteur. Photo E. Ladrey, Paris. By courtesy of Prof. Henri Fredericq, Liège.

Plate 68. Arthur Harden. By courtesy of the *Biochemical Society.*

Plate 69. Alexander Nicolaevich Lebedev. By courtesy of Prof. E. M. Kreps, Leningrad.

Plate 70. Sergei Pavlovich Kostychev. By courtesy of Prof. Kretovich, Moscow.

Plate 71. Carl Alexander Neuberg, from *Advances in Carbohydrate Chemistry*, vol. 13 (ed. by M. L. Wolfrom), New York, 1958.

Plate 72. Jacob Karol Parnas. By courtesy of Prof. T. Mann, Cambridge, England.

Plate 73. Otto Meyerhof. By courtesy of Prof. D. Nachmansohn, New York.

Plate 74. Gustav Embden. By courtesy of Prof. Erich Heinz, Frankfurt a/M.

Plate 75. Carl Ferdinand Cori and Gerty Therese Cori (1947).

Plate 76. Ragnar Nilsson.

Plate 77. Karl Lohmann.

Plate 78. Cyrus Hartwell Fiske (1932). By courtesy of Prof. J. T. Edsall, Boston, Mass.

Plate 79. Yellapragada Subbarow. By courtesy of Prof. J. T. Edsall, Boston, Mass.

Plate 80. Hans von Euler-Chelpin. By courtesy of Prof. U. S. von Euler, Stockholm.

Plate 81. Joseph Needham and Dorothy Needham. Photo Cambridge News. The scroll behind is the *Nei Ching Thu*, a classical representation of taoist physiology.

Plate 82. Otto Warburg. By courtesy of Prof. D. Nachmansohn, New York.

Plate 83. Theodor Bücher.

Plate 84. Einar Lundsgaard. By courtesy of Prof. Chr. Crone, Copenhagen.

Volume 30

Errata and Corrigenda

General Preface, line 21, instead of *Comjounds*, read *Compounds*.
Page XVI, line 1, instead of *Hermann Boerhaave*, read *Herman Boerhaave*.
Page XVIII, line 8 from bottom, instead of *A History of Biochemistry* read *A History of Chemistry*.
Page 39, line 12 from bottom, instead of *sifled* read *sifted*.
Page 119, line 15, instead of *Huenefeld*, read *Hünefeld*.
Page 130, lines 13 and 15, instead of *Thénard*, read *Thenard*.
Page 131, line 10, instead of *Keilin*, read *Keilin*[8].
Page 153, line 14, instead of *excreted*, read *exerted*.
Page 158, line 18, instead of *rinary*, read *urinary*.
Page 165, line 4 from bottom, instead of *de Liquides*, read *des Liquides*.
Page 182, line 19, instead of *by Delhoume for his Principes*, read *by Delhoume, for his Principes...*
Page 184, line 17 from bottom, instead of *contributors*, read *contributions*.
Page 188, line 15, instead of *limiting the field*, read *limited the field*.
Page 227, line 14 from bottom, instead of *Claussius*, read *Clausius*.
Page 253, line 12 from bottom, instead of *organic matter*, read *inorganic matter*.
Page 257, line 19, instead of *provisions*, read *previsions*.
Page 273, line 18; p. 277, ref. 41; p. 336, column 1, line 7 from bottom, instead of *Kostytschev*, read *Kostychev*.
Page 275, line 4, instead of *saccharose*, read *saccharase*.
Page 333, column 2, line 5 from bottom; instead of *Braconnot, 114*, read *Braconnot H., 114*.
Page 334, column 1, line 9, instead of *Buchner, H.E.A., 9, 12, 273*, read *Buchner, E., 9, 273* and on the following line, *Buchner, H., 12*.
Page 334, column 2, line 21, instead of *Desmazières, 131*, read *Desmazières, J. B., 131*.
Page 337, column 1, line 13 from bottom, instead of *Nägeli, C.V.*, read *Nägeli, C. von*.
Page 343, column 1, line 10 from bottom, instead of *Primary qualitis*, read *Primary qualities*.

Introduction

1. The emergence of biochemistry

In Hall's book on *Ideas of Life and Matter*[1] we encounter a number of very useful analyses of biological concepts but run the risk of being seduced by the idea that the philosophers of antiquity had once and for all grasped the important themes of the science of life so that their themes remain our themes. By contrast, Part II of this *History* (following Part I devoted to Protobiochemistry) showed how the myths of antiquity were replaced by the — previously entirely unsuspected — concepts of biochemistry. In this Part, the author underlines the importance of the "chemical revolution", of the work of Lavoisier and its follow-up in the accomplishment of what had been Liebig's unrealized intention (but one which, as such, proved a powerful incentive for his contemporaries). Also in Part II, the author shows where and how the concepts of biochemistry began to reveal a break with the past and how the molecular approach to biological concepts unfolded in the course of time. The intracellular location of metabolic changes, the enzymatic theory of metabolism, the emergence of bioenergetics, the recognition of proteins as definite chemical compounds and the identification of subcellular organelles appear to the author to exemplify both the removal of the roadblocks that had come into being during the protobiochemistry phase and the opening of the pathway leading to the developments in classical biochemistry described in Parts III and IV.

Part III, contained in this Volume, offers a historical survey of the sources of free energy.

The primary energy-transducing systems of cells are mitochondria and chloroplasts.

In biological systems there are three main sources of free energy. These sources are now recognized in light (photophosphorylation), disproportionation of $C_xH_yO_z$ compounds (intramolecular oxidation), and oxidation

References p. 19

(oxidative phosphorylation), besides less important systems involving the oxidation of simple inorganic compounds of the environment. Disproportionation of a $C_xH_yO_z$ compound appears as the most primitive mechanism. The nature of this mechanism may be said to be understood.

The systems of oxidative phosphorylation and of photophosphorylation remain unsolved problems. While the study of intramolecular oxidation has been spurred on by the fermentation industries. the recognition of the existence of oxidative phosphorylation has resulted in part from the progress made in microbiological and animal biochemistry. Photosynthetic phosphorylation studies have fallen behind and in many ways have relied on the results of biological oxidation studies. The limited means available to researchers specialized in photosynthesis are symptomatic of the fact that the importance to human life of a proper understanding of the biological sources of free energy is not sufficiently realized.

It has repeatedly been said that when we travel by plane around the world, we are bound to recognize the regions of Western Europe, of the northern part of the United States and of Japan as strikingly fertile. There is a connection between this fact and the historical development of technology and prosperity, though the scope of this Volume does not allow an elaboration of this point. A wise human community would have directed scientific research toward the chloroplast, and not beer brewing.

But scientific progress travelled by different routes; for example, it was studies on anaerobic phosphorylation which provided the historical setting in which biochemistry was able to develop as an autonomous discipline.

Two views have been expressed concerning this development. According to Teich[2], the science of biochemistry derived from physiological chemistry *via* a feedback process arising out of the progress made in the organic chemistry of natural substances. Kohler[3], on the other hand, developed arguments (as we shall show), in favour of the idea that biochemistry owes its beginnings to a number of different disciplines (chemistry, immunology, microbiology, zoology, physiological chemistry, *etc.*). The author believes that the historical developments outlined in the present volume substantiate Kohler's view.

The first community of biochemists was opposed to the speculative theory of "protoplasm" as a physical entity. This theory was widely believed by physiological chemists, though not by all of them. Supporting the opposition was Lavoisier's point that there is only one chemistry.

As mentioned in Chapter 16, the theory which held "protoplasm" to be a physical entity or the birthplace of the properties of living matter, was first formulated in a lecture by T. H. Huxley[4] held in 1868 (published 1869). This is the proper place to describe this theory in some detail, so that the development, in opposition to it, of biochemistry may become clear. Huxley's title *On the Physical Basis of Life* greatly contributed to the success of his theory. Huxley's views owed a great deal to a number of his predecessors who had endeavoured to isolate a basic living substance common to organisms. As Geison[5] rightly remarks:

"Huxley's lecture was in reality only a popular exposition of results gained from a long series of studies by other observers".

The same author notes that

"the first clear statement of the identity between plant and animal protoplasm was made in 1850 by a twenty-two-year-old botanist, Ferdinand Cohn[6] (1828–1898) of Breslau".

It may be said that the theory of integrated protoplasm is part of the endeavour to define the unity of life which has always haunted the mind of biologists. Generally speaking, Cohn was much indebted to the French protozoologist Félix Dujardin, author of the "sarcode" concept, but it was certainly Cohn who introduced the concept of the unity of the "protoplasm" (a term used by botanists) and of the sarcode (a term used by zoologists). This unity of protoplasm was again stated by Unger, a botanist, who concluded in 1855 that

"protoplasm must be regarded not as a fluid but as a half-fluid contractile substance, which is... comparable to the sarcode of animals, if indeed it does not coincide in identity with the latter"[7] (translation by Baker[8]).

Baker[8] and Geison[5] have retraced the elaboration of Cohn's views by several authors in the 1850's.

Schulze[9], in 1861, defined a cell as

"ein nacktes Protoplasmaklümpchen mit Kern"

in opposition to the old botanical concept of the cell as a

"bladderlike structure with membrane, contents and nucleus".

References p. 19

It is sometimes said that Schulze took this position in opposition to that of Schwann. But Schwann also defined the cell as "a layer (Zellenschichte) around a nucleus"[10]. That a differentiation of the surrounding membrane frequently occurred was, according to Schwann, due to biochemical factors. Like Cohn[6], Schulze[9] emphasized the similarity between plant "protoplasm" and animal "sarcode" and he, too, insisted on using a single word, protoplasm, for both of them.

The concept introduced by Schulze in the context of his studies on muscle was, as stated above, meant to emphasize the importance, to the cell's life, of cell content as against cell membranes. It is on this issue that Schulze opposed Schwann who had stressed (as we do) the vital importance of the cell membranes to the life of the cell.

E. W. von Brücke[11], who had not always observed the existence of a nucleus in cryptogams, went even further and opposed Schulze's stress on the importance of the nucleus. Brücke, and several others in the 1860's, developed the notion formulated by Haeckel[12] in 1869 that

"protoplasm is the original active substratum of all vital phenomena".

Understood in terms of the spirit of the age, this phrase asserted that the essential substrate of life was neither the nucleus nor the cell membrane, but protoplasm. This emphasis on protoplasm gave rise to many studies on protoplasm "structure" as described in Chapter 16; these were dropped as knowledge concerning cell organelles increased. This tendency to locate "vitality" at the protoplasm level and to deny first place to cell membrane and nucleus met with opposition from several biologists who (particularly when Schwann's views were revived and cell formation was better understood) explained the emphasis on protoplasm as due to an inadequate understanding of cell division. Cell division was first discovered by Remak[13] who did not agree with Schulze's denial of the essentiality of the cell membrane. Neither did Reichert[14] who stressed the equal importance of membranes, contents and nucleus.

So far the debate on "protoplasm" remained within the scope of technical and specialized biological problems. What changed with the publication of Huxley's lecture in 1869 was the philosophical emphasis.

In his classical studies on cell theory, Baker[8] states that Huxley's contribution to the protoplasm theory

"was first opposition, and then a phrase".

In 1853, Huxley had maintained that the cell membrane was more important than the cell's contents; in 1868, however, he joined the opposite camp. Yet Huxley's later view, which won the attention of a vast public, represented the protoplasm theory as a victory of the mechanistic interpretation over the vitalist one. As noted by Coleman[15] and by Geison[5], as early as 1853 Huxley had concluded that the faculty for manifesting vital phenomena

"resides in the matter of which living bodies are composed... the 'vital forces' are molecular forces".

In 1868 he shifted the emphasis from membranes to protoplasm, which now became "the physical basis of life". Here again we find the concept of the unity, indeed the chemical unity, of life expressed in Huxley's vivid prose.

The reader will obtain much benefit from Geison's[5] analysis of Huxley's lecture. At its centre is the concept of matter as the source of vitality ("vital forces are molecular forces"). He recognizes that at first sight organisms may appear to be widely different:

"...what community of form, or structure, is there between the animalcule and the whale; or between the fungus and the fig tree?... What hidden bond can connect the flower which a girl wears in her hair and the blood which courses through her youthful veins?"

Yet:

"a three-fold unity, namely a unity of power, or faculty, a unity of form, and a unity of substantial composition does pervade the whole living world".

The powers of organisms (the heterotrophy of animals and the autotrophy of plants excepted) differ only in degree. The unity of form is revealed by the cell theory. The unity of composition is demonstrated by the presence, in all forms of protoplasm, of the four elements carbon, hydrogen, oxygen and nitrogen. This analysis subsequently took a more precise form (Baldwin[16], Florkin[17]).

To quote Huxley:

"If the properties of water may be properly said to result from the nature and disposition of its component molecules, I can find no intelligible ground for refusing to say that the properties of protoplasm result from the nature and disposition of its molecules".

"... all vital action may... be said to be the result of the molecular forces of the protoplasm which displays it."

References p. 19

Here the root of the concept of protoplasm as a whole is added to the chemical unity of life.

Referring to the emergent properties of water, Huxley writes

> "... we live in the hope and in the faith that, by the advance of molecular physics, we shall by and by be able to see our way as clearly from the constituents of water to the properties of water, as we are now able to deduce the operations of a watch from the form of its parts and the manner in which they are put together."

Huxley's statement, "vital forces are molecular forces" pleased the materialists but incensed the vitalists who thought it preposterous to locate the source of life in a proteinaceous material; one might as well locate it in a particle of slime! The theory of protoplasm thus became the subject of heated controversies during the 70's. One of the most violent attacks against it was launched in a book by the histologist Beale[18], analysed by Geison[5]. Yet there was enough in Huxley's theory to offer the decaying fortress of vitalism the possibility of retrenchment. As Beck[19] put it:

> "... to some of the metaphysically inclined, ... protoplasm was just their dish; it had wonderful overtones of mystical music of the spheres, and conveyed poetic images of quivering disembodied life, flowing and swirling in a cloud of unknown unknowables. For those who wished to see it in that way ... the protoplasm theory may have heaped extra fuel on the fires of vitalism by implying anew a degree of complexity that only vitalism could account for while simultaneously arousing fervent opposition to this new outrage of materialistic science".

Though avoiding any profession of vitalist faith, a tendency inherited from Bichat and Barthez (see Chapter 11) did exist in the circles of physiological chemists to consider that the "vital properties" (as Claude Bernard called them), though they are the manifestation of chemical and physical forces, disappeared at death as well as in analysis.

The view which became fashionable among physiologists and physiological chemists found expression in such definitions as that of Beale's pupil Drysdale[20], who wrote in 1874, concerning protoplasm, that

> "... the elements are in a state of combination not to be called chemical at all in the ordinary sense but one which is utterly *sui generis*."

As Geison[5] states:

> "In general, exponents of this theory carefully avoided Beale's appeal to a non-material "vital force"; but they also maintained a notable skepticism about how much could really be learned from the chemical analysis of protoplasm, for they believed that living matter was killed and rendered fundamentally different by the very process of analysis."

The late Marjory Stevenson[21], alluding to the theory current at the end of the 19th century, writes in reference to F. G. Hopkins:

"When he started his work the study of chemical changes occurring within the cells, as distinct from that of substances separated from it, was seldom attempted or even thought of; these happenings were wrapped around with mystery which at that time it was considered useless and even irreverent to try and penetrate. The chemistry of the cell was thought of as something different from laboratory chemistry; the substance of living matter was protoplasm; having entered this complex, molecules of food and oxygen were believed to lose their identity and to become incorporated in the living molecule or "biogen", where they lost their characteristics and underwent mysterious and indefinable changes to reappear as end-products once more in a recognizable form such as urea and CO_2."

Hopkins, who was one of the most fervent opponents of the protoplasm myth makes repeated references, in a number of texts, to the time when this myth was prevalent. His writings are so important that they deserve to be quoted at length.

Here for instance is a passage from the address titled *The Influence of Chemical Thought on Biology*[22], given at the Harvard Tercentenary Conference of Arts and Sciences in 1936:

"It was the genius of Liebig that started modern organic chemistry on a triumphant career, and Liebig's great desire and one which directed his own efforts was to see chemistry render full service to animal physiology and to agriculture. This desire, in satisfactory measure, was not fulfilled during Liebig's own lifetime, and it is, I think, of some historical interest to decide why ..., during years when scientific minds were so alert, so promising a field was cultivated by so few. At first I think that personal attributes in leaders of thought contributed to the separation of chemistry from biology. Liebig for instance, though so brilliant a chemist, lacked biological training and, as I have always felt, a biologist's instincts. When with great enthusiasm he came to apply his chemical knowledge to the living plant and animal his thought often went obviously astray, and much of his theoretical teaching was instinctly and rightly rejected in biological thought. What was really so valuable in that teaching lost therefore some of its influence. Strange as it may seem, the influence of that other dominant mind of the time, that of Pasteur, did not altogether favour an approach between chemist and biologist. If Liebig remained too much the chemist, Pasteur, once he entered, with such immense profit to science, the biological field, became almost too much a biologist, at least in so far as he favoured the current belief that the activities of a living organism could be understood only by thinking in terms of that organism as a whole. Any analysis of its totality he held to be of little avail ...

"Apart from the divorce between chemical and biological thought, there was a tendency in the latter which in itself discouraged attempts to probe the secrets of living cells by chemical methods. Most biologists were content to ascribe the internal events of metabolism to the elusive properties of an entity insusceptible of profitable analysis; to the influence of protoplasm as a whole. There was, as I well remember, a widespread feeling that chemical studies which interfere with the full integrity of protoplasm could at most have chemical interest and must remain without bearing on the realities of biology."

References p. 19

In a report to the British Association delivered at Leicester in 1933, Hopkins[23] wrote:

"So long as the term 'protoplasm' retains a morphological significance as in classical cytology, it may be even now convenient enough, though always denoting an abstraction. In so far however, as the progress of metabolism, with all the vital activities which it supports, was ascribed in concrete thought to hypothetical qualities emergent from a protoplasmic complex in its integrity, or when substances were held to suffer change only because in each living cell they were at first built up, with loss of their own molecular structure and identity into this complex, which is itself the inscrutable seat of cyclic change, then serious obscurantism was involved.

"Had such assumption been justified the old taunt that when the chemist touches living matter it immediately becomes dead matter would also have been justified. A very distinguished organic chemist, long since dead, said to me in the late eighties: 'The chemistry of the living? That is the chemistry of protoplasm; that is superchemistry; seek, my young friend, for other ambitions.'

"... In 1895, Michael Foster, a physiologist of deep vision, dealing with the respiration of tissues, and in particular with the degree to which the activity of muscle depends on its contemporary oxygen supply, expounded the current view which may be thus briefly summarized. The oxygen which enters the muscle from the blood is not involved in immediate oxidations, but is built up into the substance of the muscle. It disappears into some protoplasmic complex on which its presence confers instability. This complex, like all living substances, is to be regarded as incessantly undergoing changes of a double kind, those of building up and those of breaking down. With activity the latter predominates, and in the case of muscle the complex in question explodes as it were, to yield the energy for contraction. 'We cannot yet trace', Foster comments, 'the steps taken by the oxygen from the moment it slips from the blood into the muscle substance to the moment when it issues united with carbon as carbonic acid. The whole mystery of life lies hidden in that process, and for the present we must be content with simply knowing the beginning and the end"...

"I must perforce limit the field of my discussion, and in what follows my special theme will be the importance of molecular structure in determining the properties of living systems. I wish you to believe that molecules display in such systems the properties inherent in their structure even as they do in the laboratory of the organic chemist. The theme is no new one, but its development illustrates as well as any other, and to my own mind perhaps better than any other, the progress of biochemistry. Not long ago a prominent biologist, believing in protoplasm as an entity, wrote: 'But it seems certain that living protoplasm is not an ordinary chemical compound, and therefore can have no molecular structure in the chemical sense of the word'. Such a belief was common."

More could be cited to describe the situation in biology at the time of the preponderance of "the conception of metabolism as comprising events determined by the obscure properties of a single biological entity (emergent properties lost with loss of integrity, Hopkins[24])" and of "a protoplasmic complex within which occurred pseudo-chemical events, mysterious and undecipherable" (Stephenson[21]).

It was in a reaction to such concepts that biochemistry took off as an autonomous science.

The present writer is convinced that the reader will recognize this rupture in the exposé offered in the following chapters. The rupture was deepened by the discovery of cell-free fermentation.

The notion that biochemistry developed as a feedback effect of organic chemistry on physiological chemistry is not, as we shall see, what the evidence shows. Physiological chemists were, by and large, believers in the protoplasm theory; active interest on the part of organic chemists in the metabolic events analyzed by biochemists came only much later. It grew into a factor of utmost importance to subsequent developments, particularly the study of metabolic pathways, engaged in since the Thirties. The individualization of biochemstry resulted from the interest shown by scientists from different fields in the influence, evident in every living tissue, of molecular structures and their inherent properties. Conducive to their interest was Lavoisier's principle that there are not two chemistries. To quote Hopkins[25] once more, the adherents of the new biochemistry believed that

"In the study of the intermediate processes of metabolism, we have to deal, not with complex substances which elude ordinary chemical methods, but with simple substances undergoing comprehensible reactions."

This confident statement was substantiated by the succession of facts and theories which followed the discovery of cell-free fermentations.

Hopkins' attitude illustrates what Bachelard[26] means when he writes

"L'histoire des sciences est l'histoire des défaites de l'irrationalisme."

Hopkins was first trained as a chemist and his intellectual frame of mind, conditioned by an era of scientific thought, could not tolerate irrationalist creations such as "protoplasm".

In Chapters 10 and 11 (Volume 30) we have referred to the views of Pflüger on the Sources of energy considered to be produced by the decomposition of high energy unstable proteins containing a labile group (cyanogen).

2. Vertical (recurrent) history and horizontal (planary) history of sciences

This volume contains, as a consequence of the very nature of its subject matter, a great deal of "internalist" points of view. Being concerned for the most part with the classical period of biochemistry, the approach to history

References p. 19

adopted in this volume is "vertical" or "recurrent", that is, we trace the process that starts with ignorance and ends with a discovery (or a truth). It certainly is legitimate for biochemists writing on the history of biochemistry to identify a particular route in the conceptual system leading to a particular concept we accept today. It is the author's conviction that a healthy situation results from the coexistence of two different attitudes toward the study of the history of science. One, which we may call the "horizontal" (or the "planary") method, describes the situation at a certain period, including all its aspects (biographical, theoretical, institutional, economical, sociological, intellectual, methodological, *etc.*) and isolates the factors leading from one horizontal plane to another.

The other method, which we may call "vertical" or, as suggested by Bachelard[26], "recurrent", traces the origins of a discovery made in our time or in an earlier age. Sources of free energy, for example, could not be identified until the concept of free energy (a subject which would well deserve a horizontal study) gained wide acceptance among biochemists; nor, as shown in this volume, could they have been identified by means of discussions concerning the whole of the living world, or organisms in their entirety. Free energy does not appear as such, going out of the mass of organisms: there we find the degraded forms of heat or light. As Hill, Meyerhof and Warburg understood so well, only the individualization of reversible steps of intermediary metabolism and the knowledge of their equilibrium constants could lead us to the sources of free energy. The history of the identification of these sources could only proceed through recurrent studies involving the isolation of the pathways of catabolism, leading to the identification of the very specific reactions in which these pathways converge. Historians should rejoice at the publication of such books by scientific experts in the field of recurrent history and familiar with the personal experience of scientific discovery. Such books as those by Keilin[27], Lipmann[28] or du Vigneaud[29], mentioned in the Introduction to Parts I and II are invaluable aids to the historian.

Since the publication of the first volume of this *History*, a new important book has been added to the list of valuable contributions made by expert biochemists to the history of biochemistry. It is Joseph S. Fruton's *Molecules and Life. Historical Essays on the Interplay of Chemistry and Biology*[30]. Fruton begins with a chapter on alcoholic fermentation and enzymes during the 19th century. Chapters 2 and 3 are concerned with protein structure and nucleic acid structure. These two chapters give the reader a vivid and detailed picture of the subject in comparison to which Chapter 15 of the present *History* (in which the history of the organic chemistry of natural substances is minimized), appears as a very sketchy picture that the reader will usefully complete by reading Fruton's presentation. The fourth chapter of Fruton's book deals with intracellular respiration, and the fifth presents a synthetic picture of the subject of metabolic pathways, which illustrates in a poignant way the importance of new methodologies, developed during the classical period of biochemistry, to the identification of catabolic and anabolic pathways. One unfortunate feature of the book (which may be corrected in future editions) is the lack of an index. This reduces its value as a reference book, despite its accuracy and reliability.

3. Methodological aspects

In the period when the methods of physiological chemistry were applied, changes taking place in the course of metabolism could only be studied at the level of the whole organism or of surviving organs. In studies on intermediary metabolism, the method consisted in measurements of changes in concentration of the reaction products. These changes were induced by varying the concentrations of suspected precursors. The complexity of the setting of the problem made success improbable, though physiological chemists were occasionally justified in boasting of some brilliant accomplishments. Such was the discovery, by Claude Bernard, of the glycogenic function of the liver and of indirect nutrition (see Chapter 10). Intermediary metabolism studies were greatly speeded up when it became possible to simplify the experimental conditions; this was done by making smaller and less complex material parts of organisms the subject of investigation. This procedure is akin to that which led Paracelsus to endeavour to reach the *Arcana* (see Chapter 2). The first far-reaching success was obtained with yeast juice for glycolysis studies. When homogeneous organelle concentrates, of mitochondria for instance, or of microsomes, were made available, the factor of complexity which had affected studies on organisms or even on isolated organs was lowered, and the practice of using minimal pieces for study and measurement reached its extreme in the case of crystallized enzymes. The reproach that such methods are inherently destructive was silenced by the introduction of isotopic tracers and of studies on inborn errors of metabolism, exacting mutants of microorganisms, induction in bacteria, *etc*.

References p. 19

(a) Application of tracer elements

This development we owe to G. Hevesy[31,32] who has reviewed it in essays to which the reader is referred.

In 1911 and 1912, Hevesy tried in Lord Rutherford's laboratory to separate radium D from lead which was combined with it and prevented the hundreds of kilograms of radiolead separated from pitchblende, presented to Rutherford by the Academy of Sciences of Vienna, to be of use in radiation studies. Failure to accomplish this separation induced Hevesy to make use of the inseparable radium D as a tracer to follow the path of lead atoms by radioactivity measurements.

Hevesy and Zechmeister[33], in 1920,

> "prepared labeled plumboacetate, dissolved equivalent amounts of this salt and of tetraphenyl lead in amyl alcohol, and separated the two compounds by crystallisation. The tetraphenyl lead was found to be inactive, demonstrating the lack of interchange between the lead atoms of tetraphenyl lead and lead ions. Other organic lead compounds investigated have shown the same absence of interchange. Between the lead atoms of formate and lead acetate or other compounds containing dissociating lead ions, prompt interchange was found to take place. These early experiments indicated the lack of interchange of atoms present in organic binding, a result which was confirmed in later experiments in which labeled phosphorus, sulfur, *etc.* were brought together, for example, in a solution of glycerophosphate, lecithin, casein or desoxyribonucleic acid with a solution of labeled phosphate; no organic labeled phosphorus was found to be present"[31].

The assumption that the tracer does not interchange with the corresponding constituents of the molecules considered was borne out by a number of experiments (literature in Hevesy[31]). The discovery of deuterium by Urey in 1931 did start a new phase of tracer studies and studies on water circulation in animal organisms were realized with the help of dilute heavy water as a tracer. Hevesy and Hofer[34], for instance, showed in 1934 that the interchange between the water of a goldfish and the surrounding water takes place at a rapid rate. This was followed by a series of studies on a number of aspects of water metabolism. The first studies on intermediate metabolism with deuterium were done in the field of fat metabolism by Schoenheimer and Rittenberg[35] in 1935 and followed by additional studies, some of which will be reviewed in Chapter 34.

After the stable isotopes ^{13}C, ^{15}N and ^{18}O were concentrated by Urey, they were used in a number of studies.

But it was after the Joliot–Curies discovered artificial isotopes that the investigation of the fate of the atoms of the common constituents of organisms could fully develop and that a large number of tracer studies could be undertaken.

While mass-spectrographic determinations of stable isotopes continued to present difficulties, the availability of radiocarbon made easy measurements possible. Discovered by Ruben and Kamen[36] in 1940, ^{14}C was first used by Barker and Kamen[37] in 1945 in a study of fatty acid catabolism in *Clostridium* and *Butyribacterium*, and in many subsequent studies as well.

(b) Inborn errors of metabolism

Alcaptonuria is the name given to a rare condition described by Marcet in 1822 and by Bödeker in 1859. In this condition the urine turns black on standing and this is a permanent feature of the individual presenting it. The very early diagnosis resulted from the blackening of the infant's napkins. The substance passed in urine was recognized as a strongly reducing catechol-like compound which received the name of homogentisic acid (for literature, see Harris[38]). In 1891, Wolkow and Baumann[39] identified the compound as 2,5-dihydroxyphenylacetic acid and showed that its excretion increased after the administration of tyrosine. These authors regarded the condition as an infectious disease. Garrod[40], in 1902, published a paper on alcaptonuria in which he called attention to the inherited differences between biochemical features. He pointed to the probability that most of the homogentisic acid is derived from tyrosine, the peculiarity of the alcaptonurics (who except for a proneness to osteoarthritis appear quite normal in other respects) being an inability to degrade the benzene ring of tyrosine.

Garrod also recognized the familial distribution of alcaptonuria and concluded that it had a genetic basis.

Bateson, consulted by Garrod, pointed to the Mendelian recessive character of the inheritance of alcaptonuria, the first example recognized in man.

Garrod deserves credit for the significant insight that in alcaptonuria as in other conditions "we are dealing with individualities in metabolism and not with the results of morbid processes".

In his Croonian Lectures[41] of 1908 and in the book[42] he published in 1909, Garrod described a number of other conditions which he called "inborn errors of metabolism": phenylketonuria, galactosemia, fructosuria, glycogen storage disease, and others. As stated by Harris[38], to whose book the reader is referred for a detailed treatment of human biochemical genetics:

References p. 19

"Thus the 'inborn errors of metabolism' can be regarded as genetically determined biochemical variations, which sharply characterize human beings. They are highly specific and represent many diverse metabolic phenomena, which result in very varied effects on the viability and fitness of the individual".

Garrod introduced the concept of an inborn metabolic block, the organism being unable to perform a particular step in the normal course of the metabolic pathway due to the lack of the enzyme effecting that step.

The importance of inborn errors of metabolism to metabolic studies was stressed by Hopkins[43] when he wrote:

"Extraordinarily profitable have been the observations made upon individuals suffering from those errors of metabolism which Dr. Garrod calls 'metabolic sports, the chemical analogues of structural malformations'. In these individuals nature has taken the first essential step in an experiment by omitting from their chemical structure a special catalyst which at one point in the procession of metabolic chemical events is essential to its continuance. At this point there is arrest, and intermediate products come to light".

Several examples of how this concept may be applied in the isolation of metabolic steps will be found in this volume. A factor which has limited the study of inborn errors of metabolism in man is the rarity of the condition; the application of concepts pertaining to biochemical genetics in the solution of problems of intermediary metabolism became widespread only after Beadle and Tatum had introduced the use of exacting mutants of microorganisms.

(c) Microbial mutants

Once biochemical geneticists had concluded that each step of a metabolic pathway is controlled by a separate gene, Beadle and Tatum conceived the idea of reversing the process.

To quote Beadle[44]:

"We reasoned that, if the one primary function of a gene is to control a particular chemical reaction, why not begin with known chemical reactions and then look for the genes that control them? In this way we could stick to our speciality, genetics and build on the work chemists had already done. The obvious approach was first to find an organism whose chemical reactions were well-known and then induce mutations in it that would block specific identifiable reactions."

Beadle and Tatum[45] chose *Neurospora crassa*, a mold which had proved itself suited to genetic studies and which was easy to grow in pure culture on a chemically defined simple medium. They obtained by X-ray irradiation a mutant which had lost the ability to synthesize pyridoxine.
Beadle and Tatum concluded that:

"... it should be possible, by finding a number of mutants unable to carry out a particular step in a given synthesis, to determine whether only one gene is ordinarily concerned with the immediate regulation of a given specific chemical reaction."

The method was widely used in the analysis of metabolic pathways by taking advantage, in a mutant, of a block of one of the enzymatic steps of the pathway. Not only have *Neurospora* mutants been used in such studies, but mutants of *Escherichia coli* and other microbial mutants as well.

4. Institutional aspects

As will be shown in this volume, biochemistry's first success came with the discovery of yeast juice capable of alcoholic fermentation. Successful experimentation was first carried out in this field by Eduard Buchner and his collaborators M. Hahn, W. Antoni and J. Meisenheimer. Eduard Buchner was stimulated by an immunologist, his brother Hans Buchner. It was another immunologist, A. Macfadyen, who engaged another chemist, A. Harden, in the field of fermentation by yeast juice. The field in which he was joined by his collaborator W. J. Young was one to which the school of Russian botanists including Lebedev, Kostychev and L. Ivanov also contributed. A number of German schools housed in academic departments participated in analyzing the pathway of fermentation and muscle glycolysis. Franz Hofmeister was one of the first who openly accepted the enzymatic theory of intracellular metabolism. Three of his pupils, J. K. Parnas, G. Embden and C. Neuberg, became leaders in the field. In Poland, J. K. Parnas developed, at Lwow, a school of biochemists among whose members we find T. Mann, T. Baranowski, P. Ostern, J. A. Guthke, J. Terzakowec, C. Lutwak-Mann and T. Karzyleski. G. Embden's school in Frankfurt am Main contributed to the development of glycolysis studies by the works of a number of collaborators: M. Oppenheimer, H. Deuticke, M. Zimmermann, K. Baldes, E. Schmitz, G. Kraft, F. Kalberlah, H. Enger, W. Griesbach, H. Jost, H. Emde, W. Wassermeyer, E. Hirsch-Kaufmann, E. Lehnartz, T. Icker.

Carl Neuberg's laboratory in Berlin contributed many important data, such as the participation of pyruvate and acetaldehyde in the glycolysis pathway; developed was also the theory of methylglyoxal, which remained current for many years. Among Neuberg's collaborators we find J. Kerb, E. Reinfurth, H. Wastenson, M. von Grab, J. Hirsch, W. Ursum, L. Larczag, M. Kobel.

References p. 19

Given the limited funds available to university departments, progress was ensured by the creation of research institutions, privately endowed and providing financial means for the acquisition of costly equipment. In these institutions the scientists were relieved of the burden of elementary instruction, while opportunities were provided for the training of limited numbers of postdoctoral students.

The Pasteur Institute was opened in Paris in 1888 and the Carlsberg Institute in Copenhagen in 1875. In London, an institute which later became the Lister Institute was founded in 1891 and we have already mentioned the pioneer work accomplished there by Harden and his collaborator Young.

The Rockefeller Institute for Medical Research was created in New York in 1901.

In Germany, the first of a series of Kaiser Wilhelm Institutes was established in 1911. The contribution of those Institutes which took part in the development of biochemical research in the Twenties was of primary importance, particularly through the contributions of schools of research under the leadership of O. Warburg and of O. Meyerhof.

O. Warburg, who had, in Berlin, been a pupil both of his father, the physicist Emil Warburg, and of Emil Fischer, was a pioneer in the analysis of biological oxidations.

Unlike the workers mentioned above, Warburg never held academic positions. After he had obtained his doctor's degree in chemistry in 1906, he worked in Krehl's laboratory at Heidelberg until 1914, meanwhile obtaining an M.D. degree in 1911. He was aged 30 in 1913 when he was appointed a member of the Kaiser Wilhelm Gesellschaft and Head of a Research Department with no administrative or teaching duties. But while he was assembling the equipment for the laboratories, war broke out on 1 August 1914. After returning to civil life in 1918, Warburg occupied part of the top floor at the Kaiser Wilhelm Institut für Biologie in Berlin-Dahlem, where he stayed until 1931 and where he secured excellent equipment, though the space allotted to him could accommodate no more than half a dozen workers.

To quote H. A. Krebs[46]:

"Throughout his life it was Warburg's policy to keep the number of research workers low and to use a relatively large fraction of his financial resources for technical equipment. His long-term collaborators throughout were technicians who in the late 1920's occupied four of the few places: E. Negelein, F. Kubowitz, E. Haas and W. Christian, to be joined later by W. Lüttgens. He recruited these technicians from skilled mechanics trained in the workshop of engineering firms. Their asset was that they knew how to deal with physical instruments, and Warburg trained them in chemistry himself. Many of these technicians, like those mentioned above, became well known through their publications."

In 1929, Warburg obtained from the Rockefeller Foundation the means to build under the aegis of the Kaiser Wilhelm Gesellschaft, a small institute for himself, the "Kaiser Wilhelm Institut für Zellphysiologie" which he occupied at the end of 1931 and where he worked until he died in 1970.

Warburg used to accept one or two postdoctoral collaborators each year. After he moved into his new Institute in 1931, he usually had 10–12 workers in his laboratory, including the couple of academics he accepted each year. Those included H. Gaffron, H. A. Krebs, W. Kempner, W. Cremer, H. Theorell, E. G. Ball, C. S. French, J. N. Davidson, Th. Bücher and D. Burk (Krebs[46]). Meyerhof, who had been Warburg's pupil in the field of cell physiology, after an academic term at the University of Kiel, joined the Dahlem institute in 1924. Meyerhof remained in Berlin-Dahlem until 1929, when he became head of the Department of Physiology at the Kaiser Wilhelm Institute for Medical Research in Heidelberg.

He developed a school of his own, mainly devoted to studies on muscle glycolysis. Among its members were K. Lohmann, J. Suranyi, S. Ochoa, R. Meier, F. Lipmann, D. McEachern, W. Kiessling, P. Schuster, K. Meyer, H. Lehmann, W. Schulz, R. McCullagh, D. Nachmansohn, P. Oesper, A. von Muralt. Meyerhof was a physiologist and his laboratory greatly benefited from K. Lohmann's chemical competence.

5. Sources of free energy

Present theory holds that ATP is the universal high-energy compound whose energy is used for the accomplishment of work.

The mechanism of ATP formation in glycolysis has been shown to consist in coupled reactions, the energy being liberated in an oxidation-reduction process. Much more important as sources of free energy are the different forms of electron-transfer chains which have been described in the case of oxidative phosphorylation, photosynthetic phosphorylation and chemoautotrophic phosphorylations. Though no satisfactory theory has been formulated as yet concerning the energetics of these electron-transfer chains, it is believed that forms of intermediate high-energy compounds other than ATP ($A \sim C$), but not so far identified, act as intermediates in

References p. 19

"energy-linked reactions", *i.e.*, energy-donating reactions involved in the electron transfer chain. In such "energy-linked reactions" ATP itself may be involved and its hydrolysis may participate as an energy-donating reaction. In current terminology, the term *energy-linked reaction* is more specially applied to a reaction requiring energy and depending on energy-donating reactions.

Reversal of electron transport in a mitochondrial respiratory chain is an example of an energy-linked process. The concept of energy-linked reversal of electron transport from reduced cytochrome *c* to pyridine nucleotide has been formulated by Chance and Hollunger[47].

6. Acknowledgements

Several colleagues and friends have, either in personal contact or in correspondence, contributed by informing or enlightening the author. Thanks are due particularly to E. G. Ball, C. Cori, H. Deuticke, J. T. Edsall, H. M. Kalckar, L. Kiesow, H. A. Krebs, R. Kohler, K. Lohmann, T. Mann, C. Martius, R. Nilsson, S. Ochoa, E. Stotz and A. Szent-Györgyi.

For permission to reproduce copyright material, the author is grateful to M. Avron, E. G. Ball, L. Kiesow, H. Theorell, J. Whiteley (née Keilin); Cambridge University Press, London; Academic Press, New York; The Biochemical Laboratory, Long Island Biological Association, Cold Spring Harbor, New York; and Annual Reviews Inc., Palo Alto, Calif., U.S.A.

The author is also indebted to those who helped him in collecting the series of Plates of Part III and whose kind contribution is acknowledged in the list of those plates.

Towards the period of completion of this volume, the author suffered from injuries resulting from a traffic accident which kept him in the hospital for a number of months. In the task of compiling the lists of references and bringing the proofs in their final state, the author was greatly aided by Dr. Ghislaine Duchâteau-Bosson who deserves his thanks for her invaluable help. In these difficult circumstances the author also derived great benefit from the efficient secretarial help of Mrs. Amy Closset without whom his task could not have been accomplished.

Liège, October 1974 Marcel Florkin

REFERENCES

1 T. S. Hall, *Ideas of Life and Matter*, 2 vols., Chicago, 1969.
2 M. Teich, *Clio Medica*, 1 (1965) 41.
3 R. E. Kohler Jr., *Isis*, 64 (1973) 181.
4 T. H. Huxley, *The Fortnightly Review*, N.S. 5 (1969) 129.
5 G. L. Geison, *Isis*, 60 (1969) 273.
6 F. Cohn, *Nova Acta Acad. Caesareae Leopoldino Carolinae*, 22 (1850) 605.
7 F. Unger, *Anatomie und Physiologie der Pflanzen*, Leipzig, 1855.
8 J. R. Baker, *Quart. J. Microsc. Sci.*, 90 (1949) 87.
9 M. Schulze, *Arch. Anat. Wiss. Med.*, (1861) 1.
10 M. Florkin, *Naissance et Déviation de la Théorie Cellulaire dans l'Oeuvre de Théodore Schwann*, Paris, 1960.
11 E. W. Von Brücke, *Sitzungsber. Kais. Akad. der Wissensch. Wien* (Math.-Naturwiss. Klasse), 44 (1861) 381.
12 E. Haeckel, *Quart. J. Microsc. Sci.*, 10 (1869) 233.
13 R. Remak, *Untersuchungen über die Entwickelung der Wirbelthiere*, Berlin, 1855.
14 K. Reichert, *Arch. Anat. Physiol. Wiss. Med.*, (1863) 86.
15 W. Coleman, *Proc. Am. Philos. Soc.*, 109 (1965) 124.
16 E. Baldwin, *An Introduction to Comparative Biochemistry*, Cambridge, 1937.
17 M. Florkin, *L'Evolution Biochimique*, Paris, 1944 (English translation by S. Morgulis).
18 L. S. Beale, *Protoplasm, or Life, Matter and Mind*, London, 1870 (A 2nd much enlarged edition appeared in the same year) (cited after Geison[5]).
19 W. S. Beck, *Modern Science and the Nature of Life*, New York, 1957 (cited after Geison[5]).
20 J. Drysdale, *The Protoplasmic Theory of Life*, London, 1874 (cited after Geison[5]).
21 M. Stephenson, in J. Needham and E. Baldwin (Eds.), *Hopkins and Biochemistry 1861–1947*, Cambridge, 1949.
22 F. G. Hopkins, *Science*, 84 (1936) 258.
23 F. G. Hopkins, Presidential Address, *Brit. Assoc., Leicester Meeting, 1933.*
24 F. G. Hopkins, *Irish J. Med. Sci.*, July (1932) 333.
25 F. G. Hopkins, Address of Sectional President to Physiological Section, *Brit. Assoc. Birmingham Meeting, 1913*; also in *Brit. Med. J.*, 2 (1913) 713 and *Lancet*, 2 (1913) 851.
26 G. Bachelard, *L'Activité Rationaliste de la Physique Contemporaine*, Paris, 1951.
27 D. Keilin, *The History of Cell Respiration and Cytochrome*, Cambridge, 1966.
28 F. Lipmann, *Wanderings of a Biochemist*, New York, 1971.
29 V. du Vigneaud, *A Trail of Research*, Ithaca, 1952.
30 J. S. Fruton, *Molecules and Life. Historical Essays on the Interplay of Chemistry and Biology*, New York, 1972.
31 G. Hevesy, *Cold Spring Harbor Symp. Quant. Biol.*, 13 (1948) 129.
32 G. Hevesy, *Minerva Nucleare*, 1 (1957) 182.
33 G. Hevesy and L. Zechmeister, *Ber. d.d. Chem. Ges.*, 53 (1920) 410.
34 G. Hevesy and E. Hofer, *Z. Physiol. Chem.*, 225 (1934) 28.
35 R. Schoenheimer and D. Rittenberg, *Science*, 82 (1935) 156.
36 S. Ruben and M. D. Kamen, *Physiol. Rev.*, 57 (1940) 599.
37 H. A. Barker and M. D. Kamen, *Proc. Natl. Acad. Sci. (U.S.)*, 31 (1945) 219.
38 H. Harris, *Human Biochemical Genetics*, Cambridge, 1959.
39 M. Wolkow and E. Baumann, *Z. Physiol. Chem.*, 15 (1891) 228.
40 A. E. Garrod, *Lancet*, 2 (1902) 1616.
41 A. E. Garrod, *Lancet*, 2 (1908) 1, 73, 142, 214.

42 A. E. Garrod, *Inborn Errors of Metabolism*, Oxford, 1909; 2d ed., 1923.
43 F. G. Hopkins, *Nature*, 92 (1913) 213.
44 G. W. Beadle, *Genetics and Modern Biology*, Philadelphia, 1963.
45 G. W. Beadle and E. L. Tatum, *Proc. Natl. Acad. Sci. (U.S.)*, 27 (1941) 499.
46 H. A. Krebs, *Biogr. Mem. Fellows Roy. Soc.*, 18 (1972) 629.
47 B. Chance and G. Hollunger, *Federation Proc.*, 16 (1957) 163.

Section I

Anaerobic Phosphorylation

Chapter 17

The Discovery of "Cell-free Fermentation"

1. Glycolysis

By the main trunk of glycolysis we now mean the degradation of glucose and of other carbohydrates into pyruvate by way of fructose diphosphate. It is along this pathway that the importance of phosphorylation in bioenergetics was first discovered in the form of anaerobic phosphorylation.

Glycolysis was first studied under two different headings: alcoholic fermentation on the one hand and muscle glycolysis on the other. Only about the year 1925 was it realized that both phenomena, while ending in different ways, were functional variations of a common biochemical mechanism. Since that time it has become clear that the glycolysis pathway is present in the great majority of cells and that, in spite of slight diversities, its unity through life is clear.

The author wishes to acknowledge his indebtedness, in the preparation of the following chapters on glycolysis, to a number of general articles[1–17b].

2. Enzymatic theories of fermentation prior to the discovery of cell-free fermentation

At the time, the theory of protoplasm considered as a single biological entity whose emergent properties were lost with loss of integrity (see Chapter 16), was generally accepted and it had been developed in several directions.

As stated in the Introduction, it was commonly believed that protoplasm had no chemical structure and that its components did not exert their chemical properties, in the sense of laboratory chemistry. Molecules of food and of oxygen entering this complex became incorporated in the huge molecule of "biogen" where their chemical properties were lost.

References p. 37

Plate 64. Moritz Traube.

The views formulated in 1875 by Pflüger[18] considered protoplasm as a giant molecule, with side chains endowed with powers of assimilation and oxidation. In the "living protein" he recognized the existence of unstable groups whose specific vibrations acted as the driving forces of the vital reactions, defined as follows by Haeckel[19]:

"This single homogeneous matter (is) the active substrate of all vital motions and of all vital activities: nutrition, growth, motion and irritability." (translation by Kohler[20])

In 1890 for example, the standard view was that oxidation, fermentation or biosynthesis were the functions of protoplasm *as a whole*. At that time, enzymology already was an established part of "physiological chemistry" (see Chapter 13) but it was agreed that the enzymes were exocellular agents performing hydrolysis. As Kohler[20] states

"No unambiguous case of a strictly intracellular enzyme was known."

Berthelot[21] isolated invertase from macerated yeast in 1860 and Schmiedeberg[22] isolated hippuric acid splitting enzymes from macerated kidney in 1881, but these observations did not appear as pressing reasons for the majority to abandon their view.

The followers of the theory of protoplasm, though forming the majority of "physiological chemists", were challenged by a small group of "enzymologists", the first of whom was Traube[23] who in 1858 suggested by analogy that definite molecules are present in yeast cells, which are capable of accomplishing fermentation. As no experimental argument was provided, this theoretical view was opposed by many physiologists, and among them Nägeli[24], who maintained that fermentation was due to the movement communicated to the fermented material by the movement of "protoplasm" and inseparable from it.

Among the first opponents of "protoplasm" we find Hoppe-Seyler[25] who, in 1881 developed a theory according to which all vital reactions were considered as combinations of hydration and dehydration, mediated by "unorganized ferments"★.

An opponent of Hoppe-Seyler, Kühne[26] attacked his speculations on "unorganized ferments" and it was on this occasion that Kühne coined

★ There is some question as to the definition of "unorganized ferments": either enzymes, or ferments in the sense of Liebig (see Chapter 6), meaning a decomposing protein.

References p. 37

the word "enzymes" the definition of which he formulated as exocellular hydrolytic agents.

After the discovery of zymase by Buchner, oxidases were also described and a theory of biosynthesis, considered as due to a reversal of the catalytic process (an aspect we shall consider in Part IV of this History) was proposed. In consequence the "protoplasmists" progressively gave way to the "enzymologists".

One of the first who openly adhered to the new faith was Hofmeister[27], who wrote in 1901:

> ... "In fact, many years ago when only the secretion ferments were known, far-sighted investigators, above all Hoppe-Seyler, expressed the opinion that such ferments were also active inside the living cell. Since then there have been innumerable cases in which such "intracellular ferments" have been brought to light from within the cell, and in many cases their significance for vital processes has been made clear. Indeed, these discoveries, increasing almost daily, have revealed so universal an occurrence of ferments in organisms and such a variety of modes of action, that we may be almost certain that sooner or later a particular specific ferment will be discovered for every vital action." (translation by Kohler[20])

This almost textually repeats a prevision of Berzelius.

Although by that time it had become unprofitable to sacrifice to the myth of protoplasm, the theory was still in being and in 1909, Halliburton[28] still expressed words of caution against "enzymologists", but the great advantage of enzymology was that it presented a new world of opportunity for biochemical work, and a program for research which led it to become the central objective of biochemical enquiry. Kohler[20] rightly stresses that:

> "The new profession of biochemistry that began to emerge about 1900 was initially composed of specialists in a variety of established fields, brought together by a common outlook on the physico-chemical nature of life, a common belief that enzymes were the key agents in life processes, and shared historical experience, and a new name, "biochemistry".

In Kohler's essay[20] referred to, the focus is on the unity of the new movement, in spite of the variety of professional backgrounds among the early biochemists; not only were there physiologists, chemists and "physiological chemists" among them, but also pharmacologists, clinicians, botanists, zoologists, and, playing a most important role, as we shall see, bacteriologists and immunologists.

One of the discoveries which gave impetus to enzymology was that of "cell-free fermentation". One of the arguments against the intracellular enzymes was that, in the case of fermentation, for example, attempts repeatedly made with a view to isolating a "fermentation unorganized ferment" had utterly failed[29]. Assays of extraction with glycerol as well as with water had given negative results[30].

Nevertheless, on the other hand, the fact that, while a number of enzymes can be extracted from cells by common solvents, other enzymes are extractable only after mechanical treatment of the cells, was established by E. Fischer[31] in 1894.

The development of the chemistry of carbohydrates, particularly under the impulsion of Fischer, also favoured the acceptance of an enzymatic concept of fermentation, as it was observed that yeast acted only on certain optical isomers and not on others, a fact which was compared to the observations on the action of enzymes on certain glycosides and not on certain of their stereoisomers.

3. Discovery of "cell free fermentation"

The concept which led to the discovery of "cell-free fermentation" was the erroneous theory, formulated by Hans Buchner, according to whom toxins and antitoxins both originate in the bacterial protoplasm. This view led Hans Buchner to endeavour to isolate the bacterial protoplasmic proteins which he considered as being of primary importance to immunity. This train of thought has been clearly retraced by Kohler[32] to whose paper the reader is referred. Hans Buchner's younger brother Eduard, was a chemist, to whom he naturally called for help in devising a method for preparing such proteins. They concluded that bacterial cells ought to be disrupted by mechanical means. Eduard Buchner had previously worked on fermentation by yeast and he selected yeast as the material for his attempts at opening cells. He found that yeast, and other microorganisms could be ground up in a mortar with the addition of fine sand, a method which had already been used as early as 1846 (for literature, see Buchner *et al.*[33]). At the time, Eduard Buchner was Privatdozent in Baeyer's laboratory in Munich. The board of this laboratory considered that the work on the contents of yeast cells was without interest. Eduard Buchner who as Privatdozent depended for his experimental work on a grant from the Board, had to put an end to his line of research. In 1893 he joined his former teacher Curtius in Kiel where he became professor in 1895. A year later he moved to Tübingen[34].

In the meantime, in 1894, Hans Buchner was appointed to the chair of Hygiene in Munich and, in 1896 he returned to his endeavour to isolate

References p. 37

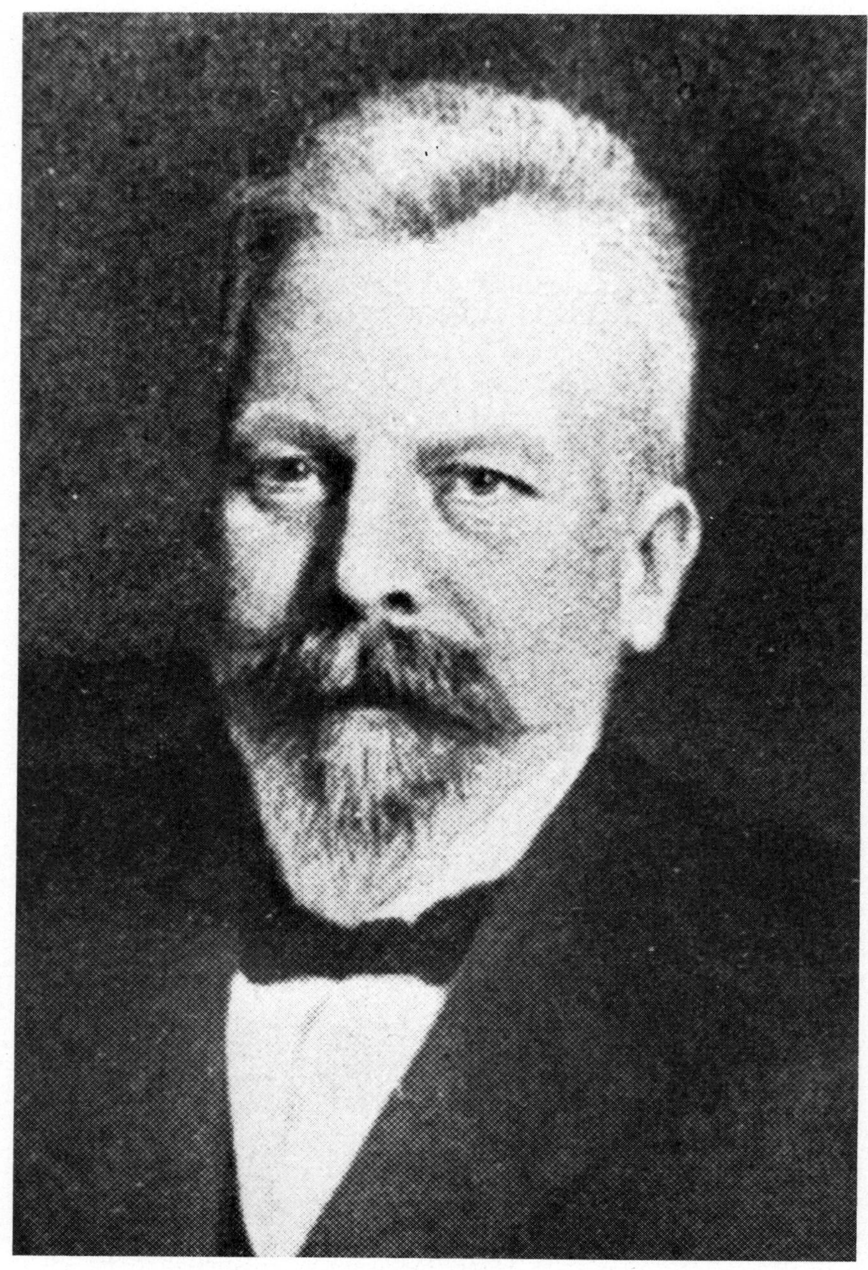

Plate 65. Eduard Buchner.

bacterial proteins. On his suggestion his assistant Hahn made further attempts to make the sand grinding method workable. He added Kieselguhr (one of the commercial forms of infusorial earth) to the mass of sand ground yeast, wrapped the whole in a cloth and pressed the mass in a hydraulic press. The Kieselguhr acted as a filter retaining the cell debris and an opalescent yellowish juice (400 ml per kg of yeast) issued from the pressed mass[35]. The juice was submitted to a study of its immunological characteristics but it was not easy to conserve. Hans Buchner, recalling the preservative effect of sugar added to fruit preserves[36], suggested the addition of 40 per cent glucose, as well as other preservatives such as glycerol, or concentrated salts. Hahn left for his vacation and Eduard Buchner arrived from Tübingen to spend his own vacation, working on press juices. He prepared the juice with an addition of 40 per cent glucose and he noticed that it generated a stream of bubbles. He recognized that he was observing the occurrence of fermentation. Hans Buchner was convinced that this was due to fragments of cells[37] but he observed that fermentation also took place in extracts treated with antiseptics which stopped the activities of living cells. Both Hans and Eduard Buchner were soon convinced after a series of experiments that cell-free fermentation had been discovered. In the spring of 1897, Eduard Buchner published the discovery[38], which came as astonishing news to the readers.

Kohler[32] notes that the theory of Hans Buchner which was at the root of the discovery of the cell-free fermentation and which made the protoplasmic proteins so important for immunity was a red herring, as the specific antigens of bacteria were recognized as polysaccharides of the cell envelope. But he rightly emphasizes that it led to the understanding that the study of intracellular proteins was of great importance and provided the method of obtaining a number of intracellular enzymes.

4. The reception of the discovery of cell-free fermentation

Buchner was particularly bold when stating the cell-free fermentation to be due to a soluble intracellular enzyme, "zymase".

Kohler[39] has analysed the reception of Buchner's discovery by the scientific circles of the time. He does not only take into account the intellectual aspects but also the social aspects. We shall here epitomize this analysis and refer the reader to Kohler's study for a more detailed treatment. As Kohler points out, what Buchner provided was more than a fact or a theory; it was the basis of a new conceptual system.

References p. 37

Zymase, opening as it did new vistas for biochemical enquiry and theory, and the zymase controversy "can best be treated as a socio-cultural event, more than a purely intellectual one"[39] and the significance it soon acquired was itself socially functional. As said above, Hans Buchner, paying tribute to the reigning protoplasmic theory and thinking along the lines of this theory, first believed that the fermentation by yeast juice was due to persisting bits of protoplasm. He added arsenate or toluene, with the purpose of inhibiting fermentation by killing the protoplasm, but the fermentation was not inhibited. This convinced him[37]:

"that cells carry out important functions, such as fermentation, not by their protoplasm, nor by their *Energiden*, but by special non-protoplasmic substances specially formed for this purpose." (translation by Kohler[39])

This citation is taken from a talk delivered by Hans Buchner to the Munich Morphological–Physiological Society on March 16, 1897, a week before the publication of the historical paper of Eduard Buchner[40]. The view adopted by Hans Buchner was rather unfavourably received by the distinguished audience and one of its members, Voit[41], raised several arguments against it, though he agreed to be prepared to consider the acceptance of substances as the causes of vital reactions if such substances were extracted from organs and showed the activity of the organ. But in the situation at hand he could not change his mind, as pieces of the fabric of protoplasm could still be active in the juice, as the enzymes so far known carried out hydrolysis, as these known enzymes could be extracted from cells without having to break them open, as yeast juice was much less active than yeast; and as the experiments with the antiseptics did not convince him, he suggested that the cell-free fermentation was not the same as the normal process. In short, he considered cell-free fermentation as an artifact. Another member present was the histologist Kupfer. As he considered protoplasm as being composed of granules (see Chapter 16) he doubted that press juice was lifeless[42]. That the concept of an intracellular enzyme was in contradiction with the spirit of the time is even confirmed by the writings of Eduard Buchner himself, who believed that zymase was not really intracellular, but acted outside the cell[38].

Eduard Buchner demonstrated by a number of experiments that fermentation by press juice[43]

"is due not to living bits of plasma but to an enzyme-like chemical substance." (translation by Kohler[39]).

In one of these experiments he carefully desiccated yeast cells and heated the residue to 100 °C without destroying the faculty of fermenting glucose, in spite of the fact that such residues were unable to grow. At the end of the same year, he concluded even more positively that[14]:

"the assumption that an enzyme in press juice causes fermentation is in accord with all the facts." (translation by Kohler[39])

A number of reports soon came in stating that their authors had been unable to repeat Buchner's experiments. Kohler[39] notes that the reception of Buchner's view was uniformly negative in the circles of the brewing technologists, an aspect he rightly considers as related to "their practical concern with whole cells and conservative Pasteurian tradition"[39].

As examples of this scepticism linked with professional commitments, Max Delbrück and Will, the leading figures in brewery technological research in Germany, may be cited[39]. Among other disbelievers was for instance the botanist J. R. Green who failed in his attempts to repeat Buchner's experiments and facetiously remarked that English yeasts, at least, did not contain an alcohol-producing enzyme[45].

Stavenhagen, a chemist whose background was mainly technological, reported that press juice filtered through a porcelain filter was inactive. He was aware of the retention of enzymes by porcelain filters but he nevertheless refused to recognize zymase, on the grounds that it "was in complete contradiction to Pasteur's theory"[46]. To this Buchner retorted that Pasteur's authority was no argument against experimental evidence[44]. The negative reports were summarized in a paper by Wehmer[47], a Dozent of the technical college at Hannover who had mainly studied Sauerkraut and pickled herrings and who concluded as follows:

"Needless to say, the hypothesis of "zymase" is still completely unsubstantiated and it deserves to be publicized no further. By all appearances this reawakened version of Traube's old fermentation theory will enjoy none too long a life." (translation by Kohler[39])

In spite of the opposition of the technologists, the opinion of the chemists, by the time of the publication of Wehmer's paper, had become favourable to zymase. This was confirmed by the fact that Buchner[48] addressed the German Chemical Society on zymase in 1898. In the light of the opposition of the technologists on the basis of Pasteurian conservatism, it appears paradoxical to note that the most active support of

References p. 37

zymase came from the disciples of Pasteur. Emile Duclaux[49], director of the Pasteur Institute, hailed the discovery with enthusiasm:

"For a long time enzymes were believed to carry out only hydrolysis... then came Rey Pailhade's hydrogenases and the oxidases of G. Bertrand. Buchner's alcoholic enzyme continues the series and is the first to cause changes in the carbon chains and rearrangement of groups." (translation by Kohler[39])

It must be stressed here that in a centre of research in immunology, as Paris was at the time, immunochemistry was regarded as a major aspect of the young science of biochemistry. The common denominator being the existence of active proteins, enzymes, toxins and antitoxins were considered to be aspects of a single scientific field ending in the study of chemical reactions. Roux[50], in a lecture delivered in November 1898, expressed once more the enthusiasm of Pasteur's disciples for zymase:

"The science of ferments and that of microbes both end up in the study of chemical reactions, due for the most part to enzymes. ... the discovery of the alcoholic enzyme is unquestionably one whose importance will never be dimmed." (translation by Kohler[39])

Among the early supporters of zymase, we also find Pfeffer, the plant physiologist.

As Kohler[39] states,

"Zymase provided a touchstone for identifying those groups, experimental botanists and zoologists, and immunologists, that would be most active in the development of the new discipline of biochemistry in the years following Buchner's discovery."

5. Enzymes versus protoplasm

Even if there were strong opponents of zymase, there were, as we have said above, a number of scientists who accepted it. But there were also many biologists who persisted in the opinion that zymase was nearer to "protoplasm" than to an enzyme as known by enzymologists of the period. This view, which derived from the early attitudes of Voit or Kupfer referred to above, was clearly expressed in December 1897 by Neumeister[51], a physiologist and a protein chemist.

"It is more likely that the activity of the press juice is due not to one substance but to a number of different proteins that even after their removal from the living cell persist in the metabolic activities characteristic of the protoplasm." (translation by Kohler[39])

Buchner agreed that zymase was in some ways different from the enzymes known so far.

Among the "protoplasmists" opposing the enzymatic nature of zymase, we may cite Abeles[53], who, in 1898 insisted that zymase was a piece of surviving protoplasm, and that it was alive, in contrast with "dead ferments". Buchner[54] answered by expressing what was to become the credo of the new community of biochemists:

"The concept of "living plasma" is indefinite and chemically indefinable; it is generally conceived as a mixture of various proteins which act as the carriers of life. Among these proteins enzymes are no doubt included. If definite substances in the living intracellular plasma can be identified by specific chemical reactions, such advances should not be denied by saying that such reactions are due to the whole protoplasm unless there are compelling reasons to do so. The presence of such mysterious agents in press-juice, these pieces of living plasma, must be demonstrated before the simpler enzyme theory, which agrees with all the facts, is discarded." (translation by Kohler[39])

An argument of the same nature as that presented by Abeles appears in a historically important paper published in 1900 by the bacteriologist and immunologist Macfadyen. He was impressed by the rate of autofermentation of the juice of top-fermenting yeast, without addition of glucose. When he added glucose, only a part of it was fermented. This he interpreted as indicating that glucose is only fermented after its inclusion in protoplasm. For him, the autofermentation when the yeast protoplasm is still present in the juice consumed its own glucose and the added glucose was only partly fermented because its inclusion in protoplasm outran its decomposition by protoplasm. For him, as for Abeles, zymase was a fragment of living protoplasm. Reporting on the views of Macfadyen, Kohler[39] rightly stresses that the model in Macfadyen's mind was undoubtedly the theory of Pflüger on intracellular metabolism which was at the time popular among bacteriologists.

Buchner was supported on this issue, by Green who was by that time converted to the views of Buchner, and who insisted that zymase, which could survive desiccation and antisepsis, could not be identified with protoplasm[55].

The protoplasmic theory of zymase put forward by Wroblewski was of a different nature. Wroblewski[56], who worked in a nutrition laboratory at Cracow and had studied invertase, confirmed Buchner's results and made an analysis of yeast juice. He found no less than twenty-one different

References p. 37

Plate 66. Frederick Gowland Hopkins.

substances, including several enzymes.*. He also made the important discovery of the stimulating effect of phosphate on zymase, which is sometimes wrongly attributed to Harden.

He considered the protoplasm of the cell as a web-like structure of proteins to which reactive side chains were attached, each catalyzing a metabolic function. Some of the side chains could be detached and secreted as the classical enzymes. Other side chains such as zymase could be expressed in their active form but still attached to remnants of protoplasm.

To all the "protoplasmic" arguments, Buchner repeatedly opposed his experimental data showing the enzymatic nature of zymase: persistence of activity in dried yeast juice after nine months; lack of sedimentation of splinters of protoplasm in centrifuged juice; reversibility of poisoning zymase by hydrogen cyanide and retention of ability to ferment in dried yeast heated to 100° and completely sterile.

The controversy over zymase was in fact the last fight between the adepts of the fading myth of "protoplasmic substance" and the adepts of the molecular viewpoint, the enzymologists.

Returning to the opposition of the technologists, we may state that, after having put Buchner into their black books, they then, after becoming convinced, made him their darling[39]. They were interested in the brewing possibilities of zymase until it soon became clear that there was no economic future for beer brewed by enzymes. In 1898, Buchner had become professor at the Agricultural College in Berlin, and Director of the Institute for the Fermentation Industry. After he received the Nobel Prize, Buchner was appointed to the chair of Physiological Chemistry in Breslau. In 1911 he moved to Würzburg where he stayed until he was mortally wounded on the Rumanian front, during the first World War, in 1917 (ref. 57).

There is no doubt that the discovery of cell-free fermentation opened new vistas in research and contributed to the development of one of the leading concepts of biochemistry, the enzymatic theory of metabolism (see Chapter 13). It introduced a far-reaching rupture with the past, as it endeavoured no less than to approach the biological whole through its molecular parts. This was clearly visualized by Buchner and also by Hopkins[58] when in his 2nd Purser Memorial lecture, delivered at Trinity College, Dublin in 1932, he said:

* According to our present knowledge, Buchner's zymase is in reality a system of several enzymes (see Table VI).

"Research during the present century has, in the minds of many, replaced the conception of metabolism as comprising events determined by the obscure properties of a single biological entity (emergent properties lost with loss of integrity) by such a view as the following. The living cell ... is to be pictured as the seat of diverse but organized chemical reactions, in which substances identifiable by chemical methods undergo changes which can be followed by chemical methods."

More observations came to the support of the enzymologic nature of cell-free fermentation. Lebedev[59,60] observed that by a cautious heating of pressed yeast at 15–35 °C, a preparation is obtained (Dauerhefe) which produces fermentation. By macerating this dried yeast in water, followed by filtration, a juice is obtained which produces fermentation. Nilsson and Alm[61] described a method for the preparation of an active maceration juice from baker's yeast. In the same paper the authors reported the production of trehalose in such juices.

Preparations (known under the name of *zymin*) were obtained by treating yeast with alcohol and ether[62] or with acetone and ether[63] and they showed the fermenting property. The active juice obtained by grinding such acetone-yeast when precipitated with alcohol and ether yielded an amorphous powder which produced sugar fermentation.

As late as 1927, a new form or objection was raised against the existence of "cell-free fermentation" by Kostychev *et al.*[64] who suggested that the fermentation by "yeast juice" was due to the persistence of a few yeast cells, the activity of which was stimulated by the presence of the substance discovered by Wildiers[65]★ and called *bios* by him, which is liberated during the preparation of the juice. This theory was refuted by Kluyver and Struyk[66] and since that time the concept of cell-free fermentation became widely accepted.

★ E. Wildiers was a student of M. Ide in Louvain who, after the publication of his work on *bios* (1901) migrated, as a doctor, to the United States where, as far as the present author knows, his trace has been lost.

REFERENCES

1 G. Embden, in Bethe's *Handbuch der Physiologie*, Vol. 8, I, Berlin, 1925.
2 O. Meyerhof, in Bethe's *Handbuch der Physiologie*, Vol. 8, I, Berlin, 1925.
3 A. Gottschalk, in *Oppenheimer's Handbuch der Biochemie*, 2nd edn., Vol. 2, Jena, 1925.
4 C. Neuberg, in *Oppenheimer's Handbuch der Biochemie*, 2nd edn., Vol. 2, Jena, 1925.
5 R. Nilsson, in H. von Euler (Ed.) *Chemie der Enzyme*, Vol. 2, Part 2, Munich, 1934.
6 O. Meyerhof, *Ergeb. Enzymforsch.*, 4 (1935) 208.
7 J. K. Parnas, *Ergeb. Enzymforsch.*, 6 (1937) 57.
8 T. Thunberg, *Ergeb. Enzymforsch.*, 7 (1938) 163.
9 F. Lipmann, *Advan. Enzymol.*, 1 (1941) 99.
10 C. F. Cori, *Biol. Symp.*, 5 (1941) 131.
11 C. F. Cori, in *Symposium on Respiratory Enzymes*, Madison, 1942, p. 175.
12 F. Lipmann, *Advan. Enzymol.*, 6 (1946) 231.
13 J. B. Sumner and G. F. Somers, *Chemistry and Methods of Enzymes*, 2nd ed., New York, 1947.
14 F. F. Nord and S. Weiss, in J. B. Sumner and K. Myrbäck (Eds.), *The Enzymes. Chemistry and Mechanism of Action*, Vol. 2, Part 1, New York, 1951.
15 F. Dickens, in J. B. Sumner and K. Myrbäck (Eds.), *The Enzymes. Chemistry and Mechanism of Action*, Vol. 2, Part 1, New York, 1951, p. 624.
16 P. K. Stumpf, in D. M. Greenberg (Ed.), *Chemical Pathways of Metabolism*, Vol. 1, New York, 1954.
17 B. Axelrod, in D. M. Greenberg (Ed.), *Metabolic Pathways*, New York and London, 1960 (2nd edn. of *Chemical Pathways of Metabolism);* 1967 (3rd edn. of same).
17a C. Oppenheimer and R. Kuhn, *Lehrbuch der Enzyme*, Thieme, Leipzig, 1927.
17b A. Fodor, *Das Fermentproblem*, Dresden, 1922.
18 E. Pflüger, *Arch. Ges. Physiol.*, 10 (1875) 251.
19 E. Haeckel, *Jenaische Z. Med. Naturw.*, 4 (1868) 64.
20 R. Kohler, *Isis*, 64 (1973) 181.
21 M. Berthelot, *Compt. Rend.*, 50 (1860) 980.
22 O. Schmiedeberg, *Arch. Exptl. Pathol.*, 14 (1881) 288.
23 M. Traube, *Theorie der Fermentwirkungen*, Berlin, 1858.
24 C. von Nägeli, *Theorie der Gärung*, Munich, 1879.
25 F. Hoppe-Seyler, *Physiologische Chemie*, Berlin, 1881.
26 W. Kühne, *Unters. aus d. physiol. Inst.*, Heidelberg, 1 (1877) 291.
27 F. Hofmeister, *Die chemische Organisation der Zelle*, Strasburg, 1901.
28 W. D. Halliburton, *Ann. Rept. Progr. Chem.*, (1909) 210.
29 A. Harden, *Alcoholic Fermentation*, London, 1911 (4th edn., 1932).
30 C. von Nägeli and O. Loewy, *Ann. Chem.*, 193 (1878) 322.
31 E. Fischer, *Ber. deut. Chem. Ges.*, 27 (1894) 3480.
32 R. Kohler, *J. Hist. Biol.*, 4 (1971) 35.
33 E. Buchner, H. Buchner and M. Hahn, *Die Zymase*, Munich, 1903.
34 C. Harris, *Ber. deut. Chem. Ges.*, 50 (1917) 1850.
35 M. Hahn, *Münch. med. Wochschr.*, 55 (1908) 515.
36 M. Gruber, *Münch. med. Wochschr.*, 55 (1908) 342.
37 H. Buchner, *Münch. med. Wochschr.*, 44 (1897) 299.
38 E. Buchner, *Ber. deut. Chem. Ges.*, 30 (1897) 117.
39 R. Kohler, *J. Hist. Biol.*, 5 (1972) 327.
40 E. Buchner, *Münch. med. Wochschr.*, 44 (1897) 322.
41 C. Voit, *Münch. med. Wochschr.*, 44 (1897) 321.

42 C. Kupfer, *Münch. med. Wochschr.*, 44 (1897) 322.
43 E. Buchner, *Ber. deut. Chem. Ges.*, 30 (1897) 1110.
44 E. Buchner and R. Rapp, *Ber. deut. Chem. Ges.*, 30 (1897) 2668.
45 J. R. Green, *Ann. Botany*, 11 (1897) 555.
46 A. Stavenhagen, *Ber. deut. chem. Ges.*, 30 (1897) 2422.
47 C. Wehmer, *Botan. Z.*, 65 (1898) 57.
48 E. Buchner, *Ber. deut. chem. Ges.*, 31 (1898) 568.
49 E. Duclaux, *Ann. Inst. Pasteur*, 11 (1897) 287.
50 E. Roux, Abstract in *Schweiz. Wochschr. Chem. Pharm.*, 37 (1899) 54.
51 R. Neumeister, *Ber. deut. chem. Ges.*, 30 (1897) 2963.
52 E. Buchner and R. Rapp, *Ber. deut. chem. Ges.*, 31 (1898) 209.
53 H. Abeles, *Ber. deut. chem. Ges.*, 31 (1898) 2261.
54 E. Buchner and R. Rapp, *Ber. deut. chem. Ges.*, 32 (1899) 127.
55 J. R. Green, *Nature*, 63 (1900) 106.
56 A. Wroblewski, *J. prakt. Chem.*, 64 (1901) 1.
57 H. Schriefers, in C. C. Gillispie (Ed.), *Dictionary of Scientific Biography*, Vol. II, p. 560, New York, 1970.
58 F. G. Hopkins, *Irish J. Med. Sci.*, 1932, p. 333.
59 A. von Lebedev, *Z. physiol. Chem.*, 73 (1911) 447.
60 A. von Lebedev, *Ann. Inst. Pasteur*, 26 (1912) 8.
61 R. Nilsson and F. Alm, *Z. Physiol. Chem.*, 239 (1936) 179.
62 R. Albert, *Ber. deut. chem. Ges.*, 33 (1900) 3775.
63 R. Albert, E. Buchner and R. Rapp, *Ber. deut. chem. Ges.*, 35 (1902) 2376.
64 S. Kostychev, G. Medvedev and H. Kardo-Systejewa, *Z. Physiol. Chem.*, 168 (1927) 244.
65 E. Wildiers, *La cellule*, 18 (1901) 311.
66 A. J. Kluyver and A. P. Struyk, *Z. physiol. Chem.*, 170 (1927) 110.

Chapter 18

Discovery of the "Coenzyme of Alcoholic Fermentation" and of the Esterification of Phosphate

1. Recognition of the by-product nature of certain substances found in fermenting mixtures

When yeast is added to a solution of sugar containing the proper nitrogen compounds and mineral salts, fermentation takes place and leads not only to a production of alcohol and carbon dioxide but also of substances such as fusel oil, succinic acid, glycerol, acetic acid, aldehyde, formic acid, esters and traces of many other non-nitrogenous compounds. These substances were considered by Pasteur[1] as transformation products of the sugar. Since the advent of experimentation with yeast juice or with zymin, some of these substances have been recognized as being derived from other metabolic processes of yeast cells and as originating, for instance, from amino acids.

(a) Fusel oil

This name, derived from the German word "Fusel", meaning rot-gut, is given to the high-boiling fraction, more concentrated in low quality spirits, obtained for instance from the fermentation of potatoes. It is a mixture of several substances among which isoamyl alcohol, $(CH_3)_2\cdot CH\cdot CH_2CH_2OH$ and D-amyl alcohol, $CH_3\cdot CH\cdot(C_2H_5)\cdot CH_2OH$ predominate. Smaller amounts of propyl alcohol and isobutyl alcohol are present, as well as traces of fatty acids, aldehydes, etc.

First thought to be derived from the sugar[2–7], these substances were considered by Felix Ehrlich, Professor at the University of Breslau, as products of what he called "the alcoholic fermentation of amino acids" and of the nitrogen metabolism of yeast cells (for literature see Harden[8]).

References p. 56

Plate 67. Louis Pasteur.

According to F. Ehrlich[9–11] the source of the D-amyl alcohol of fusel oil is D-isoleucine along to the following pathway:

$$\begin{matrix}H_3C\\H_5C_2\end{matrix}\!\!>\overset{*}{C}H\cdot CHNH_2\cdot COOH+H_2O=CO_2+NH_3+\begin{matrix}H_3C\\H_5C_2\end{matrix}\!\!>\overset{*}{C}H\cdot CH_2OH$$

Neubauer and Fromherz[12], in the course of the fermentation of phenylaminoacetic acid by yeast in the presence of sugar, isolated phenylglyoxylic acid and suggested that the alcohols derived from the amino acids during the fermentation process are produced by way of the formation of α-ketonic acids. Neuberg and Steenbock[13,14] obtained amyl alcohol from valeraldehyde in maceration juices of yeast.

It is now assumed that most of the fusel oils are by-products of the biosynthesis of amino acids. Some of the α-ketonic acids which are a stage of the biosynthesis of amino acids are decarboxylated instead of being aminated and converted into the alcohols.

(b) Glycerol

Pasteur[1] had demonstrated that, in alcoholic fermentation, glycerol and succinic acid arise. He considered them as products of sugar fermentation. Brefeld[15,16] and von Udranszky[17] opposed Pasteur's views and proposed that glycerol was not a product of alcoholic fermentation. Delbrück[18] suggested that fats and oils were the source of the glycerol produced in alcoholic fermentation, but Buchner and Meisenheimer[19] put this theory aside on a quantitative basis, as the amounts of fat present could not explain the amounts of glycerol produced.

The paper of Buchner and Meisenheimer[19] also refuted the view proposed tentatively by Carracido[20] and by F. Ehrlich[21] concerning the possible derivation of glycerol from proteins. Buchner and Meisenheimer[19], in contradiction to a previous paper of Buchner and Rapp[22], found that glycerol was produced in fermentation by yeast juice as well as in fermentation by living yeast. Buchner and Meisenheimer[19] considered that glycerol derives from sugar "nicht als direktes Nebenprodukt der Zuckerspaltung in Alkohol und Kohlensäure, sondern durch einen gesonderten Vorgang".

Buchner and Meisenheimer[23] shared the opinion that glycerol was a product of the reduction of the first triose sugars (dihydroxyacetone and glyceraldehyde) resulting from the scission of the six-carbon glucose chain.

References p. 56

Plate 68. Arthur Harden.

An experimental study of the question was accomplished by Embden, Schmitz and Baldes[24] who reached the conclusion that in minced organs and in artificially perfused liver, glyceraldehyde was transformed into glycerol. The possibility of reduction of glyceraldehyde to glycerol appeared in the light of the previous work of Parnas[25].

On the other hand, Neuberg and Kerb[26] proposed the view according to which methylglyoxal, an intermediate in Neuberg's scheme of fermentation, undergoes a Cannizzaro reaction with production of pyruvic acid and glycerol.

Oppenheimer[27] (in Embden's laboratory) confirmed the production of glycerol from glucose in yeast juice and showed that in maceration juice of yeast, both dihydroxyacetone and glyceraldehyde are actively transformed into glycerol. He therefore considers that glycerol is derived from dihydroxyacetone in a side reaction of glucose fermentation.

(c) Succinic acid

This is also a product of the metabolism of amino acids, as demonstrated by F. Ehrlich[21]. The pathway of this metabolism has been the subject of a number of studies (for literature, see Harden[8]).

2. Discovery of the "coenzyme of alcoholic fermentation"

We have noted, while dealing with the discovery of cell-free fermentation, how much immunochemistry was determinant in the period of the individualization of the new science of biochemistry.

Another example is afforded by the discovery of the "coenzyme of alcoholic fermentation" by Harden.

In one of his penetrating studies on the history of alcoholic fermentation, Kohler[28] has narrated how Macfadyen, whose objections to the enzymatic nature of zymase have been reported above, and Harden became associated at the Lister Institute in London. In 1897, Macfadyen became Director of the Institute in which, during the same year, Harden was appointed as chemist. Harden had been trained as an organic chemist and, as an assistant of H. E. Dixon in Manchester, had been more involved in teaching and textbook-writing than in research.

Harden, who was in charge of routine analysis of water and of foodstuffs at the Lister Institute, did not leave Manchester with the purpose of finding

References p. 56

new ways of research, but his association with the Lister Institute brought these ways, and their rewards, to him.

After the discovery of antibodies, Macfadyen, who was five years Harden's senior, had been interested in the intracellular enzymes and toxins of bacteria.

As many others at the time, Macfadyen was impressed by the similarities between enzymes and toxins: both were heat-sensitive and both were specific functional proteins. He started looking for enzymes inside the bacterial protoplasm and in fact he obtained by the current procedure of glycerol extraction, preparations with proteolytic activity, from cholera and other bacilli. He thought[29] that

"it seemed probable that the study of the formation of these enzymes would also throw further light on the mode of production of the toxalbumins."

He found that the extracts of cholera bacilli were toxic to animals[28]:

"the soluble cell proteins which possess these toxic properties are closely related in constitution to the ferments and may be identical with them, or a modification of them."

In 1898, at the suggestion of Macfadyen, Harden started a study of the chemical metabolic products of *B. coli* as a possible way of diagnosing varieties. Harden accomplished a thorough chemical study in which he identified several new products and proposed a theory of glucose metabolism which involved the simultaneous cleavage of two glucose molecules by a coupled decomposition[30]. Macfadyen himself turned to alcoholic fermentation and accomplished the work we referred to above, concluding that zymase was a piece of protoplasm and that sugar molecules were first incorporated in the large molecules of living protoplasm, which then broke down into alcohol and carbon dioxide. But his interest in endotoxins was revived and he left the work on zymase to Harden. In 1901 Harden and Rowland[31], who had been Macfadyen's assistant, published a study of the autoliquefaction of pressed yeast. They showed that this spontaneous digestion was a true alcoholic fermentation analogous to Macfadyen's auto-fermentation of yeast juice. They also showed that it was not the yeast protoplasm which was fermented, but glycogen. Harden studied yeast glycogen with Young, who had just joined the Lister Institute[32].

As he wrote later, Harden[8] inherited from Macfadyen another subject, the study of the effect of blood on zymase:

"In the course of some preliminary experiments (commenced by the late Allen Macfadyen but subsequently abandoned) on the production of antiferments by the injection of yeast juice into animals, the serum of the treated animals was tested for the presence of such antibodies both for the alcoholic and proteoclastic enzymes of yeast juice, and it was then observed that the serum of normal and treated animals alike greatly diminished the autolysis of yeast juice."

The main argument against the enzymatic nature of zymase, derived from Macfadyen's experiments with yeast juice, was the "disappearing glucose". Harden and Young[33] confirmed the loss of reducing glucose but showed that it could be accounted for by a non-reducing compound from which glucose could be liberated by treatment with boiling acids.

Harden then tackled the question of the inhibition of yeast protease by blood serum.

The fact that the fermentative power of yeast juice fades out on standing was considered as being due to the action of proteolytic enzymes. Harden observed that, in the presence of serum, 60 to 80 per cent more sugar was fermented[34]. As this was attributed to an antitryptic action, other methods were tried to obtain the same effect. The idea of accomplishing it by an inhibition of the proteolytic enzyme by the products of its activity led to the use of autolyzed juices, previously boiled and filtered. In fact, Harden and Young observed that this solution had the effect of increasing the fermentation[35]. But boiled fresh juice had the same effect, as had already been observed by Buchner and Rapp[36].

Amino acid inhibition of protease was thus not the explanation of the increase of fermentation by yeast juice deprived of zymase activity, and Harden concluded that a specific heat-stable factor present in yeast juice promoted this activity[8].

In the Proceedings of the meeting of the Physiological Society, November 12, 1904, Harden and Young[35] reported that the coenzyme was heat-stable, dialyzable, precipitable by alcohol, and absolutely necessary for the action of zymase.

The concept of coferment was not new: several examples had been discovered since Bertrand formulated the concept in 1897 in the case of laccase (see Chapter 26).

But the case of the coenzyme of alcoholic fermentation was unusual. When they published their first detailed paper on the coenzyme of alcoholic fermentation in 1906, Harden and Young[37] showed that part of the stimulating effect of boiled yeast juice was due to phosphate. As we recalled above, the fact that phosphates stimulate

References p. 56

fermentation by yeast juice had already been described by Wroblewski and by E. Buchner, H. Buchner and Hahn[38].

Harden and Young observed that the effect is lost if the boiled and filtered juice is dialyzed in a parchment tube. But when Harden had separated by filtration the coenzyme from the residue left on the filter, he could not restore the activity of the residue by an addition of phosphate[37], though he could restore this activity by adding the filtrate.

In performing these justly famous experiments, Harden and Young were lucky to be able to use, at the Lister Institute, a Martin gelatin filter and it may be noted that it was C. J. Martin who communicated their paper to the Royal Society.

To quote Harden and Young[37]:

> "This method of rapid dialysis was chosen because the yeast-juices at our disposal lost their activity too rapidly to permit of the ordinary process of dialysis through parchment being carried out. Either a 10 or a 7.5 per cent solution of gelatin was used to impregnate the Chamberland filter and the filtration was carried out under a pressure of 50 atmospheres.
>
> Only a portion of the juice placed in the filter was actually filtered, the remainder being simply poured out of the case as soon as a sufficient quantity of filtrate had passed through. The residue adhering to the candle, which consisted of a brown viscid mass, was dissolved in water and made up to the volume of the juice filtered. Glucose was then added and one portion incubated at 25° with an equal volume of sugar solution and a second portion with an equal volume of the filtrate or of boiled juice, containing an equal amount of glucose. Before incubating the carbon dioxide was pumped out of all the solutions.
>
> The filtrate was invariably found to be devoid of fermenting power, none of the enzyme having passed through the gelatin. The results show that in this way an almost inactive residue can be obtained which is rendered active by the addition of the filtrate or a boiled juice."

Harden was too careful an experimenter to decide on a single negative result in a complex experiment. Harden and Young performed a variety of experiments in which they could never restore the activity by phosphate without also adding boiled yeast juice. They therefore concluded that a second cofactor was required, besides phosphate[39].

The final proof was obtained when Harden devised a method for a continuous study of the kinetics of the fermentation, by a volumetric measurement of the CO_2 released. The addition of phosphate or of boiled juice caused a sharp initial rise in rate and also a prolongation of the normal fermentation. It appears that the fact that boiled juice generally gave a longer-lasting effect than phosphate convinced Harden that yeast juice, besides phosphate, contained a second cofactor. One of the most striking results was that the initial burst of CO_2 was strictly proportional to the amount[37] whether phosphate or boiled yeast juice was added. Harden and Young had announced their discovery of the "coenzyme of alcoholic fermentation"⋆ in two preliminary notes[40,41]. Buchner and Antoni[42] repeated their experiments and confirmed them, but the lack of details in the preliminary publications led them to believe that in the experiments of Harden and Young the concentration of glucose and of enzyme had not been kept constant and they partly attributed the effect produced by boiled juice to a diminution of concentration of glucose and of alcohol, which are always present. The details given in the full publication[37] show that these influences have no share in the interpretation. It is

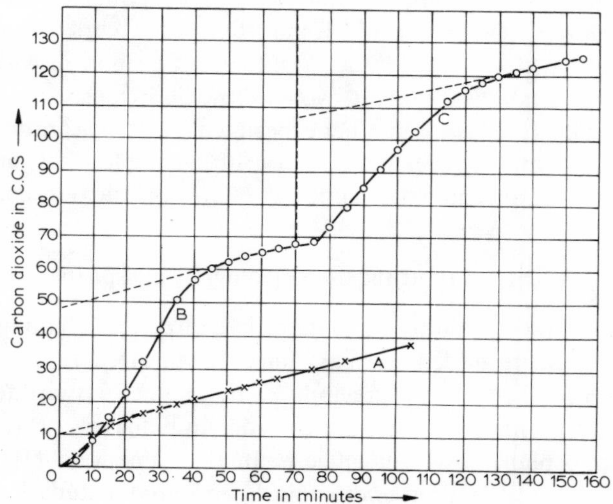

Fig. 5. Harden and Young[37]. Explanation in text.

⋆ This "coenzyme" was to be recognized later (see Chapter 23) as a system composed of ATP, NAD(P), cocarboxylase, and Mg^{2+} as cofactor.

References p. 56

worth recalling here one of the masterly experiments of Harden and Young retraced in the diagram of Fig. 5. In this experiment, 25 ml of an aqueous solution of glucose are added (curve A) to 25 ml of yeast juice. To a second 25 ml quantity of yeast juice are added 5 ml of a 0.3 molar solution of the mixed primary and secondary sodium phosphates, and 20 ml of a solution containing 5 g of glucose (curve B). Curve A shows the normal course of fermentation of glucose by yeast juice. During the first twenty minutes, due to the phosphate being present, there is a slight acceleration followed by a steady rate of about 1.4 ml in five minutes. Extrapolating the line of steady state to zero time, the authors find that 10 ml of extra carbon dioxide are evolved. The effect of added phosphate is shown by curve B. The rate first rises and then gradually falls until after an hour it remains steady at about 1.4 ml per five minutes. Extrapolating the steady state line to the ordinate axis we find that the extra amount of carbon dioxide is 48 ml. Subtracting the 10 ml due to the juice alone (curve A) we find that an addition of 38 ml results from the addition of phosphate. The amount calculated from the phosphate added corresponds to 38.9 ml. After seventy minutes another equal amount of phosphate is added. As curve C shows, the whole phenomenon recurs, the rate again becoming steady at 1.4 ml. The extra amount of carbon dioxide evolved calculated graphically is 39 ml (107 − 68) shown previously.

The conclusion (confirmed in other experiments with maceration extracts, dried yeast or zymin) is that the addition of a soluble phosphate causes the production of an equivalent amount of carbon dioxide and alcohol.

3. Discovery of the esterification of phosphate

Another important observation was made by Harden and Young[37,41]: the inorganic phosphate added to yeast juice disappeared rapidly and was replaced by a form no longer precipitable by magnesium citrate mixture; as indicated by preliminary tests they considered that it was probably in the form of a phosphoric ester of glucose. In the same year (1905) as the preliminary paper of Harden and Young[41] appeared a study by Ivanov[43] who showed that living yeast converted inorganic phosphate into organic derivatives. During the following year (1906) Ivanov[44] also found, as Harden and Young had observed with yeast juice, that inorganic phosphate added to zymin became unprecipitable by uranium acetate. He was the first to isolate the compound in the form of a copper salt and he correctly determined his empirical formula as $C_3H_5O_2PO_4H_2$, which he erroneously identified as a triosephosphate.

That the ester isolated from fermenting solutions prepared with yeast juice was a hexosephosphate, was recognized, in 1907, by Young[45], who isolated it in the form of its copper salt.

In 1909, Ivanov[46], following on his erroneous interpretation of the nature of the accumulating phosphoric esters, proposed a theory of fermentation. His conclusion was that triosephosphates take part in the pathway of alcoholic fermentation which he conceives as consisting of three phases:

1. a depolymerisation of hexose;

2. a combination of the products with phosphate in the presence of a hypothetical easily soluble enzyme called by him "synthetase";

3. a breakdown of the product in the presence of another enzyme (alcoholase, not easily soluble) with a production of CO_2 and ethanol.

Ivanov[47] found a stimulating effect of phosphate on the fermentation by living yeast as well as by germinated wheat and peas. The same effect was observed by N. N. Ivanov[48] on killed-etiolated terminal stalks of *Vicia faba*.

A confirmation of the introduction of added phosphate into an organic derivative not precipitable by magnesia was afforded by von Euler and Johansson[49] in experiments with dried yeast. They showed that after the early stages, the amount of phosphate esterified was equivalent to the carbon dioxide evolved. This was confirmed by Boyland[50] in 1929.

That the Harden and Young ester is of the nature of a hexosediphosphate was shown by Young[51,52,52a].

4. Hexosephosphates

As said above, the Harden and Young ester was identified by Young as a hexosediphosphate and not as a triosephosphate as had been claimed by Ivanov. But the nature of the hexosediphosphate, fructose-1,6-diphosphate, was established by Levene and Raymond[52b] in 1928 only.

In 1914, Harden and Robison[53] detected a hexosemonophosphate in fermenting mixtures, but it was only in 1931 that Robison and King[54] isolated this hexosemonophosphate and recognized it as glucose-6-phosphate (Robison ester, G-6-P).

In 1918, Neuberg[55] prepared a new hexosemonophosphate (fructose-6-

References p. 56

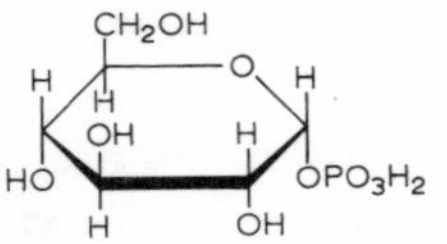

α-D-Glucose-1-phosphoric acid (Cori ester)

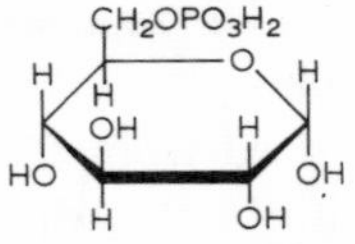

α-D-Glucose-6-phosphoric acid (Robison ester)

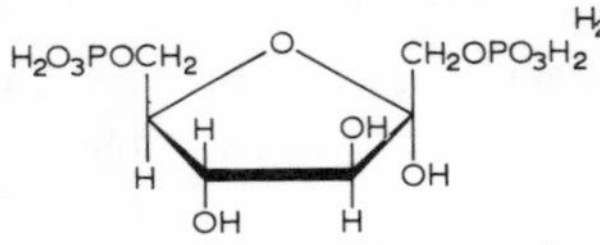

α-D-Fructose-1,6-diphosphoric acid (Harden and Young ester)

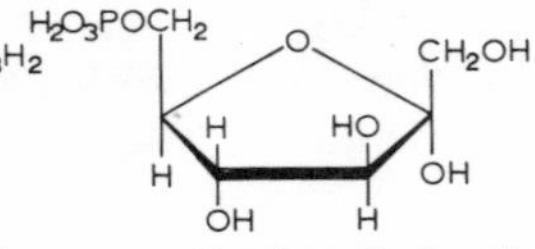

α-D-Fructose-6-phosphoric acid (Neuberg ester)

phosphate, F-6-P, Neuberg ester) by partial hydrolysis of fructose diphosphate. That this ester is also present in the hexosemonophosphate fraction of yeast juice fermentation was shown by Robison[56] in 1932.

In 1927, Embden obtained from muscle a hexosemonophosphate fraction which, at the time, was called 'Embden ester', but which was later found (by Lohmann[56b]) to be a mixture of 2/3 G-6-P and 1/3 F-6-P.

Glucose-1-phosphate, the Cori ester, was isolated in 1937 by Cori, Colowick and Cori[56c] as a result of the action of frog muscle enzymes on glycogen and inorganic phosphate. The synthesis of this ester was accomplished by the authors. The ester is alkali-stable and non-reducing, and liberates glucose on acid hydrolysis.

The esters of Robison, of Neuberg and of Harden and Young have been recognized, as we shall see, as intermediates on the path from glucose to pyruvic acid. G-1-P (Cori ester) has also been recognized as an intermediate, on the path from glycogen to G-6-P (Robison ester).

One of the factors, besides the lack of chemical knowledge, which delayed for such a long time the recognition of phosphoric acid esters as intermediates in glycolysis was the fact that (with the exception of G-6-P) they are metabolized as rapidly as produced and consequently are not found when living cells are analyzed. Another aspect was, as will be shown, the concept according to which an intermediate in the pathway of glycolysis should not be metabolized more slowly than glucose is in tissues, a concept which did ignore the alterations of composition resulting from the methods of obtention of the different preparations from yeast or from muscle, the two main materials used in studies on glycolysis.

5. Interpretation of the Harden and Young reaction

This name has become attached to the experimental conditions in which, in the case of yeast juice as shown in Fig. 5, a break occurs in the curve between a rapid phase and a slow phase when either glucose or phosphate is exhausted and there is an accumulation of hexose esters (later shown as mostly hexosediphosphate). This experiment, as was shown later, corresponds to an artifact, the destruction of ATPase in the course of yeast juice preparation (see Chapter 22). In the stationary state, living yeast ferments free hexoses much more slowly than free trioses, phosphoric esters not at all. If the yeast is dried, a metabolic alteration sets in and at a certain point the speed of fermentation slows down, inorganic phosphate disappears and hexosephosphates accumulate.

Harden's[8] interpretation is given in his book on *Alcoholic fermentation*[8]. He concludes from experiments with yeast juice such as that shown in Fig. 5 that the inorganic phosphate is used up in the reaction.

This he expresses by the equation:

$$2\,C_6H_{12}O_6 + 2\,PO_4HR_2 = 2\,CO_2 + C_2H_5OH + 2\,H_2O + C_6H_{10}O_4(PO_4R_2)_2 \qquad (1)$$

a balance sheet which has been confirmed repeatedly.

According to equation (1) two molecules of sugar are involved, one being split and the other phosphorylated. The carbon dioxide and the alcohol are equivalent in weight to one of the sugar molecules used, and the hexosephosphate and water represent the other half.

If a fermenting mixture of yeast juice, sugar and phosphate is examined at different intervals and the amounts of inorganic phosphate are estimated, it is found that during the first period of accelerated fermentation following the addition of the inorganic phosphate, the latter rapidly diminishes and at the close of this initial period, the inorganic phosphate concentration becomes very low. As the speed slackens, a marked and rapid rise of free phosphate occurs at the expense of the hexosephosphate, which diminishes. The following equation represents this aspect

$$C_6H_{10}O_4(PO_4R_2)_2 + 2\,H_2O = C_6H_{12}O_6 + 2\,PO_4HR_2 \qquad (2)$$

According to Harden, the diminution of inorganic phosphate during the rapid phase corresponds to equation (1). During the whole of the two phases the hydrolysis of the hexosephosphate by a phosphatase proceeds according to equation (2). The phosphate thus produced enters into equation

(1) up to the end of stage 2 and reacts with the sugar which is present in excess and is reconverted into hexosephosphate.

As long as the fermentation proceeds, no mineral phosphate can accumulate but, as soon as the fermentation ceases when the sugar is exhausted, it is impossible for the phosphate to pass back into hexosephosphate and it accumulates. As soon as all the free phosphate is exhausted, the supply of phosphate depends on the rate of its liberation from hexosephosphate by an enzyme considered by Harden to be a hexosephosphatase. Two entirely different reactions therefore rule the rate of the process during the two stages. This explains why it is the extra carbon dioxide evolved during the initial rapid stage which is equivalent to the added phosphate.

The theory according to which equations (1) and (2) represent the nature of the action of phosphate was formulated by Harden and Young[57] in 1908. In this theory, fermentation and phosphorylation appear as two separate, coupled reactions, the formation of the hexosephosphate providing, in the phosphorylation of a molecule of glucose, the driving force for the decomposition without phosphorylation of another molecule of glucose to CO_2 and alcohol. The remaining phosphorylated molecule is hydrolyzed, liberating a molecule of glucose and two molecules of phosphate which is recycled and accomplishes at catalytic function. This mechanism was akin to the one proposed in Harden's paper of 1901 on *B. coli*[30]. When it was realized that in the reaction of Harden and Young, hexosediphosphate was accompanied by a smaller amount of hexosemonophosphate, Harden[58] modified the equations expressing his theory as follows:

$$\text{(a)} \quad 3\,C_6H_{12}O_6 + 2\,Na_2HPO_4 = 2\,CO_2 + 2\,C_2H_5OH + 2\,C_6H_{11}O_5(Na_2PO_4) + 2\,H_2O$$

$$\text{(b)} \quad 2\,C_6H_{12}O_6 + 2\,Na_2HPO_4 = 2\,CO_2 + 2\,C_2H_5OH + C_6H_{11}O_5(Na_2PO_4)_2 + 2\,H_2O$$

$$\text{(a+b)} \quad 5\,C_6H_{12}O_6 + 4\,Na_2HPO_4 = 4\,CO_2 + 4\,C_2H_5OH + 2\,C_6H_{11}O_5(Na_2PO_4) + C_6H_{10}O_4(Na_2PO_4)_2 + 4\,H_2O \quad (3)$$

According to Harden's theory, the molecule which provides the CO_2 and the alcohol is not phosphorylated and the phosphoric esters are not to be considered as intermediate products in the pathway but as taking place in an initial, preparatory reaction making glucose able to enter the pathway in the form of free glucose.

The theory proposed by its discoverers in 1908 to explain Harden and Young's reaction was a logical product of imagination derived from the discovery of the proportionality of the production of one mole of CO_2 and of the disappearance of one molecule of inorganic phosphate. They believed this phosphate was combined with glucose.

In the present state of our knowledge, the interpretation is quite different. We no longer consider the reason for the chemical balance of one phosphate used and one CO_2 produced as the consequence of a fixation of inorganic phosphate by glucose, as we know that the molecules of phosphate associated with glucose in the hexosephosphates are derived from ATP. But, on the other hand, we know where inorganic phosphate enters the metabolic pathway (see Fig. 7). It is at the level of the phosphorylation of glyceraldehyde-3-P into 1,3-diphosphoglyceric acid, followed by a transphosphorylation to ADP with a formation of 3-phosphoglyceric acid and ATP. This gives the explanation of the ratio of phosphate used and CO_2 produced, as observed by Harden and Young, as one mole of phosphate for each mole of triose, from which one mole of CO_2 is derived: one more example of the possibility of the deceiving nature of well-planned, accurately performed and logically analyzed experiments.

In the conditions of Harden and Young's reaction performed with yeast juice, two more or less distinct phases are recognized and are the basis of the theory which was generally accepted until the thirties.

For Nilsson[59] the break between the rapid and the slow phase confirms his theory of the decomposition of the hexose into two halves, one phosphorylated and one non-phosphorylated, fermented with equal speed in living yeast. As the result of the alteration of the conditions due to the extraction of yeast juice, only the un-phosphorylated half ferments while the other half "abnormally" condenses to hexosediphosphate. This is based on a theory which, while recognizing a phosphoric ester as an intermediate, still refused this role to hexosediphosphate.

In the frame of the accepted theory, Meyerhof[60], during the last period of his life, when he was working in Philadelphia, brought more arguments in favour of an explanation he had suggested a number of years before[61] and according to which the slow phase is the result of the lack of the enzyme ATPase, damaged in the process of the preparation of yeast juice or of the drying of yeast, or of its maceration. The fermentation rate of hexosediphosphate during the slow phase, which is about 1/10th that of glucose is raised to the level of free glucose fermentation if ATPase, prepared for example from potatoes, is added.

References p. 56

Meyerhof[62] has afforded further arguments in favour of his interpretation when he showed that fermentation by quickly dried brewer's yeast proceeds without any break in the curve until as much CO_2 is produced as corresponds to the complete fermentation of the sugar. If ATPase inhibitors (toluene, octyl alcohol, ph nylurethane, etc.) are added, the break occurs if either free phosphate or free sugar is exhausted, and hexose esters, mostly hexosediphosphate, accumulate. It is therefore generally admitted on the basis of these experiments of Meyerhof that a removal, destruction or inhibition of ATPase is responsible for the Harden–Young effect in a variety of fermenting preparations derived from yeast. When phosphate, for example, is exhausted, the ATP formed as described above can be split if ATPase is present and the reaction kept going at normal speed.

Though they reached different conclusions, there is a close relation between the experiments of Nilsson on "intake Trockenhefe" and those of Meyerhof. Some questions remain unanswered and Nilsson[59] points out some of them. He stresses the fact that preformed HDP is only slowly fermented by "intakte Trockenhefe" in comparison with glucose. It may be noted that in the same paper, Nilsson discusses the formation of HDP from galactose. Nilsson[63] had previously shown that, in the case of a galactose-adapted yeast, the HDP formed from galactose is the same as in the case of glucose, fructose and mannose. But the hexosemonophosphate is different in each case.

During the second phase of glucose fermentation with intact dried yeast, Nilsson observed the formation of trehalose. This, at the time, appeared a curiosity. It has since been shown that in the frame of glucidic metabolism of insects, trehalose plays an essential role.

Nilsson[64,65], after the publication of the views of Meyerhof, recalled a number of facts which could not be explained on the basis of Meyerhof's theory, and which should be reconsidered.

In the course of the long controversy about the mechanism of glycolysis, Harden never recognized phosphorylated compounds as intermediates in the pathway.

When he received the Nobel Prize in 1929, at a time when the nature of the hexosephosphates as intermediates was widely recognized, Harden[66] still clung to his opinion according to which the phosphate esterification of one sugar molecule by inorganic phosphate induced the decomposition of another molecule which is not phosphorylated:

"It appears to me that the fundamental idea expressed in the original equation of Harden and Young is nearer to the truth than any alternative that has yet been suggested. A coupled reaction of some kind occurs, as the result of which the introduction of two phosphate groups into certain sugar molecules induces the decomposition of another molecule".

There is a striking parallelism between the careers of Buchner and of Harden[28]. Neither man would have been expected *a priori* to have achieved fame*. Both received the seed of this fame from an immunologist, Eduard Buchner from Hans Buchner and Harden from Macfadyen.

Hopkins and Martin[67], in their obituary notice on Harden, write

"Harden's outstanding qualities as an investigator were clarity of mind, precision of observation, and a capacity to analyse dispassionately the results of an experiment and define their significance. He mistrusted the use of his imagination beyond a few paces in advance of the facts. Had he exercised less restraint, he might have gone further; as it was he had little to withdraw."

Not only did the studies on yeast juice stimulate the development of interest in biochemical studies but the classical experiments of Harden and Young, even if performed under abnormal conditions and wrongly interpreted was of a great seminal value as it gave way to the method of progressive simplification which under the impact of Warburg did tend to reach the level of the purified enzyme. The reaction of Harden and Young was the first step along this methodological path.

On the other hand, Harden's views, popularized by his widely-read book[8], remained for many years an obstacle to a correct interpretation of the meaning of phosphorylation.

As rightly stated by Lipmann[68]:

"for a long time after its discovery by Harden and Young, phosphorylation of hexose to alcoholic fermentation was thought to be significant only as a means of modeling the hexose molecule to fit for fermentation breakdown."

---

* Nevertheless Buchner has contributed in the field of preparative organic chemistry, a valuable work on the synthesis in the cycloheptane series.

REFERENCES

1 L. Pasteur, *Ann. Chim. Phys.*, 58 (1960) 323.
2 O. Emmerling, *Ber. deut. chem. Ges.*, 37 (1904) 3535.
3 O. Emmerling, *Ber. deut. chem. Ges.*, 38 (1905) 953.
4 H. H. Pringsheim, *Ber. deut. chem. Ges.*, 38 (1905) 486.
5 H. H. Pringsheim, *Biochem. Z.*, 3 (1907) 121.
6 H. H. Pringsheim, *Biochem. Z.*, 10 (1908) 490.
7 H. H. Pringsheim, *Biochem. Z.*, 16 (1909) 243.
8 A. Harden, *Alcoholic Fermentation*, London, 1911 (4th ed., 1932).
9 F. Ehrlich, *Biochem. Z.*, 2 (1907) 52.
10 F. Ehrlich, *Ber. deut. chem. Ges.*, 40 (1907) 1030.
11 F. Ehrlich, *Ber. deut. chem. Ges.*, 40 (1907) 2551.
12 O. Neubauer and K. Fromherz, *Z. Physiol. Chem.*, 70 (1911) 326.
13 C. Neuberg and H. Steenbock, *Biochem. Z.*, 52 (1913) 494.
14 C. Neuberg and H. Steenbock, *Biochem. Z.*, 59 (1914) 188.
15 O. Brefeld, *Landwirthsch. Jahrb.*, 3 (1874) (after von Udranszky[17]).
16 O. Brefeld, *Landwirthsch. Jahrb.*, 4 (1875) (after von Udranszky[17]).
17 L. von Udranszky, *Z. physiol. Chem.*, 13 (1889) 539.
18 M. Delbrück, *Wochschr. Brau.*, 20 (1903) 66.
19 E. Buchner and J. Meisenheimer, *Ber. deut. chem. Ges.*, 39 (1906) 3201.
20 R. Carracido, *Ref. Biochem. Zentralbl.*, 3 (1904–5) 439.
21 F. Ehrlich, *Biochem. Z.*, 18 (1909) 391.
22 E. Buchner and R. Rapp, *Ber. deut. chem. Ges.*, 34 (1901) 1521.
23 E. Buchner and J. Meisenheimer, *Ber. deut. chem. Ges.*, 43 (1910) 1773.
24 G. Embden, E. Schmitz and K. Baldes, *Biochem. Z.*, 45 (1912) 174.
25 J. Parnas, *Biochem. Z.*, 28 (1910) 274.
26 C. Neuberg and J. Kerb, *Biochem. Z.*, 58 (1913) 158.
27 M. Oppenheimer, *Z. physiol. Chem.*, 89 (1914) 63.
28 R. Kohler, *Bull. Hist. Med.*, 48 (1974) 22.
29 A. Macfadyen, *J. Anat. Physiol.*, 26 (1892) 409.
30 A. Harden, *J. Chem. Soc.*, 79 (1901) 610.
31 A. Harden and S. Rowland, *J. Chem. Soc.*, 79 (1901) 1227.
32 A. Harden and W. J. Young, *J. Chem. Soc.*, 81 (1902) 1224.
33 A. Harden and W. J. Young, *Ber. deutl chem. Ges.*, 37 (1904) 1052.
34 A. Harden, *Ber. deut. chem. Ges.*, 36 (1903) 715.
35 A. Harden and W. J. Young, *J. Physiol. (London)*, 32 (1905) 1.
36 H. Buchner and R. Rapp, *Ber. deut. chem. Ges.*, 32 (1899) 2086.
37 A. Harden and W. J. Young, *Proc. Roy. Soc. (London), Ser. B*, 77 (1906) 405 (this paper is reprinted in Kalckar[69]).
38 E. Buchner, H. Buchner and M. Hahn, *Die Zymasegärung*, Munich, 1903.
39 A. Harden and W. J. Young, *Proc. Soc. (London), Ser. B*, 78 (1906) 369.
40 A. Harden and W. J. Young, *J. Physiol., (London)*, 32 (1904), Proc. Nov. 12.
41 A. Harden and W. J. Young, *Proc. Chem. Soc.*, 21 (1905) 189.
42 E. Buchner and W. Antoni, *Z. physiol. Chem.*, 46 (1905) 136.
43 L. Ivanov, *Trav. Soc. Natur. St. Petersbourg*, 34 (1905).
44 L. Ivanov, *Z. physiol. Chem.*, 50 (1906) 281.
45 W. J. Young, *Proc. Chem. Soc.*, 23 (1907) 65.
46 L. Ivanov, *Centr. Bakt.*, 24 (1909) 1.
47 L. Ivanov, *Biochem. Z.*, 25 (1910) 171.

48 N. N. Ivanov (cited after L. Ivanov[44]).
49 H. von Euler and D. Johansson, *Z. physiol. Chem.*, 85 (1913) 192.
50 E. Boyland, *Biochem. Z.*, 23 (1929) 219.
51 A. Harden and W. J. Young, *Proc. Roy. Soc. (London), Ser. B*, 80 (1908) 299.
52 W. J. Young, *Proc. Roy. Soc. (London), Ser. B*, 81 (1909) 528.
52a W. J. Young, *Biochem. Z.*, 32 (1911) 178.
52b P. A. Levene and A. L. Raymond, *J. Biol. Chem.*, 80 (1928) 323.
53 A. Harden and R. Robison, *Proc. Chem. Soc. (London), Ser. B*, 30 (1914) 16.
54 R. Robison and E. J. King, *Biochem. J.*, 25 (1931) 323.
55 C. Neuberg, *Biochem. Z.*, 88 (1918) 432.
56 R. Robison, *Biochem. J.*, 26 (1932) 2191.
56a G. Embden and M. Zimmermann, *Z. Physiol. Chem.*, 167 (1927) 114.
56b K. Lohmann, *Biochem. Z.*, 262 (1933) 137.
56c C. F. Cori, S. P. Colowick and G. T. Cori, *J. Biol. Chem.*, 21 (1937) 465.
57 A. Harden and W. J. Young, *Proc. Roy. Soc. (London), Ser. B*, 80 (1908) 299.
58 A. Harden, *Erg. Enzymforsch.*, 1 (1932) 113.
59 R. Nilsson, *Naturwissenschaften*, 31 (1943) 25.
60 O. Meyerhof, *J. Biol. Chem.*, 157 (1945) 105.
61 O. Meyerhof, *Erg. Physiol.*, 39 (1937) 10.
62 O. Meyerhof, *J. Biol. Chem.*, 180 (1949) 575.
63 R. Nilsson, *Svenska Vetesk. Ark. Kemi*, 10A (1930) 132.
64 R. Nilsson and F. Alm, *Acta Chem. Scand.*, 3 (1949) 213.
65 R. Nilsson, *Schweiz. Z. Allg. Path. Bakteriol.*, 13 (1950) 672.
66 A. Harden, *Nobel Lectures, Chemistry 1922–1941*, Amsterdam, 1966, p. 131.
67 F. G. Hopkins and C. J. Martin, *Obit. Not. F.R.S.*, 4 (1942) 3.
68 F. Lipmann, *Advan. Enzymol*, 1 (1941) 99.
69 H. M. Kalckar, *Biological Phosphorylations. Development of Concepts*, Englewood Cliffs, N.J., 1969.

Chapter 19

The Search, Before 1930, for Metabolic Intermediates Between Glucose and Ethanol, in Alcoholic Fermentation

During the period considered, the study of the pathway of alcoholic fermentation consisted mainly of the search for the successive intermediates in the catabolism. Those who pursued this quest for intermediates generally considered phosphorylation either as an accessory preliminary phenomenon, or even as an artifact.

Buchner and his collaborators were among the active researchers in the field, as well as a number of Russian botanists and microbiologists such as Lebedev and Kostychev. A first concept of a metabolic pathway involving pyruvate and acetaldehyde as intermediates spread and came to be generally recognized. To this theory, Neuberg who had been a collaborator of Hofmeister in Strassburg, and who had recognized pyruvate and acetaldehyde as intermediates, accounted for the formation of pyruvate by considering methylglyoxal as an intermediary product between glucose and pyruvate.

1. Chemical schemes of alcoholic fermentation derived from non-biochemical experiments

An experiment accomplished *in vitro* by Buchner and Meisenheimer[1] in 1905 showed the production of small quantities of alcohol by the action on cane sugar of boiling concentrated caustic soda solutions. In the same experiment lactic acid was also produced in larger proportions.

As far back as in 1886, Duclaux[2] exposed to sunlight a mixture of glucose and caustic potash and obtained alcohol and carbon dioxide. When he used weaker alkalis instead of caustic potash, no alcohol was

References p. 77

formed but 50 per cent of the glucose was transformed into inactive lactic acid[3,4]. Duclaux concluded from these experiments that alcohol and carbon dioxide were secondary products of the action of concentrated alkali on lactic acid.

The mechanism of the formation of lactic acid during the action of alkalis on sugar was the subject of other experiments, the results of which pointed to the presence of an aldehyde or ketone containing three carbon atoms among the intermediate products.

In the presence of phenylhydrazine, the action of alkali on glucose[5–7] produces the osazone of methylglyoxal, $CH_3 \cdot CO \cdot CHO$. But this osazone may also be formed from acetol $CH_3 \cdot CO \cdot CH_2 \cdot OH$ or from lactic aldehyde[8], $CH_3 \cdot CH(OH) \cdot CHO$. Furthermore, methylglyoxal may be derived, by a process of molecular dehydration, from glyceraldehyde $CH_2(OH) \cdot CH(OH) \cdot CHO$ or from dihydroxyacetone $CH_2(OH) \cdot CO \cdot CH_2(OH)$, and on the other hand methylglyoxal, in the presence of alkalis, goes into lactic acid by the addition of a molecule of water:

$$CH_3 \cdot CO \cdot CHO + H_2O = CH_3 \cdot CH(OH) \cdot COOH$$

The theory of the action of alkalis on sugar, as it was accepted in the beginning of the present century, considered that the first product is glyceraldehyde, which passes into methylglyoxal and finally into lactic acid, these changes taking place at ordinary temperature in the presence of a catalyst:

$$C_6H_{12}O_6 \rightarrow 2\,CH_2(OH) \cdot CH(OH) \cdot CHO$$
$$CH_2(OH) \cdot CH(OH) \cdot CHO \rightarrow CH_3CO \cdot CHO + H_2O$$
$$CH_3 \cdot CO \cdot CHO + H_2O \rightarrow CH_3 \cdot CH(OH) \cdot COOH$$

In 1870, Baeyer[9] had proposed a scheme of successive removals and additions of water. In this scheme, the series of intermediate stages derives from the hydrated aldehyde formula of glucose by additions and removals of water. Other schemes have been proposed (for literature, see Harden[10]). From these studies *in vitro*, Wohl[11] derived a chemical scheme of the succession of steps in the process of alcoholic fermentation:

I. Glucose		II. Enol		III. Ketone
CHO		CHO		CHO
|		|		|
CH(OH)		C(OH)		CO
|		‖		|
CH(OH)	$\xrightarrow[OH]{H}$	CH	$\rightleftharpoons$	CH_2
|		|		|
CH(OH)		CH(OH)		CH(OH)
|		|		|
CH(OH)		CH(OH)		CH(OH)
|		|		|
$CH_2(OH)$		$CH_2(OH)$		$CH_2(OH)$

(ketone III has the constitution of a condensation product of methylglyoxal and glyceraldehyde)

$$\underset{\text{Methylglyoxal}}{\begin{matrix}CHO\\|\\CO\\|\\CH_3\end{matrix}} \xrightarrow[OH]{H} \begin{matrix}COOH\\|\\CH(OH)\\|\\CH_3\end{matrix} + \begin{matrix}CO_2\\ \hline CH_2OH\\|\\CH_3\end{matrix}$$

$$\underset{\text{Glyceraldehyde}}{\begin{matrix}CHO\\|\\CH(OH)\\|\\CH_2(OH)\end{matrix}} \xrightarrow[OH]{H} \begin{matrix}CHO\\|\\C(OH)\\\|\\CH_2\end{matrix} \rightleftharpoons \underset{\text{Methylglyoxal}}{\begin{matrix}CHO\\|\\CO\\|\\CH_3\end{matrix}} \xrightarrow[OH]{H} \underset{\text{Lactic acid}}{\begin{matrix}COOH\\|\\CH(OH)\\|\\CH_3\end{matrix}} \longrightarrow \underset{\text{Alcohol}}{\begin{matrix}CO_2\\ \hline CH_2OH\\|\\CH_3\end{matrix}}$$

Such a scheme postulates as intermediate products three different compounds containing chains of three carbon atoms: glyceraldehyde, methylglyoxal, and lactic acid.

2. Lactic acid

Buchner and Meisenheimer[1,12–14] having frequently, but not regularly, identified the presence of small amounts of lactic acid (0.2 per cent), proposed considering lactic acid as an intermediate product of fermentation. According to their theory, in the presence of an enzyme which they call zymase or yeast-zymase, sugar goes into lactic acid, which is then broken down by a second hypothetical enzyme, lactacidase, into alcohol and CO_2. This theory was consistent with the chemical scheme proposed by Wohl but it immediately met severe criticism. It was stressed that the amounts of lactic acid found were extremely small[15], and proof was afforded that the lactic acid added to the juice, and which sometimes disappears, is transformed[14] into alcohol and CO_2. If lactic acid was an intermediate product, the reaction by which it is fermented should proceed as rapidly as it is formed or it would accumulate. That this is not the case has been shown by Slator[16–19]. Buchner and Meisenheimer[14,20] themselves, from very careful quantitative experiments in which the proper precautions had been taken against bacterial contamination, concluded that lactic acid is neither formed nor fermented by pure yeast. There were still a few objections to this final proof that lactic acid is not an intermediate (literature, see Harden[10]) but the theory was abandoned. Slator had already come to the conclusion that lactic acid is probably a side product of the alcoholic fermentation.

Buchner and Meisenheimer[14], looking for a product of fermentation

References p. 77

Plate 69. Alexander Nicolaevich Lebedev.

from which lactic acid could be derived, considered methylglyoxal, glyceraldehyde and dihydroxyacetone as possible intermediates. As an argument they proposed the fact that in order to be recognized as an intermediary step, the substance should be identified among the products of fermentation or at least should be fermented by living yeast or by yeast juice. Buchner and Meisenheimer confirmed Mayer[21] and Wohl[22] on this issue and found that methylglyoxal was not fermented, and therefore discarded it as a possible precursor of lactic acid.

This argument was not considered as of any value by Neuberg[23] who, in his review published in 1913, retorted that there was a difference between the fate of an added substance to living material and a substance naturally produced. But, said Oppenheimer[24], why should not the same argument apply to lactic acid? It is true that Lebedev and Griaznoff[25] had announced that they would publish data showing that methylglyoxal can be fermented, but this demonstration never appeared.

Neuberg[26] in support of his methylglyoxal theory showed that a production of lactic acid from methylglyoxal is observed in experiments with "obergäriger lebender Hefe" and with maceration juices of the same. On the other hand no such reaction is observed in Lebedev's maceration juice.

Therefore Buchner and Meisenheimer[1], in their search for lactic acid precursors, only considered dihydroxyacetone and glyceraldehyde as intermediary products of fermentation able to play that role. They observed a definite fermentability of the first substance and confirmed the fact, already observed by Bertrand[27] that dihydroxyacetone was fermented by yeast juice.

Buchner and Meisenheimer[1] concluded that the great fermentability of dihydroxyacetone compared with that of the two other possible precursors of lactic acid was in favour of dihydroxyacetone as intermediary product in fermentation. A confirmation was found by them in the symmetry of dihydroxyacetone which made it clear why in the cell-free fermentation the racemic form of lactic acid was always produced.

Embden, Baldes and Schmitz[28] showed in animal experiments that glyceraldehyde is more readily transformed in blood red cells and in the perfused liver than dihydroxyacetone. They conclude as follows[28]:

"wir möchten auf Grund unserer Erfahrungen glauben dass als Zwischenprodukt bei dem zu optisch aktiver Milchsäure führenden Traubenzuckerabbau im tierischen Organismus praktisch weitaus am erster Stelle optisch aktiver Glycerinaldehyd in Betracht kommt. Im

References p. 77

Zusammenhalt mit den Ergebnissen von Buchner und Meisenheimer liegt es nahe, ganz allgemein als Muttersubstanz der *d*-Milchsäure optisch aktiven Glycerinaldehyd, als Muttersubstanz der *d, l*-Milchsäure Dioxyaceton anzunehmen."

The authors insist on the possibility of the conversion of glyceraldehyde to dihydroxyacetone and the reverse. Embden and Oppenheimer[29] have again insisted on the same views and presented them as a conjecture.

There was another 3-carbon compound which could be considered to be the source of lactic acid: pyruvic acid. It can be easily transformed by reduction to lactic acid in the animal body as shown by Mayer[30] by injection of the whole animal and by Embden and Oppenheimer[31] by liver perfusion. Oppenheimer[24], carefully avoiding contamination with lactic bacteria, showed that lactic acid is a side-product of fermentation by yeast maceration juice. This production is greatly increased by the addition of glyceraldehyde and of dihydroxyacetone which the author considers as the main precursor. On the other hand, contrary to what obtains in animal tissues, pyruvic acid is not the precursor of lactic acid in fermentation. Furthermore why does lactic acid not appear in the experiments with living yeast? Oppenheimer[32] answers this by accepting the fact that the enzymes involved in lactic acid production by yeast are more resistant than the other components of the system and that the fermentation by zymase, in which the "force" of fermentation is not as strong, permits the production of lactic acid.

Neuberg and Kerb[33] strongly opposed the two papers by Oppenheimer on the formation of glycerol and of lactic acid, but Oppenheimer soon answered this criticism[34].

A very extensive history of the ancient work on glycerol formation by yeast is to be found in Neuberg and Reinfurth's work[35].

The question was studied again in considerable detail by von Fürth and Lieben[36] who found that when added in the form of its sodium or lithium salts, lactic acid is readily destroyed by yeast in the presence of a plentiful supply of oxygen. Under these conditions lactic acid disappears from the medium with simultaneous evolution of CO_2. The authors recognized that the carbon dioxide came from the carbohydrate of the yeast itself. Several authors have confirmed the destruction of lactic acid by yeast[37,38].

In order to determine the fate of lactic acid, which could not be accounted for by the formation of CO_2, nor of easily hydrolysable carbohydrate of yeast, von Fürth and Lieben[39] performed new experiments and concluded that a non-hydrolysable carbohydrate must be formed in yeast from the lactic acid or else that proteins with a higher carbohydrate content are synthetized. This theory has been rejected by Hoffert[40]. She nevertheless confirmed that lactic acid is actually destroyed by yeast.

3. Formic acid

A theory devised by Schade[41] has proposed considering that in alcoholic fermentation under the influence of catalytic agents, glucose is decomposed into acetaldehyde and formic acid. In this theory, alcohol is considered as resulting from a reduction of acetaldehyde accomplished by the formate produced at the same time.

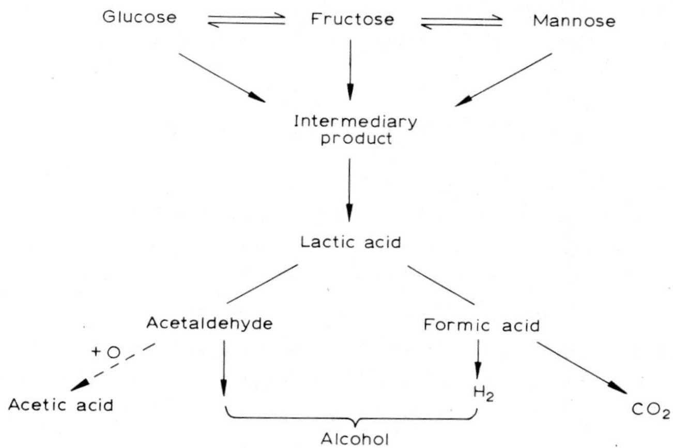

Fig. 6. Schade's theory of alcoholic fermentation.

Another version of formate theory was proposed by Ashdown and Hewitt[42]. According to this theory the sugar is first decomposed to two triose molecules and these react with formamide to produce alanine and formic acid. The alanine enters into reaction with formic acid, producing alcohol, carbon dioxide and formamide.

Both these theories postulated by Schade and by Ashdown and Hewitt have been recognized to be inadequate, as a mixture of alanine and formate is not fermented by yeast juice, and as the addition of a mixture of alanine and formate, or of formamide, to glucose fermenting in the presence of yeast juice, does not affect the fermentation.

References p. 77

4. Glycogen

It has been suggested by several authors[43,44] that all the sugars, before being fermented by yeast, pass through the stage of glycogen.

The formation of glycogen has been observed in yeast juice but the rate of autofermentation of yeast is very low and the fermentation of added glycogen is lower than that of glucose (for literature, see Harden[10]). These data render the theory improbable.

5. Formaldehyde

Another theory proposed by Löb[45–55] suggests considering as intermediates of alcoholic fermentation those substances which, in the other direction appear as intermediates of the synthesis of sugar from formaldehyde. But Löb himself has been unable to identify the postulated intermediates.

6. Glyceraldehyde and dihydroxyacetone

Fischer and Tafel[56,57] made the very important observation that a mixture of glyceraldehyde and dihydroxyacetone (called glycerose because it resulted from the oxydation of glycerol) was readily fermented by yeast.

Pure glyceraldehyde $CH_2(OH) \cdot CH(OH) \cdot CHO$ was tried with negative results by Wohl[58] and by Emmerling[60] using different yeasts. Piloty[59] and Emmerling[60] also obtained negative results with pure dihydroxyacetone but careful experiments by Buchner[61] and by Buchner and Meisenheimer[20] demonstrated that glyceraldehyde as well as dihydroxyacetone are fermentable.

Confirming the fermentability of dihydroxyacetone, Lebedev[62] observed that during the fermentation of dihydroxyacetone, the same hexosephosphate is formed as when hexoses are fermented. On the basis of this, Lebedev[62–64] proposed the following scheme of alcoholic fermentation:

hexose →2 mol. of triose→2 triosephosphates

→hexosephosphate→alcohol (phosphate + readily fermentable form of hexose?)

(*1*) $C_6H_{12}O_6 = 2\ C_3H_6O_3$

(*2*) $2\ C_3H_6O_3 + 2\ RHPO_4 = 2\ C_3H_5O_2RPO_4 + 2\ H_2O$

(*3*) $2\ C_3H_5O_2RPO_4 = C_6H_{10}O_4(RPO_4)_2$

(*4*) $C_6H_{10}O_4(RPO_4)_2 + H_2O = C_2H_5OH + CO_2 + C_3H_5O_2RPO_4 + RHPO_4$

(*5*) $C_6H_{10}O_4(RPO_4)_2 + 2\ H_2O = 2\ C_2H_5OH + 2\ CO_2 + 2\ RHPO_4$

The basis of the theory is the fact that a yeast-juice obtained by maceration in water of air-dried yeast ferments dihydroxyacetone and, in the presence of phosphate, produces a hexosephosphate identical with that obtained in the fermentation of glucose, fructose or mannose.

As stated above, a similar theory had been proposed by Ivanov who postulated that the sugar is first transformed into triose and the latter into triosemonophosphate, fermented directly to alcohol, carbon dioxide and phosphate, but it must be pointed out that Ivanov's "triosemonophosphate" introduction results from a wrong interpretation of the nature of the ester he had isolated and which, in reality, was a mixture of FDP and G-6-P.

The hypothetical views according to which trioses are intermediates in fermentation were opposed with reference to experiments related to the influence of phosphates on fermentation (for literature, see Harden[10]). But Lebedev[65,66,25], extending his experiments to glyceraldehyde, showed that fermentation of glyceraldehyde takes place without any change in free phosphate, contrary to the fermentation of sugar. He therefore concluded that phosphate plays no role in the fermentation of glyceraldehyde. In the new theory he proposed[65], he considered that:

(*1*) the sugar is split into equimolecular proportions of glyceraldehyde and dihydroxyacetone:

$$(a) \qquad C_6H_{12}O_6 \rightleftharpoons C_3H_6O_3 + C_3H_6O_3$$

(*2*) the dihydroxyacetone passes through the stages he postulated in his first theory:

$$(b)\ 4\ C_3H_6O_3 + 4\ R_2HPO_4 \rightarrow 4\ CO_3H_5O_2PO_4R_2 + 4\ H_2O$$

$$(c)\ 4\ C_3H_5O_2PO_4R_2 \rightarrow 2\ C_6H_{10}O_4(R_2PO_4)_2$$

$$(d)\ 2\ C_6H_{10}O_4(R_2PO_4)_2 + 4\ H_2O \rightarrow 2\ C_6H_{12}O + 4\ R_2HPO_4$$

(*3*) glyceraldehyde is converted to pyruvic acid, and then into CO_2 and acetaldehyde, which is reduced into alcohol by the hydrogen liberated along with the pyruvic acid. This scheme corresponds to the theory suggested by Kostychev (see below), but Lebedev[67,68] proposed a primary formation of glyceric acid which is later converted into pyruvic acid in the presence of a hypothetic enzyme he calls dehydratase:

$$(1)\ CH_2(OH)\cdot CHOH\cdot CHO \xrightarrow{+H_2O} CH_2(OH)\cdot CH(OH)\cdot CH(OH)_2 \rightarrow$$
$$\rightarrow CH_2(OH)\cdot CHOH\cdot COOH + 2\ H$$

$$(2)\ CH_2(OH)\cdot CH(OH)\cdot COOH \rightarrow CH_2\cdot CO\cdot COOH + H_2O$$

References p. 77

The experimental basis of this theory is the observation that glyceric acid is fermented by dried yeast and by maceration juice, but it neglected the possibility of the substance being transformed into an intermediate in the course of the experiment.

In the theory of Lebedev, the role of phosphate is to remove dihydroxyacetone, therefore preventing it from inhibiting further conversion of hexose into triose. The main objection to the theory first came from the fact, demonstrated by Harden and Young[69] in 1912 that dihydroxyacetone does not inhibit glucose fermentation even in high amounts. On the other hand, alcoholic fermentation does not proceed in the absence of phosphate, a point which was also made against Lebedev's views, for if phosphate had only the role of eliminating an intermediate which does not in fact inhibit the process, its absence should not prevent fermentation. Dihydroxyacetone had been shown to be formed in the course of the fermentation of glucose by yeast. The reaction was accomplished by Boysen-Jensen[70,71] in the presence of sodium sulphate or chlorhydrate of hydroxylamine in order to stop it at the stage of dihydroxyacetone. (The observation could not be repeated by Chick[72]).

7. Pyruvic acid and acetaldehyde

The recognition of the role of pyruvic acid as an intermediary step in alcoholic fermentation was of an indirect nature. As we have said, F. Ehrlich recognized in 1907 that the production of higher alcohols by yeast fermentation has its origins in the metabolism of amino acids. Neubauer and Fromherz[73] admitted that when alanine was metabolized with the production of ethyl alcohol, the following intermediary reactions took place:

CH_3–C(H)(NH_2)–COOH	$\xrightarrow{(+O)}$	CH_3–C(OH)(NH_2)–COOH	$\xrightarrow{(-NH_3)}$	CH_3–CO–COOH	$\xrightarrow{(-CO_2)}$	CH_3–CHO + CO_2	$\xrightarrow{+H_2}$	CH_3–CH_2OH
Alanine		Hydrate of α-iminopropionic acid		Pyruvic acid		Acetaldehyde		Ethyl alcohol

This scheme recognized the reality of the fermentation of pyruvic acid by yeast and this fact had been applied to the fermentation of glucose by several authors.

At the beginning of 1911, Neuberg and Wastenson showed that pyruvate is fermented by yeast[74]. Neubauer[75] also showed that pyruvic acid was easily fermented by yeast. This fact came as a surprise, as A. Mayer, in a well-known treatise[76], had classified pyruvic acid among the substances which, for chemical reasons, could never be fermented by yeast.

The step of the production of pyruvic acid from hexose, was confirmed by the experiment of Fernbach and Schoen[77] who isolated pyruvic acid in the form of its calcium salt, from fermentation mashes containing living yeasts. Yeast rapidly decomposes α-ketonic acids with an evolution of CO_2[73,78,79] (this gives an explanation, so far non-existent, of the evolution of CO_2, accounted for by the action of carboxylase).

According to Kostychev[79], alcoholic fermentation proceeds in three steps:

(*1*) a production of pyruvate from the hexoses, with a loss of hydrogen

$$C_2H_{12}O_6 \rightarrow 2\ CH_3 \cdot CO \cdot COOH + 4[H]$$

(*2*) the decomposition of pyruvic acid into acetaldehyde and CO_2 in the presence of carboxylase

$$2\ CH_3 \cdot CO \cdot COOH \rightarrow 2\ CH_3 \cdot CHO + 2\ CO_2$$

(*3*) the reduction of the acetaldehyde to ethyl alcohol by the active hydrogen produced in a previous phase

$$2\ CH_3 \cdot CHO + 4\ H \rightarrow 2\ CH_3 \cdot CH_2OH$$

Step *2* has been demonstrated by Neuberg and Karczag[80] and step *3* by Kostychev[82] and Hübbenet[81] according to Kostychev who, nevertheless, in 1914 considered that he had introduced the hydrogen atoms into an earlier step of the pathway. A priority contestation against Kostychev was raised by Lebedev[83] about this scheme.

The demonstration of the production of pyruvic acid as an intermediate in alcoholic fermentation has been accomplished using a "fixing agent". The condensation with β-naphthylamine of the pyruvic acid formed in the fermentation of sugar by yeast-juice was accomplished by von Grab[86] in Neuberg's laboratory.

The product of the condensation reaction is α-methyl-β-naphthocinchoninic acid:

$$2\ CH_3 \cdot CO \cdot COOH + H_2N \cdot C_{10}H_7 = 2\ H_2O + CO_2 + H_2 + \begin{array}{l} CH_3 \cdot C{:}N \diagdown \\ \quad | \qquad\qquad C_{10}H_6 \\ HC \cdot C \cdot COOH \diagup \end{array}$$

References p. 77

Plate 70. Serguei Pavlovich Kostychev.

The second step of Kostychev's scheme was at that time well documented[75,81,87]. The introduction of acetaldehyde into Kostychev's scheme was the consequence of the fact that during fermentation with different kinds of yeast preparations in the presence of $ZnCl_2$, important amounts of acetaldehyde are found, while only very small amounts are produced in normal conditions.

An objection was raised against these experiments on the basis that $ZnCl_2$ possibly modified the pathway by producing modified forms of enzymes. Kostychev and Scheloumoff[88] showed that after a few days of action of $ZnCl_2$, the ratio CO_2/alcohol is modified. From this they concluded that alcohol and CO_2 are not produced simultaneously in fermentation but in different steps, and that $ZnCl_2$ modifies step *3* of Kostychev's scheme by the fact that only a part of the acetaldehyde is reduced into alcohol while the step corresponding to the evolution of CO_2 is not modified.

Neuberg and Kerb[89] at the time opposed Kostychev's observations on $ZnCl_2$ with the argument that the amounts of acetylaldehyde produced were too small to show that the substance was an intermediate in the pathway. Kostychev[90] answered this by saying that it was impossible in the study of intermediate steps to look for a quantitative balance as is possible in complete reaction systems.

Important evidence for the third step was furnished by several authors and notably by Kostychev and his collaborators (for literature see Harden[10]). The last step of Kostychev's scheme, the reduction of acetaldehyde to alcohol, remained the most controversial aspect, and was far from being generally recognized (see Harden[10]). It had frequently been considered possible before that acetaldehyde[91–95] is an intermediate step of alcoholic fermentation. Traces of acetaldehyde are always produced in alcohol fermentation, but it had been suggested that they might result from the oxidation of alcohol[96–99]. Nevertheless, even in anaerobic conditions, or under water saturated with CO_2, the autolysis of yeast produces acetaldehyde[99,100].

A technique first introduced by Neuberg and preserving acetaldehyde from reduction[101–103] finally gave the proof of its role as an intermediary step in alcoholic fermentation. When sodium sulphite, Na_2SO_3, is added to a fermenting mixture of yeast and sugar, a decrease of yield in alcohol and CO_2 is observed, with an increase in glycerol and acetaldehyde production. The principle of this observation was applied on a large scale

References p. 77

Plate 71. Carl Alexander Neuberg.

during the first world war for the production of glycerol in Germany.

As the aldehyde produced is equivalent to the glycerol[33,104,106] and as this equivalence persists through the fermentation[105], the explanation given by Neuberg is that the aldehyde combines to form the stable bisulphite complex as rapidly as to escape reduction to alcohol. As a consequence another substrate is reduced to provide glycerol.

It is true that pyruvic acid also combines with sodium bisulphite, but the compound is readily fermented[106–108]. Sulphite can be replaced by dimedone, forming aldomedone with aldehyde, without combining with glucose or pyruvic acid[109].

Fermentation in the presence of sulphite is called a second form of fermentation by Neuberg, while he calls a third form of fermentation the fermentation in the presence of alkalis producing equivalent amounts of alcohol and acetic acid[110–113]. In this third form of fermentation, aldehyde is converted into equal molecular proportions of acetic acid and alcohol in the course of a Cannizzaro reaction:

$$2\ C_2H_4O + H_2O = C_2H_6O + C_2H_4O_2$$

8. Methylglyoxal

What is the mechanism of the production of pyruvic acid from glucose? It is clear that it could have been derived from lactic acid by oxidation, but at that time lactic acid was already recognized as not belonging to the intermediates of alcoholic fermentation. Glyceraldehyde as well as glyceric acid could be considered, from chemical evidence, as possible intermediates[114]. As we have recalled above, schemes based on purely chemical evidence had also considered the possibility of methylglyoxal as an intermediate.

Kostychev[84,115,116] has repeatedly expressed the view that the mystery of alcoholic fermentation resided mainly in the introduction of the methyl. This may have favoured the adoption (without any direct proof) of the methylglyoxal theory.

Methylglyoxal had been tested without success with living yeast[21,22,117], with acetone yeast[13] and with yeast juice[20]. But an objection was always possible to such addition experiments, namely that they may have inhibited the fermentation. Neuberg and Kerb[89] proposed in 1913 a new theory of the metabolic sequence of fermentation based on the production

References p. 77

of methylglyoxal. It must be noted that in 1912 Lebedev[65] had stated that under certain circumstances methylglyoxal was fermentable by yeast. He retracted this in 1914[118] but proposed it again in 1918[119]. Neuberg and Kerb suggested the following pathway:

(*1*) The sugar is split up into two molecules of methylglyoxal:

$$C_6H_{12}O_6{-}2\,H_2O = C_6H_8O_4 = 2\,CH_2{:}C(OH){\cdot}CHO$$

methylglyoxal aldol

or

$$2\,CH_3{\cdot}CO{\cdot}CHO$$

methylglyoxal

(*2*) A portion of the methylglyoxal is converted into glycerol and pyruvic acid (Cannizzaro transformation):

$$\begin{array}{lcl} CH_2{:}C(OH){\cdot}CHO + H_2O & H_2 & CH_2(OH)CH(OH){\cdot}CH_2OH \\ & +\,|\;= & \\ CH_2{:}C(OH){\cdot}CHO & O & \text{glycerol} \\ \text{methylglyoxal} & & + \\ & & CH_2{:}C(OH){\cdot}COOH \\ & & \text{pyruvic acid} \end{array}$$

(*3*) The pyruvic acid, in the presence of carboxylase, is split into acetaldehyde and CO_2:

$$CH_3{\cdot}CO{\cdot}COOH = CH_3{\cdot}CHO + CO_2$$

(*4*) A molecule of methylglyoxal and a molecule of aldehyde undergo a Cannizzaro reaction, yielding alcohol by reduction of the acetaldehyde and pyruvic acid by oxidation of the methylglyoxal

$$\begin{array}{lccc} CH_3{\cdot}CO{\cdot}CHO & O & & CH_3{\cdot}CO{\cdot}COOH \\ & +\;| & = & + \\ CH_3{\cdot}CHO & H_2 & & CH_3{\cdot}CH_2(OH) \end{array}$$

and pyruvic acid goes through stage *3*, being again decarboxylated. The cycle was supposed to continue as follows:

Sugar → methylglyoxal → pyruvic acid → acetaldehyde + CO_2

Methylglyoxal + acetaldehyde → pyruvic acid + alcohol

As stated above, the insertion of methylglyoxal as an intermediate of alcoholic fermentation had first been proposed on the basis of chemical evidence.

Biochemical arguments in favour of the theory were also presented, such as the identification of a methylglyoxalase in animal tissues (Neuberg[123]; Dakin and Dudley[124]), besides the repeated identification of methylglyoxal among the products of alcoholic fermentation. On the other hand it was recognized by Meyerhof[125] to be transformed at a faster rate than sugar, to lactic acid by slices of liver and of other organs.

The methylglyoxal theory, formulated in 1913, was to keep general recognition for twenty years, and many great experts praised it in their writings. In his book on muscle chemistry[126] (1930), and in the French translation[127] which appeared in 1932, Meyerhof still did characterize Neuberg's scheme "as the one which is best demonstrated experimentally".

Even in 1934, after the publication of Embden, Deuticke and Kraft (1933), Nilsson[128], who had discovered 3-phosphoglyceric acid in glycolyzing mixtures, but did not consider it as an intermediate in the pathway, still adheres to the methylglyoxal theory.

Another pathway of methylglyoxal metabolism had been proposed by Lebedev[66]:

$$(1)\quad CH_3COCOH + H_2O = CH_3COCH\begin{matrix} \diagup OH \\ \diagdown OH \end{matrix}$$

$$(2)\quad CH_3COCH\begin{matrix} \diagup OH \\ \diagdown OH \end{matrix} - H_2 = CH_3COCOOH$$

$$(3)\quad CH_3COCOOH = CH_3COH + CO_2$$

$$(4)\quad CH_3COH + H_2 = CH_3CH_2OH$$

This same mechanism was again proposed by Kluyver, Donker and Visser 't Hooft[120] as a variant of Neuberg's scheme and was the occasion of a priority claim by Lebedev[121].

Neuberg's theory made no reference to the intervention of phosphate in the normal course of fermentation, as he believed that the phosphorylation reaction represents a phenomenon which takes place in preparations derived from yeast but not in the living cells (for a detailed presentation of the extensive literature which developed around this theory, see Neuberg[122]).

Referring to the view of Neuberg, according to which the formation of hexosephosphates in yeast is a phenomenon which does not occur in normal yeast but only when it has been injured in some way, by drying, grinding, action of toluene, etc., Hemmi[129] examined the effect of the presence of phosphate on the production of aldehyde by yeast and found that the addition of phosphate in no way interfered with its production.

References p. 77

The fermentation of potassium hexosephosphate itself also yields aldehyde in the presence of sulphite. Further, the presence of sulphite does not prevent the esterification of the phosphoric acid so that the two processes in no way interfere. She concludes that the phosphate is concerned in the early stages of the decomposition of sugar, according to Harden and Young's view. It is clear that in the period considered, the positive knowledge of the intermediate compounds involved in the fermentation of glucose by yeast was mainly due to the explorative studies of the British and of the Russian Schools.

From the work accomplished before 1929, related in this section, the acquisitions which were to survive are contained in a scheme, according to which glucose is decomposed into pyruvate which, in the presence of carboxylase gives acetaldehyde and CO_2 while the acetaldehyde is reduced into ethanol.

REFERENCES

1 E. Buchner and J. Meisenheimer, *Ber. deut. chem. Ges.*, 38 (1905) 620.
2 E. Duclaux, *Compt. Rend.*, 103 (1886) 881.
3 E. Duclaux, *Ann. Inst. Pasteur*, 7 (1893) 751.
4 E. Duclaux, *Ann. Inst. Pasteur*, 10 (1896) 129.
5 G. Pinkus, *Ber. deut. chem. Ges.*, 31 (1898) 31.
6 J. U. Nef, *Ann. Chem., Liebigs*, 335 (1904) 247.
7 J. U. Nef, *Ann. Chem., Liebigs*, 357 (1907) 214.
8 A. Wohl, *Ber. deut. chem. Ges.*, 41 (1908) 3599.
9 A. Baeyer, *Ber. deut. chem. Ges.*, 3 (1870) 63.
10 A. Harden, *Alcoholic Fermentation*, London, 1911 (4th ed., 1932).
11 A. Wohl, in E. von Lippmann (Ed.), *Die Chemie der Zuckerarten*, Brunswick, 1904.
12 E. Buchner and J. Meisenheimer, *Ber. deut. chem. Ges.*, 37 (1904) 417.
13 E. Buchner and J. Meisenheimer, *Ber. deut. chem. Ges.*, 39 (1906) 3201.
14 E. Buchner and J. Meisenheimer, *Z. wiss. Landwirtschaft*, 38 Ergänzungs-band V (1909) 265.
15 A. Harden, *J. Inst. Brewing*, 11 (1905) No. 1.
16 A. Slator, *J. Chem. Soc.*, 89 (1906) 128.
17 A. Slator, *Ber. deut. chem. Ges.*, 40 (1907) 123.
18 A. Slator, *J. Chem. Soc.*, 93 (1908) 217.
19 A. Slator, *Chem. News*, 98 (1908) 175.
20 E. Buchner and J. Meisenheimer, *Ber. deut. chem. Ges.*, 43 (1910) 1773.
21 P. Mayer-Karlsbad, *Biochem. Z.*, 2 (1907) 435.
22 A. Wohl, *Biochem. Z.*, 5 (1907) 45.
23 C. Neuberg, *Die Gärungsvorgang und der Zuckerumsatz der Zelle*, Jena, 1913.
24 M. Oppenheimer, *Z. physiol. Chem.*, 89 (1914) 45.
25 A. von Lebedev and N. Griasnoff, *Ber. deut. chem. Ges.*, 45 (1912) 3256.
26 C. Neuberg, *Biochem. Z.*, 51 (1913) 484.
27 G. Bertrand, *Ann. Chim. Phys.*, 3 (1904) 256.
28 G. Embden, K. Baldes and E. Schmitz, *Biochem. Z.*, 45 (1912) 108.
29 G. Embden and M. Oppenheimer, *Biochem. Z.*, 45 (1912) 186.
30 P. Mayer, *Z. physiol. Chem.*, 40 (1912) 444.
31 G. Embden and M. Oppenheimer, *Z. physiol. Chem.*, 55 (1913) 335.
32 M. Oppenheimer, *Z. physiol. Chem.*, 89 (1914) 63.
33 C. Neuberg and J. Kerb, *Biochem. Z.*, 62 (1914) 489.
34 M. Oppenheimer, *Z. physiol. Chem.*, 93 (1914/15) 262.
35 C. Neuberg and E. Reinfurth, *Biochem. Z.*, 92 (1918) 234.
36 O. von Fürth and F. Lieben, *Biochem. Z.*, 128 (1922) 144.
37 E. Kayser, *Bull. Soc. Chim. Biol.*, 6 (1924) 345.
38 I. Smedley Maclean and D. Hoffert, *Biochem. J.*, 17 (1923) 720.
39 O. von Fürth and F. Lieben, *Biochem. Z.*, 132 (1922) 165.
40 D. Hoffert, *Biochem. J.*, 20 (1926) 358.
41 H. Schade, *Z. physikal. Chem.*, 57 (1906) 1.
42 O. E. Ashdown and J. T. Hewitt, *J. Chem. Soc.*, 97 (1910) 1636.
43 J. Gruss, *Z. ges. Brauw.*, 27 (1904) 689.
44 F. G. Kohl, *Ber. deut. botan. Ges.*, 25 (1907) 74.
45 W. Löb, *Landwirtsch. Jahrb.*, 35 (1906) 541.
46 W. Löb, *Chem. Z.*, 42 (1906) 540.
47 W. Löb, *Z. Elektrochem.*, 12 (1906) 282.

48 W. Löb, *Biochem. Z.*, 12 (1908) 78.
49 W. Löb, *Biochem. Z.*, 12 (1908) 466.
50 W. Löb, *Biochem. Z.*, 17 (1909) 132.
51 W. Löb, *Biochem. Z.*, 20 (1909) 516.
52 W. Löb, *Biochem. Z.*, 22 (1909) 103.
53 W. Löb and G. Pulvermacher, *Biochem. Z.*, 23 (1909) 10.
54 W. Löb and G. Pulvermacher, *Biochem. Z.*, 17 (1909) 343.
55 W. Löb, *Biochem. Z.*, 29 (1910) 311.
56 E. Fischer and J. Tafel, *Ber. deut. chem. Ges.*, 21 (1888) 2634.
57 E. Fischer and J. Tafel, *Ber. deut. chem. Ges.*, 22 (1889) 106.
58 A. Wohl, *Ber. deut. chem. Ges.*, 31 (1898) 1796.
59 O. Piloty, *Ber. deut. chem. Ges.*, 30 (1897) 3161.
60 O. Emmerling, *Ber. deut. chem. Ges.*, 32 (1899) 542.
61 E. Buchner, *Bull. Soc. Chim. de France*, [4] 7 (1910) 1.
62 A. von Lebedev, *Ber. deut. chem. Ges.*, 44 (1911) 2932.
63 A. von Lebedev, *Compt. Rend.*, 153 (1911) 136.
64 A. von Lebedev, *Bull. Soc. Chim.*, 9 (1911) 678.
65 A. von Lebedev, *Biochem. Z.*, 46 (1912) 483.
66 A. von Lebedev, *Bull. Soc. Chim.*, 11–12 (1912) 1039.
67 A. von Lebedev, *Ber. deut. chem. Ges.*, 47 (1914) 660.
68 A. von Lebedev, *Ber. deut. chem. Ges.*, 47 (1914) 965.
69 A. Harden and W. J. Young, *Biochem. Z.*, 40 (1912) 458.
70 P. Boysen-Jensen, *Dissertation*, Copenhagen, 1910.
71 P. Boysen-Jensen, *Ber. deut. botan. Ges.*, 26a (1908) 666.
72 F. Chick, *Biochem. Z.*, 40 (1912) 479.
73 O. Neubauer and K. Fromherz, *Z. physiol. Chem.*, 70 (1910–11) 326.
74 C. Neuberg and H. Wastenson, *Sitzber. Berlin physiol. Ges. von 20–1–1911.*
75 O. Neubauer, *Z. physiol. Chem.*, 70 (1911) 350.
76 A. Mayer, *Lehrbuch der Gärungschemie*, 4th ed., Heidelberg, 1895.
77 A. Fernbach and M. Schoen, *Compt. Rend.*, 157 (1913) 1478.
78 C. Neuberg and J. Kerb, *Z. Gärungsphysiol.*, 1 (1912) 114.
79 S. Kostychev, *Z. Physiol. Chem.*, 79 (1912) 359.
80 C. Neuberg and L. Karczag, *Biochem. Z.*, 36 (1911) 68.
81 S. Kostychev and E. Hübbenet, *Z. physiol. Chem.*, 79 (1912) 359.
82 S. Kostychev, *Z. Physiol. Chem.*, 89 (1914) 367.
83 A. von Lebedev, *Z. physiol. Chem.*, 84 (1913) 308.
84 S. Kostychev, *Z. physiol. Chem.*, 79 (1912) 130.
85 A. von Lebedev, *Ber. deut. chem. Ges.*, 45 (1912) 3267.
86 M. von Grab, *Biochem. Z.*, 123 (1921) 69.
87 S. Kostychev, *Z. Physiol. Chem.*, 83 (1913) 93.
88 S. Kostychev and A. Scheloumoff, *Z. physiol. Chem.*, 85 (1913) 493.
89 C. Neuberg and J. Kerb, *Biochem. Z.*, 58 (1913) 158.
90 S. Kostychev, *Biochem. Z.*, 64 (1914) 237.
91 A. Magnus-Levy, *Arch. Anat. Physiol.*, 3/4 (1902) 365.
92 J. B. Leathes, *Problems in Animal Metabolism*, London, 1906, p. 81.
93 E. Buchner and J. Meisenheimer, *Ber. deut. chem. Ges.*, 41 (1908) 1410.
94 A. Harden and D. Norris, *Proc. Roy. Soc. (London), Ser. B*, 84 (1912) 492.
95 E. C. Grey, *Biochem. J.*, 7 (1913) 359.
96 E. Buchner, K. Langheld and S. Kraup, *Ber. deut. chem. Ges.*, 47 (1914) 2550.
97 C. Neuberg and J. Kerb, *Ber. deut. chem. Ges.*, 47 (1914) 1308.

98 C. Neuberg and J. Kerb, *Biochem. Z.*, 64 (1914) 251.
99 C. Neuberg and J. Kerb, *Ber. deut. chem. Ges.*, 47 (1914) 2730.
100 C. Neuberg and E. Schwenk, *Biochem. Z.*, 71 (1915) 126.
101 C. Neuberg and E. Reinfurth, *Biochem. Z.*, 89 (1918) 365.
102 W. Connstein and W. Ludecke, *Ber. deut. chem. Ges.*, 52B (1919) 1385.
103 K. Schweizer, *Helv. Chim. Acta*, 2 (1919) 167.
104 C. Neuberg and E. Reinfurth, *Ber. deut. chem. Ges.*, 52B (1919) 1677.
105 C. Neuberg and J. Hirsch, *Biochem. Z.*, 98 (1919) 141.
106 C. Neuberg and E. Reinfurth, *Ber. deut. chem. Ges.*, 53B (1920) 462.
107 C. Neuberg and E. Reinfurth, *Ber. deut. chem. Ges.*, 53B (1920) 1039.
108 E. Zerner, *Ber. deut. chem. Ges.*, 53B (1920) 324.
109 C. Neuberg and E. Reinfurth, *Biochem. Z.*, 106 (1920) 281.
110 C. Neuberg and J. Hirsch, *Biochem. Z.*, 96 (1919) 175.
111 C. Neuberg and J. Hirsch, *Biochem. Z.*, 100 (1919) 304.
112 C. Neuberg, J. Hirsch and E. Reinfurth, *Biochem. Z.*, 105 (1920) 307.
113 C. Neuberg and W. Ursum, *Biochem. Z.*, 110 (1920) 193.
114 E. Erlenmeyer, *Ber. deut. chem. Ges.*, 14 (1881) 320.
115 S. Kostychev, *Ber. deut. chem. Ges.*, 45 (1912) 1289.
116 S. Kostychev, *Z. physiol. Chem.*, 92 (1914) 402.
117 A. Wohl, *Z. angew. Chem.*, 20 (1907) 1169.
118 A. von Lebedev, *Ber. deut. chem. Ges.*, 47 (1914) 967.
119 A. von Lebedev, *Chem. Zentr.*, 2 (1918) 52.
120 A. J. Kluyver, H. J. L. Donker and F. Visser 't Hooft, *Biochem. Z.*, 161 (1925) 361.
121 A. von Lebedev, *Biochem. Z.*, 166 (1925) 407.
122 C. Neuberg, in *Oppenheimer's Handbuch der Biochemie*, 2nd edn., Vol. 2, Jena, 1925.
123 C. Neuberg, *Biochem. Z.*, 49 (1913) 502.
124 H. D. Dakin and H. W. Dudley, *J. Biol. Chem.*, 14 (1913) 155.
125 O. Meyerhof, *Biochem. Z.*, 159 (1925) 432.
126 O. Meyerhof, *Die chemische Vorgänge im Muskel und ihr Zusammenhang mit Arbeitsleistung und Wärmebildung*, Berlin, 1930.
127 O. Meyerhof, *Chimie de la Contraction Musculaire* (translated by L. Genevois), Paris, 1932.
128 R. Nilsson, in H. von Euler, *Chemie der Enzyme*, 2. Teil, 3 Abschnitt, Munich, 1934.
129 F. Hemmi, *Biochem. J.*, 17 (1923) 327.

Chapter 20

Muscle Glycolysis Before 1926

It was during the second half of the 19th century that the concept prevailed that the energy used by muscle in contraction was derived from chemical reactions taking place in the muscle cells. This view was one of the aspects of a concept alluded to by Robert Mayer[1] in a letter of 1844 to his friend Griesinger:

"For some time a logical instinct has led physiologists to the axiomatic proposition: No action without metabolism." (translation by Rosen[2])

1. Lactic acid formation from glycogen

E. du Bois Reymond[3] recognized that fatigue and *rigor mortis* were accompanied by a production of lactic acid. Several authors confirmed this observation, but others denied it (for literature, see von Fürth[4]).

With respect to the origin of lactic acid, several authors, including Liebig, du Bois Reymond and Bernard, considered it as being derived from glycogen (for literature see refs. 5–7). In careful quantitative experiments Fletcher and Hopkins[8] in 1907 clearly showed that fatigue and *rigor mortis* are really accompanied by production of lactic acid. They succeeded in this by reducing to a minimum the stimulation of the muscle fibres during the experimental manipulations, while in many cases some of their predecessors had, by mistreating the tissues, produced the maximal possible concentration of lactic acid. Parnas and Wagner[9] also adopted the same caution in their manipulations of the tissues and showed that the lactic acid is derived from the glycogen of the muscle. Their demonstration, whilst clear in a number of cases, did not show the same quantitative agreement in others, and it was left to Meyerhof to afford the demonstration that in all the various circumstances studied by him[10–12] a complete equivalence and time coincidence was observed between carbohydrate

References p. 90

Plate 72. Jakób Karol Parnas.

disappearing and lactic acid appearing (for a number of accessory works on the same subject, see Embden[22]).

2. Lactacidogen

After it was recognized that in many organs or tissues the degradation of carbohydrates led to the production of lactic acid (for literature see Embden[22]), Embden, Kalberlah and Engel[23] tried to demonstrate a production of lactic acid from carbohydrates added to muscle press juice, an experiment inspired by the success of Buchner with yeast juice.

Embden observed that in a press juice of dog muscle to which some sodium bicarbonate had been added to ensure buffering, an increase in lactic acid concentration took place but the addition of glucose or of glycogen produced no increase in formation. Fletcher and Hopkins[8] had already observed this when, in an attempt to account for the increase in lactic acid in the frog muscles which they were studying, carbohydrates were added to minced rabbit muscle without any formation increase. These facts led Embden, Kalberlah and Engel[23] to postulate the presence in the dog muscle juice obtained through the press, of a still unknown precursor of lactic acid, which they called "lactacidogen."

Various authors had observed that muscular work produces an increase in urinary excretion of phosphoric acid and the increase in phosphoric acid (presently recognized as mainly due to creatine phosphate splitting) had also been detected at the time of contraction. But these observations were sometimes contradictory (for literature see Embden[22]). The convincing demonstration of the fact that the formation of lactic acid in muscle was accompanied by a liberation of phosphoric acid was provided by Embden, Griesbach and Schmitz[24]. After a number of different experimental trials, these authors recognized conditions in which, in spite of different amounts produced, lactic acid and phosphoric acid appeared in the muscle juice in equimolecular proportions. From this they concluded that their "lactacidogen" was at the same time the precursor of lactic acid and of phosphoric acid. They added a number of organic phosphoric derivatives to the muscle juice but obtained negative results with deoxyribonucleic acid, inosinic acid and phytin. At that time it had already been demonstrated by Young that alcoholic fermentation by yeast juice was accompanied by a production of hexosediphosphate. By adding this compound to the press juice of muscle, the authors obtained an increased production of lactic acid and of phosphoric acid.

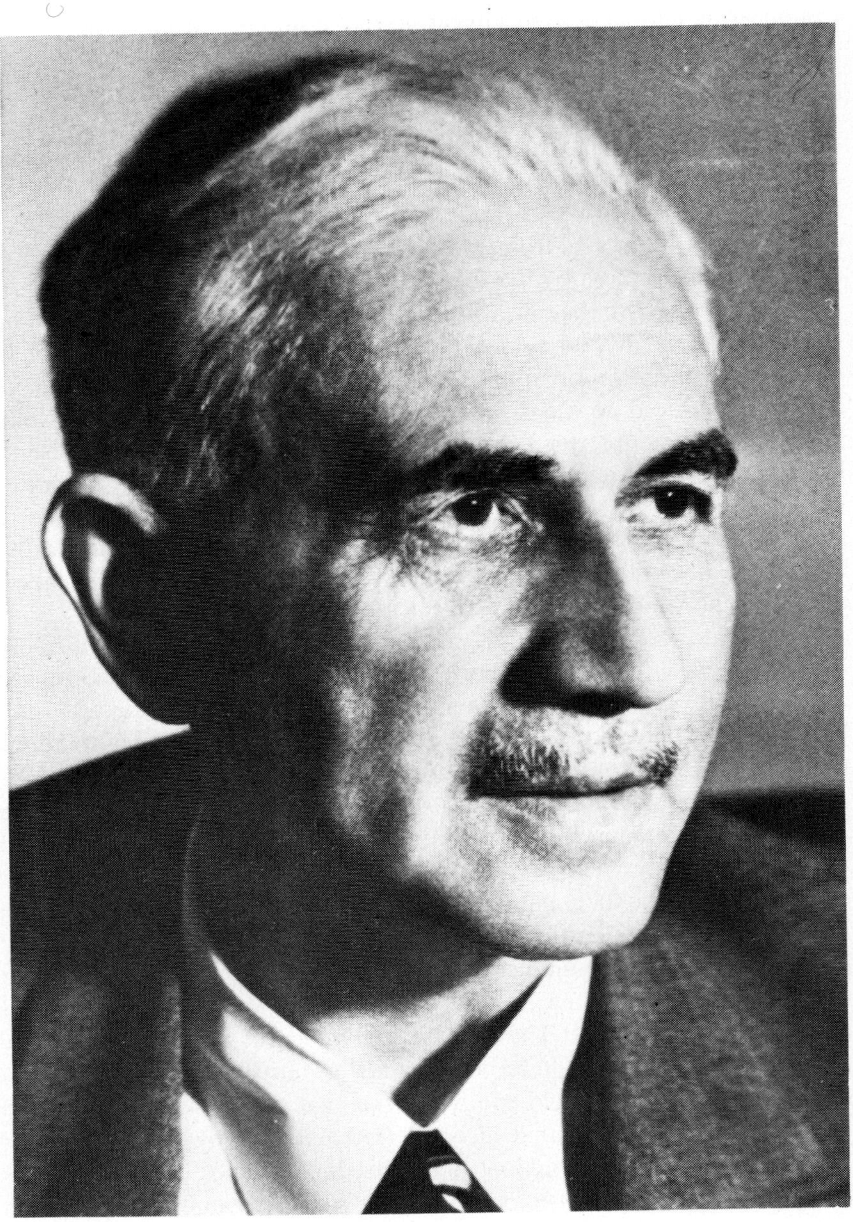

Plate 73. Otto Meyerhof.

From this Embden concluded that his "lactacidogen" was related to hexosediphosphate. It was found that muscle press-juice contained a reducing phosphate compound. From dog muscle Embden and Laquer[25] had by 1921 prepared an osazone compound identical to that derived from yeast hexosediphosphate. The conclusion was (at least for some time) that "lactacidogen" was in fact hexosediphosphate. Later, hexosediphosphate was prepared from muscle in the form of its brucine salt by Embden and Zimmerman[26], by a process in which the authors added sodium fluoride to the press juice of muscle, as this had been found to cause a diminution of free phosphate when added to a muscle juice together with glycogen. In another set of experiments, without addition of fluoride, Embden and Zimmerman[27,28] obtained a hexosemonophosphoric ester.

Later on it was shown[29] that living muscle does not contain hexosediphosphate but that this substance can be detected in the press juice of muscle in the presence of fluoride. Embden and Zimmerman[26] considered the supply of "lactacidogen" as being replenished from the glycogen stores in the muscle. This rested on the observation that in the process of incubation of the chopped muscle (2 hours at 37°C in 2 per cent bicarbonate solution) glycogen disappeared in amounts equivalent to the lactic acid increase. Embden never gave up his belief that a phosphorylated intermediate in carbohydrate metabolism yielded, in the course of a same reaction lactic acid and phosphoric acid, both contributing, according to his belief to an acidification of the muscle. Of course, at the time, little was known of the esters of carbohydrates and their dissociation constants had not been determined. The esters were later demonstrated to be stronger acids than phosphoric acid[30,31] and consequently, when a carbohydrate ester is split up, an alcalinisation results. In the physiological range of pH values in muscle (6–7.5), phosphagens do not buffer, but as a consequence of phosphagen splitting, phosphate, a powerful buffer, is liberated. As, in the course of muscle contraction, more phosphagen is split than lactic acid is produced, the latter is buffered and the result is not an acidification, but an alcalinisation.

In the presence of creatine phosphate, the concentration of ATP is maintained in muscle and, therefore an acidifying effect which could result in contraction from the dephosphorylation of ATP to ADP or of ADP to AMP and from the resulting liberation of an acid group of the strength of the second dissociation constant of phosphoric acid (Lohmann[32,33]) is not to be considered.

References p. 90

Plate 74. Gustav Embden.

The lactacidogen theory was discarded (as a consequence of the discovery of ATP) after Cori and Cori[33a] confirmed, in 1931, that the increase of inorganic P during incubation (corresponding to Embden's lactacidogen) was mostly derived from ATP (as Lohmann[33b] had suggested), very little coming from hexosemonophosphate.

As stated above, one of the origins of the long-lasting concept of "lactacidogen" was the increased urinary excretion of phosphate during muscular activity. The main origin of this phosphate is creatine phosphate and its excretion is probably the expression of the fact that it diffuses from the muscle more rapidly than does creatine phosphate.

3. Fate of the lactic acid

In 1914, A. V. Hill[34] showed that heat measurement excluded a combustion of lactate, to account for the removal of lactic acid during the recovery phase of muscle contraction. From these data, the theory of the resynthesis of glycogen from lactate was derived. As the breaking down of glycogen into lactic acid was considered, as said above, to be associated with the liberation of the energy of contraction (Hill–Meyerhof's theory) a supply of energy was needed for the reverse process and this energy was thought to be provided by combustion of carbohydrate. According to this theory, glucose is the fuel of the muscle.

For Fletcher and Hopkins[35], lactic acid was removed by complete combustion, with the production of CO_2 and a consumption of oxygen, the greater part of the energy being retained in the muscle for the restoration of the original physico-chemical state. This theory considers lactic acid as the fuel of the muscle. On the other hand, Parnas[9], in his paper with Wagner published in 1914, had concluded that there is no resynthesis of glycogen during the phase of muscle recovery in oxygen.

When he introduced in 1920 what was accepted for several years as the "Meyerhof cycle", Meyerhof stated that, contrary to the views of Parnas, a resynthesis of glycogen took place in muscle during recovery. He claimed that the extra oxygen consumption of fatigued muscles is equivalent to one-fifth of the amount of lactic acid disappearing. According to him, one-fifth of the lactic acid is completely oxidized, the rest being rebuilt into glycogen. Meyerhof considered that the work done is proportional to the lactic acid production[12,13]. The theory according to which lactic acid was the agent of muscle contraction persisted until Lundsgaard showed the existence of contraction without glycolysis, an aspect which will be dealt with in Chapter 25.

From his Nobel lecture we can derive the views held by Meyerhof[36] in 1924 on the degradation of glycogen in muscle:

"Während der anaeroben Arbeitsphase häuft sich annährend proportional mit der geleisteten Arbeit Milchsäure in Muskel an. Gleichzeitig verschwindet eine entsprechende Menge Glykogen, während die Menge niedriger Kohlehydrate, vor allem freier Glucose und der von Embden im Muskel entdeckten Hexosephosphorsäure sich nicht deutlich ändert. In der zweiten oxydativen Phase dagegen schwindet die gebildete Milchsäure, während eine ganz bestimmte Menge Extrasauerstoff aufgenommen wird, und zwar geht der Schwund der Milchsäure während dieser Periode genau proportional dem Mehrverbrauch an Sauerstoff."

The Meyerhof cycle remained a current theory for many years, until the formulation of the Cori cycle, in which it is considered that the lactic acid produced in the first minute of work is completely removed during the next 15 minutes by diffusion into the blood and not by increased oxidation or by resynthesis into glycogen. In the Cori cycle, lactic acid is resynthesized to glycogen in the liver, liberating the glucose from which the muscles can synthesize glycogen.

Meyerhof, Lohmann and Meier[37] recorded some 10 per cent increase in glycogen in hind legs of frogs perfused with lactate. This could not be confirmed by Eggleton and Lovatt Evans[38] who found no evidence of glycogen synthesis in the muscles of cats injected intra-arterially or intravenously with lactate.

Sacks and Sacks[39], in rabbit muscles recovering from contraction, could find no evidence of glycogen synthesis, but they observed a loss of lactic acid by diffusion from muscle.

Cori and Cori[40], when feeding or subcutaneously injecting lactate to rats found at least 70 per cent retained as glycogen in the liver and defined what is now known as the Cori cycle, attributing to the liver the main part of the lactic acid resynthesis into glycogen

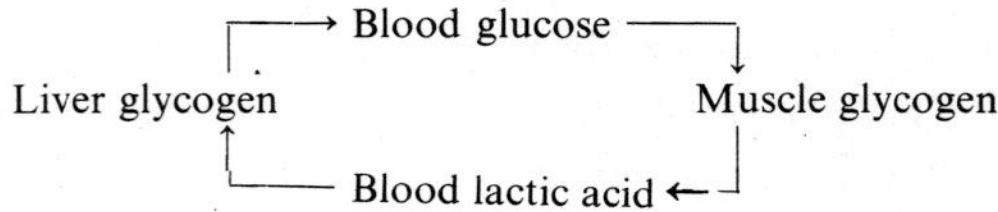

It is clear that a gluconeogenesis from lactate in muscle must be considered as negligible. The enzymes responsible for gluconeogenesis are very weak or are completely lacking in muscle, particularly in the case of the conversion of pyruvate into phosphopyruvate (Krebs and Woodford[41]).

Recently, Bendall and Taylor[42] have reconsidered the matter and have concluded that

"the total amount of lactate disappearing from frog muscle during aerobic recovery from tetanus or a period of anoxia is 5–6 times the amount that is oxidized"

and

"that the missing lactate reappears in muscle in the form of glycogen."

On the other hand, Houghton[43] has found no uptake by muscle, of lactate added during rest or recovery after severe exercise, even with a concentration of 5 mM of lactate in the perfusion fluid.

Krebs[44] underlines the abnormal conditions of the experiments of Bendall and Taylor[42] and the fact that the rates of glycogen formation observed in their experiments, if measurable,

"would be negligible in relation to overall events *in vivo*"

It appears that the experiments of Meyerhof were misleading due to the material used and the conditions of the experiments, and that in the case of active muscle or in recovery, the Cori cycle is highly significant in an equilibrium of flux involving a number of different pathways.

References p. 90

REFERENCES

1 W. Preyer, *Robert von Mayer und die Erhaltung der Energie*, Berlin, 1889.
2 G. Rosen, in C. McC. Brooks and P. F. Cranefield (Eds.), *The Historical Development of Physiological Thought*, New York, 1959.
3 E. du Bois Reymond, *Gesammelte Abhandlungen*, Vol. 2, Freiberg/Sa., 1859.
4 O. von Fürth, *Ergeb. Physiol.*, 1 (1903) 574.
5 A. Gottschalk, in *Oppenheimer's Handbuch der Biochemie*, 2e Aufl., Jena, 1925.
6 H. Fries, *Biochem. Z.*, 35 (1911) 368.
7 D. M. Needham, *Machina Carnis. The Biochemistry of Muscular Contraction in its Historical Development*, Cambridge, 1971.
8 W. M. Fletcher and F. G. Hopkins, *J. Physiol. (London)*, 35 (1907) 247.
9 J. Parnas and R. Wagner, *Biochem. Z.*, 61 (1914) 387.
10 O. Meyerhof, *Arch. ges. Physiol.*, 175 (1919) 20.
11 O. Meyerhof, *Arch. ges. Physiol.*, 175 (1919) 88.
12 O. Meyerhof, *Arch. ges. Physiol.*, 182 (1920) 232.
13 O. Meyerhof, *Arch. ges. Physiol.*, 182 (1920) 284.
14 O. Meyerhof, *Arch. ges. Physiol.*, 185 (1920) 11.
15 O. Meyerhof, *Arch. ges. Physiol.*, 188 (1921) 114.
16 O. Meyerhof, *Arch. ges. Physiol.*, 191 (1921) 128.
17 O. Meyerhof, *Arch. ges. Physiol.*, 195 (1922) 22.
18 O. Meyerhof, *Klin. Wochschr.*, 1 (1922) 230.
19 O. Meyerhof, *Klin. Wochschr.*, 3 (1924) 292.
20 O. Meyerhof, *Ergeb. Physiol.*, 22 (1923) 328.
21 O. Meyerhof, *Arch. ges. Physiol.*, 204 (1924) 295.
22 G. Embden, in *Bethe's Handbuch der n. und path. Physiol.*, Vol. 8/1 Berlin, 1925.
23 G. Embden, F. Kalberlah and H. Engel, *Biochem. Z.*, 45 (1912) 45.
24 G. Embden, W. Griesbach and E. Schmitz, *Z. physiol. Chem.*, 93 (1914–15) 1.
25 G. Embden and F. Laquer, *Z. physiol. Chem.*, 93 (1914) 94; 113 (1921) 1.
26 G. Embden and M. Zimmerman, *Z. physiol. Chem.*, 141 (1924) 225.
27 G. Embden and M. Zimmerman, *Z. physiol. Chem.*, 167 (1927) 114.
28 G. Embden and M. Zimmerman, *Z. physiol. Chem.*, 167 (1927) 137.
29 J. Pryde and E. T. Waters, *Biochem. J.*, 23 (1929) 573.
30 P. A. Levene and H. S. Simms, *J. Biol. Chem.*, 70 (1926) 319, 327.
31 O. Meyerhof and J. Suranyi, *Biochem. Z.*, 178 (1926) 427.
32 K. Lohmann, *Biochem. Z.*, 254 (1932) 381.
33 K. Lohmann, *Biochem. Z.*, 28 (1935) 120.
33a C. F. Cori and G. T. Cori, *J. Biol. Chem.*, 94 (1931) 581.
33b K. Lohmann, *Biochem. Z.*, 227 (1930) 39.
34 A. V. Hill, *J. Physiol. (London)*, 48 (1914) X.
35 W. M. Fletcher and F. G. Hopkins, *Proc. Roy. Soc. (London), Ser. B*, 89 (1917) 444.
36 O. Meyerhof, *Naturwissenschaften*, 12 (1924) 137.
37 O. Meyerhof, K. Lohmann and R. Meier, *Biochem. Z.*, 157 (1925) 459.
38 M. G. Eggleton and C. Lovatt Evans, *J. Physiol. (London)*, 70 (1930) 269.
39 J. Sacks and N. C. Sacks, *Am. J. Physiol.*, 122 (1935) 565.
40 C. Cori and G. T. Cori, *J. Biol. Chem.*, 135 (1940) 733.
41 H. A. Krebs and M. Woodford, *Biochem. J.*, 94 (1965) 436.
42 J. R. Bendall and A. A. Taylor, *Biochem. J.*, 118 (1970) 887.
43 C. R. S. Houghton, *Ph.D. Thesis*, Oxford, 1971 (cited after Krebs[44]).
44 H. A. Krebs, *Essays Biochem.*, 8 (1972) 1.

Chapter 21

The Unified Pathway of Glycolysis, and the Process of Phosphate Esterification

1. The unified pathway of glycolysis

As emphasized in Chapter 17, the discovery of cell-free fermentation did stimulate the molecular approach to intermediary metabolism, and it favoured the growth of a community of biochemists sharing the views expressed by E. Buchner and by F. G. Hopkins, and considering that the cell is the seat of chemical reactions in which substances identifiable by chemical methods undergo changes which can be followed by chemical methods (Hopkins[1]).

Studies that merged into a common trend were performed on the subjects of alcoholic fermentation and of muscle glycolysis. These studies fostered the evolution of "physiological chemistry" into biochemistry, as they led to the first formulation of a metabolic pathway, the pathway of glycolysis. Physiological chemistry had been oriented by physiological methodology. In the first decades of the present century, the molecular viewpoint was more and more often adopted by a number of leaders. After Buchner's contribution and its consequences were developed, physiological chemists more and more explored enzymological patterns of metabolism. This trend was at the origin of the individualization, in a few German Universities, of departments of physiological chemistry where the molecular aspects of metabolism were studied. One of these departments in which the new biochemistry developed was the Institute of Franz Hofmeister (1850–1922) in Strasburg. Hofmeister was a precise and imaginative mind, and at the same time a man of broad culture, who created in Strasburg a brilliant school to which Parnas, Embden and Neuberg belonged. Rotschuh[2] calls him "ein ideenreicher Fanatiker der exakten Forschung".

The first metabolic pathway which was clarified was the pathway of glycolysis, which was recognized as a common pattern subserving two different cellular phenomena, alcoholic fermentation and muscle glycolysis.

References p. 103

Cells can also take advantage of respiration as a more efficient source of metabolic energy. We shall show (Chapter 26) how it came to be recognized that it is through essentially similar ways that the utilization of metabolic energy occurs in respiration and in glycolysis. The unravelling of the glycolytic pathway was the first clue to the understanding of the sources of the free energy available in cells. Buchner and Harden were chemists. While Buchner progressively turned the major part of his activity towards biological problems, Harden remained limited in his achievements by his exclusive taste for facts and his aversion for theories. Such great architects of the first periods of biochemistry as Embden, Parnas, Warburg or Meyerhof had, in addition to their chemical knowledge, developed in the course of medical studies an understanding of biological problems which favoured the broadening of their viewpoints, to the benefit of the new discipline.

The recognition of a unified pathway of glycolysis represents a major progress as it provided the first example of the opposition to diversity at the cellular level (alcoholic fermentation in one case, muscle contraction in the other) in favour of a unity at the molecular level.

That the course of the reactions in the animal cell is very similar to that in yeast cells, and that many intermediates were the same in alcoholic fermentation as in muscle glycolysis, had already been stressed in 1913 by Neuberg[3]. This similarity was illustrated by the experiments of Embden, Griesbach and Schmitz[4] who, by adding to muscle juice the ester of Harden and Young isolated from yeast juice, observed an increase of lactic acid and of phosphoric acid production.

After the first world war another striking argument in favour of the unified pathway was provided when Meyerhof showed in 1924 that the same "coenzyme", later recognized as the same complex system, was active in fermentation and in muscle glycolysis[5]. A year later (1925), von Euler, Myrbäck and Karlsson[6] expressed the concept of the unity of the pathway in the following terms:

"Die Klarlegung des Zusammenhanges zwischen dem biochemischen Kohlehydratabbau bei den anaeroben Spaltung in höheren und niedrigen Pflanzen, besonders in der Hefe einerseits und im tierischen Muskel anderseits, scheint uns von prinzipieller Bedeutung für die Einsicht in die Zellfunktionen überhaupt zu sein und wir haben uns seit einiger Zeit damit beschäftigt, die Analogie, welche zwischen der Gärungspaltung in der Hefe und der anaeroben Kohlehydratspaltung in dem Muskel bestehen, heraus zu arbeiten."

Many other aspects, as we shall see, confirmed the unity of the pathway and, in 1939, two collaborators of Embden, Jost and Emde[7], proposed the use of the single term, glycolysis, to designate the common pathway in both instances.

"In Anbetracht der weitgehend gleichartigen Abbauvorgänge sollte die Bezeichnung Glykolyse für alle Formen des biologischen anaeroben Kohlenhydratabbaues angewendet werden, sowohl für die Milchsäurebildung in tierischen und pflanzichen Zellen, als auch für die Alkoholbildung der Hefe. Gärung ist dann die spezielle Bezeichnung für diese Formen der Glykolyse, die mit Gasentwickelung einhergehen."

The notion that glycolysis was running along the same lines in yeast and in muscle evolved from the accumulation of knowledge, with the proviso of a number of differences, the main being that pyruvic acid was transformed into acetaldehyde in the anaerobic metabolism of yeast, with a reduction to ethanol, and gave lactic acid in the anaerobic metabolism of animal cells.

2. Reactions leading to the esterification of inorganic phosphate

These were very difficult to identify. One of the reasons was that the esterification taking place at the hexose level is not the only one, and that a second kind of phosphorylation is accomplished during the oxido-reduction phase (see Chapter 24).

Before an understanding of the process of inorganic phosphate esterification in glycolysis could be set on a firm footing, it was necessary to know more about the cyclic process and about the reversibility of certain reactions. It was also necessary to know that different enzymes were active in the case of the esterification of glucose and in the esterification of polysaccharides, and to purify the enzymes in order to individualize the different steps involved.

Another difficulty was due to the ignorance of the precise differences between the enzymatic system at work in cells and the enzymatic system in the juice or in other preparations used in experimentation and derived from yeast or muscle. It therefore took a long time to reach the present stage of our knowledge of phosphate esterification, which was recognized as belonging to the pathway only in the early thirties.

When considering the fate of glucose in alcoholic fermentation, we now teach that it is phosphorylated at the expense of ATP and that the resulting G-6-P is converted into F-6-P which is phosphorylated by ATP to give FDP (F-1, 6-PP; Harden and Young ester).

In muscle, the glycogen is phosphorylated by inorganic phosphate into G-1-P which goes into G-6-P entering the path just described, and it is also through the G-6-P stage that other hexoses enter the pathway. The first stage of glycolysis, from carbohydrate to FDP, a prelude to the scission of phosphorylated hexose molecules into triosephosphates, appears quite clear and simple to the many students of our time who memorize it in preparing their examinations. In fact it was the subject of bitter and protracted polemics.

The prestige of Harden had widely spread his theory according to which the phosphorylation of hexose molecules by inorganic phosphate was a method for rendering other hexose molecules more labile and apt to undergo a scission into alcohol and CO_2 without formation of phosphorylated intermediates, a theory which was liquidated as late as the early thirties.

According to the methylglyoxal theory of Neuberg, the introductory reaction of alcoholic fermentation

$$\text{Hexose} \rightarrow \text{2-methylglyoxal} + 2\ H_2O$$

did not involve phosphorylated intermediates. Neuberg considered the formation of hexosediphosphate (Harden–Young ester) as an unphysiological process obtaining in preparations derived of yeast, but not in living cells.

On the other hand, Lebedev thought the phosphate had the function of removing dihydroxyacetone, preventing it from inhibiting further conversion of hexose to trioses, as described in his theory of fermentation[8]. But, as stated above, Harden and Young[9] showed that dihydroxyacetone does not inhibit fermentation. The introduction of the concept according to which hexosephosphates act as intermediary steps in glycolysis was in contradiction with the ruling theories and it therefore met strong opposition.

3. The concept of "stabilization stages"

The accumulation of hexosediphosphate in Harden and Young's reaction was by all means an attractive subject of experimentation. Von Euler and Nilsson[10], while considering that the accumulation of hexosediphosphate was an artifact, nevertheless suggested that from its study "ein Rückschluss auf die normalen Vorgänge" could be done, as well as the great accumulation of acetaldehyde and other substances in experiments which were realized in conditions differing from those obtaining in living cells, had led to important information.

In experiments devised to study the reaction of Harden and Young and the methods of dialyzing the juices and of adding inhibitors such as fluoride or iodoacetate to stop the process before the oxido-reduction stages of the pathway, were repeatedly used. There were logical arguments (later shown to be based on artifacts) to discard hexosediphosphate as an intermediate in the glycolysis pathway. Neuberg and Kobel[11], for instance, had observed that hexosediphosphate is not fermented by living yeast and that when fermented by yeast juice it was at a much lower speed than glucose which was later considered as a result of the alteration of ATPase (see Chapter 22).

Nilsson[12] observed that the Robison ester is fermented by living yeast and more easily than hexosediphosphate by yeast juice.

But here again the fact that it is fermented more slowly than glucose by yeast remained a stumbling block against its insertion in the pathway.

Should it perhaps be admitted that a hexosemonophosphate produced in fermentation is not in the same state as if introduced in the fermenting mixture? The concept was proposed by Meyerhof[13] at the Physiological Congress of Stockholm in 1926. Meyerhof considered that all molecules of glucose go through the stage of hexosephosphate, a concept which embodied a first rupture with Harden's views. According to the theory developed by Meyerhof and Lohmann[14] on the basis of experiments with hexosephosphates added to muscle extracts, in muscle glycolysis and alcoholic fermentation a compound of the nature of a monoester, an "active form", is an intermediate in the pathway and it transfers a phosphate residue to another molecule which becomes the "stabilized form" of hexosediphosphate, while a free molecule of triose is liberated and undergoes further degradation.

$$2\,C_6H_{12}O_6 + 2\,HR_2PO_4 \rightarrow 2\,C_6H_{11}O_5(R_2PO_4)^* + 2H_2O$$
$$\rightarrow 2\,C_3H_6O_3 + C_6H_{10}O_4(R_2PO_4)_2 + 2\,H_2O$$

(the asterisk designates an "active form").

The concepts of the active form and of the stabilized form was presented by Meyerhof[15] as an outcome of biochemical experiments and not as the result of the identification of chemical properties.

References p. 103

The concept is mentioned by D. M. Needham[16] among

> "the rather curious ideas held in the early nineteen-thirties by some workers on fermentation and glycolysis".

The background of this concept, which lasted until the publication of Cori and Cori's work on glycogen phosphorolysis, in 1936, is considered in an appendix (p. 439).

In the laboratory of Von Euler in Stockholm, another theory was formulated to account for the formation of hexosediphosphate, in which it was also considered, though through another pathway than that proposed by Meyerhof[16a], as a stabilisation product. To quote Nilsson[17]

> "The conversion of hexosemonophosphate follows by means of an oxido-reduction activated by cozymase, two molecules of hexosemonophosphate entering at the same time into the reaction, whereby results a splitting of the 6-carbon chain of both ester molecules into two 3-carbon chains. The two triosephosphate residues unite to give hexosediphosphate. Through the oxido-reduction a redistribution of energy takes place, so that on the one hand an energy-poor stable compound (hexosediphosphate) is formed, on the other hand there is formation of an energy-rich phosphate free residue, ready to break down." (translation by D. M. Needham[16])

Nilsson proposes the following scheme of the glycolytic degradation of hexoses and of polysaccharides:

```
                                          G-6-P
Polysaccharide                             ⇅
              ↘
               Hexose enol form ⇌ Hexose enol → oxido-reduction
              ⇗                     form-6-P
n-Hexose                                   ⇅
                                          F-6-P
```

According to this concept, G-6-P, F-6-P and FDP are considered as "stabilization products", a labile hexosemonophosphate being the real intermediary, a concept which was in contradiction to Harden's equation as it recognized the participation, as intermediate, of a phosphorylated compound, but which maintained hexosediphosphate outside the metabolic pathway.

4. Meyerhof's work on glycolysing extracts of muscles (1926–1927)

By grinding frog skeletal muscles at -1 to -2°C with water or isotonic KCl solutions, Meyerhof obtained cell-free filtered glycolysing extracts of muscle (glycolytic enzyme, myozymase), the action of which lasted only a few hours, even when maintained at $-3°C$. Meyerhof performed a number of experiments with such "myozymase extracts"[17–21]. Phosphate is necessary, as is the case for alcoholic fermentation by yeast juice.

Laquer[22] had observed that, in chopped muscles incubated in diluted bicarbonate, only 60 per cent of the glycogen was converted into lactic acid. Meyerhof[20] observed that, under the same conditions, the addition of phosphate produces a complete conversion of the glycogen and increases the yield of lactic acid up to threefold. He observed the formation of lactic acid from starch or glycogen as well as from other polysaccharides: amylopectin, amylose, dihexosan and trihexosan. During a very short time the extracts did act on glucose, fructose, or mannose, but not saccharose or galactose, to form lactic acid. Meyerhof showed that a coenzyme was present in boiled extracts of fresh muscle, which could replace coenzyme in alcoholic fermentation. In an "aged" muscle extract having lost glycolytic activity, the addition of boiled extract of fresh muscle restored this activity.

On the other hand, Meyerhof showed that a yeast maceration extract or plasmolyzed yeast contained a factor (later named by him hexokinase) which conferred on muscle extracts the ability to attack fermentable hexoses at a rate up to twice that of the conversion of glycogen into lactic acid by these extracts.

Hexokinase was a thermolabile substance inactivated by acids or alkalis.

When a limited amount of glucose and an excess of hexokinase is added to a muscle extract, there is a rapid formation of lactic acid and an esterification of phosphate at almost an equimolecular rate until the added glucose is used up. From this point on, the production of lactic acid diminishes and esterified phosphate is hydrolyzed in proportion to the slower rate of glycolysis which obtains[20]. These impressive experiments of Meyerhof, paraphrasing the reaction of Harden and Young with substituted components extracted from muscle were the turning point in the history of glycolysis, as they greatly emphasized the resemblances between alcoholic fermentation and muscle glycolysis. The preparation of Meyerhof's muscle extract which had been preceded by the preparation of suspensions of chopped muscle in ice-cold distilled water (Meyerhof and Lohmann[23]) put the glycolysis studies on a new experimental plane, as it allowed a study of the formation of lactic acid by the rapid determination allowed by the manometric method of Warburg's apparatus[24]. Yeast

References p. 103

hexokinase was recognized later as catalyzing the phosphorylation of glucose according to the following reaction[25–28]

$$\text{Glucose} + \text{ATP} \rightarrow \text{G-6-P} + \text{ADP}$$

This concept put an end to the theory of the direct phosphorylation of glucose by inorganic phosphate as had been accepted since Harden.

But it must be recalled that, until 1932, Meyerhof maintained that glucose phosphorylation went on by direct transfer from ATP to an active hexose (resulting from the action of hexokinase). It was first believed that in the phosphorylation of glucose or fructose by ATP in the presence of hexokinase, ATP was transformed into AMP in the course of the process. Later (1941) Colowick and Kalckar[29] showed that, in yeast, not two but one phosphate group is transferred to glucose or fructose, according to the equation:

$$\text{ATP} + \text{hexose} \rightarrow \text{ADP} + \text{hexose-6-P}$$

It was observed that only the fructofuranose form of D-fructose is converted by yeast into F-6-P. When myokinase from muscle is added to the yeast extract, a second phosphate of ATP can be transferred, as shown by Colowick and Kalckar[29] and by Kalckar[32,33]:

$$\text{ADP} + \text{glucose} \rightarrow \text{AMP} + \text{G-6-P}$$

Colowick and Kalckar[29] and Kalckar[32,33] have studied the system and have found that, while the hexokinase only catalyses the reaction:

$$\text{Glucose} + \text{ATP} \rightarrow \text{G-6-P} + \text{ADP}$$

the myokinase catalyses a dismutation of the ADP to give ATP

$$2\ \text{ADP} \rightleftharpoons \text{ATP} + \text{AMP}$$

ATP reacting again with the hexose to give more phosphosugar.

5. The phosphorolysis of glycogen

Until late in the 1930's the mechanism of the rapid esterification taking place, for instance, in fresh muscle press-juice with glycogen or with hexosemonophosphate as substrate, remained a controversial subject.

In 1935, the school of Parnas[34] still considered hexosediphosphate as being formed by a direct action of ATP on glycogen. But in that year,

Parnas and Baranowski[35] observed that in the absence of ATP, dialyzed extracts and inorganic phosphate, after an addition of Mg, could phosphorylate glycogen to a difficultly hydrolysable ester. They suggested that this ester was the Embden ester, *i.e.* an equilibrium mixture of G-6-P and F-6-P. This was confirmed by Ostern and Guthke[36] who also showed that only in the presence of ATP could the hexosemonophosphate be phosphorylated to hexosediphosphate.

As shown by Lundsgaard[37], the esterification of glycogen is completely prevented by phlorizin in muscle *brei* treated with fluoride. Ostern, Guthke and Terzakowec[38] showed that *M*/200 phlorizin had no effect on the formation of FDP from hexosemonophosphate in autolyzed and dialyzed muscle extract, while the phosphorylation of glycogen by inorganic phosphate is, in these conditions, inhibited up to 75 per cent.

Working with intact frog muscle, Cori and Cori[39] in 1936 found that if the muscles were soaked in solutions of adrenaline and iodoacetate, there was an accumulation of hexosemonophosphate and a decrease of inorganic phosphate, while the concentration of creatine phosphate was only slightly decreased. But if iodoacetate alone was used, there was an accumulation of FDP. No inorganic phosphate was liberated, in spite of a decrease in creatine phosphate.

In the same report[39], Cori and Cori showed that, in a mince of frog muscle, washed with water and consequently unable to form lactic acid and containing only traces of acid-soluble inorganic phosphate, the hexosemonophosphate increased in the presence of isotonic phosphate buffer and that this increase was enhanced by small amounts of ATP (considered at the time as transferring phosphate to glycogen). The authors found that the ester first formed was not G-6-P. It was not reducing and it was readily hydrolysable in acid, properties which pointed to G-1-P. Cori, Colowick and Cori[40] prepared the synthetic G-1-P and found that it was converted to G-6-P when added to dialyzed frog muscle extract. Parnas[41] pointed out that the reaction between glycogen and inorganic phosphate could be considered as an addition of one atom of hydrogen to the glycogen residue, and of the phosphate residue to the esterified glucose unit in the presence of "phosphorylase", and did not involve water. He called the process phosphorolysis.

In the presence of phosphoglucomutase, G-1-P is converted into G-6-P, as found by Cori and Cori[42,43]. The enzyme has been isolated and crystallized from rabbit muscles by Najjar[44]. It has been found in yeast

References p. 103

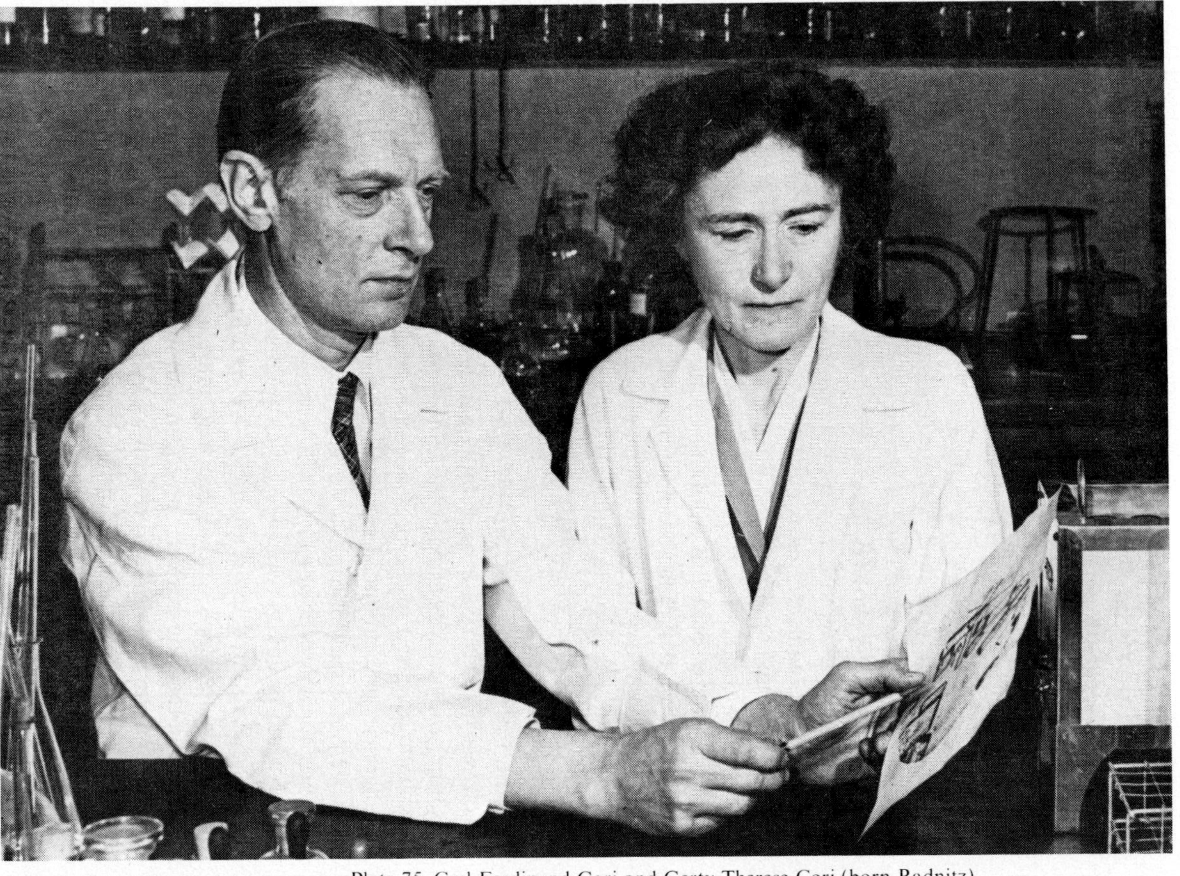

Plate 75. Carl Ferdinand Cori and Gerty Therese Cori (born Radnitz).

and in a number of tissues[45]. It requires the presence of G-1,6-PP and of Mg, Mn or Co ions[46].

$$\text{G-1,6-PP} + \text{G-1-P} \rightarrow \text{G-6-P} + \text{G-1,6-PP}$$

6. From G-6-P to F-6-P

In 1933, Lohmann[47] identified an enzyme (glucosephosphate isomerase) catalyzing the establishment of an equilibrium between G-6-P and F-6-P, when it was added either to G-6-P or to F-6-P.

On the other hand, while acid hydrolysis transforms FDP into F-6-P, in the presence of an active juice deprived by dialysis of its coenzyme, and to which Mg ions are added, FDP is not transformed into F-6-P, but into G-6-P. When the organic components composing the "coenzyme", and Mg ions are present, no hydrolysis takes place, but in the presence of Mg ions, *one* phosphate residue is released. But why is the resulting product G-6-P and not F-6-P? When F-6-P is added to a muscle extract, even deprived of Mg, it is transformed into G-6-P or more exactly, into a mixture of G-6-P (70 per cent) and F-6-P (30 per cent) in equilibrium. If it is G-6-P which is added, the same form of equilibrium of G-6-P and F-6-P is reached. From these experimental data, Meyerhof[48] concluded that, in yeast maceration juice, G-6-P is formed first and that it is converted into F-6-P. The enzyme catalyzing the conversion of G-6-P to F-6-P (glucosephosphate isomerase) has since been found in yeast, in plants and in muscle extracts[49,50]. G-6-P and F-6-P can also be formed directly from glucose or fructose in the presence of hexokinase, as stated above.

7. From F-6-P to FDP (Phosphofructokinase).

It was found by Ostern, Guthke and Terszakowec[51] that F-6-P, originating either from fructose or from G-6-P, was converted into FDP. This reaction, confirmed by von Euler and Adler[52] and by Lutwak-Mann and Mann[26], takes place in the presence of Mg^{2+}, of ATP and of the enzyme phosphofructokinase[51] transferring the terminal group of ATP to F-6-P.

In a paper published in 1938, Parnas[53] gave an account of all the work accomplished in his laboratory in Lwow. Soon afterwards, the activities of this great center of research were to come to a complete

References p. 103

halt due to the war. In this paper, Parnas formulates the reactions involving hexose-6-P phosphorylation as:

$$\text{Hexose-6-phosphate} + \text{ATP} + H_2O \rightarrow \text{fructose-1,6-diphosphate} + \text{adenylic acid}$$

To quote from a letter of Th. Mann, who was one of the collaborators of Parnas in Lwow, to the author:

"Although one might say that Parnas recognized the role of ADP in P-transfer reactions, he looked upon that nucleotide as a P-donor rather than a P-acceptor. It was only later when the intermediary enzymes were purified [and in particular, when diphosphoglyceric acid was also recognized as an intermediary] that the right course was established, *i.e.*

$$\text{Hexose-6-phosphate} + \text{ATP} \rightarrow \text{Fructose-1,6-diphosphate} + \text{ADP"}$$

The whole process of the Harden and Young reaction, through the series of studies mentioned above, was therefore recognized as being accomplished through the sequence of G-6-P, F-6-P and FDP. In the case of polysaccharides, G-1-P was first formed and transformed into G-6-P which followed the same pathway in which the function of ATP as phosphorylation coenzyme (glucose → G-6-P; G-6-P → G-1,6-PP) is identified.

In 1931, a chemist, Ohle, had published[54] a theoretical study on the possible intermediates in the first phase of glycolysis, and pointed to the notion according to which enzymes were mere catalysts, unable to accomplish what is not an inherent property of the chemical nature of substrates. The first stage of glycolysis was regarded by Ohle as a protection of the aldehyde function. Glucose is transformed into G-6-P and thereby prepared for its isomerization into F-6-P which is again phosphorylated into FDP. This discussion certainly favoured the adoption of the pathway leading from glucose to hexosediphosphate which was considered by Ohle as an intermediate in the pathway of glycolysis.

The progress reported in this chapter afforded a knowledge of the mechanism of Harden and Young reaction, but from the point of view of the understanding of glycolysis, the theory reigning at the end of the 1920's was an "active hexosemonophosphate" theory, as introduced by Meyerhof in 1926. More than twenty years after its discovery and that of the esterification of phosphates, hexosediphosphate was still considered as a mere "stabilization product" playing no role in the glycolysis pathway.

REFERENCES

1 F. G. Hopkins, *Irish J. Med. Sci.*, (1932) 333.
2 K. E. Rotschuh, *Geschichte der Physiologie*, Berlin, 1953.
3 C. Neuberg, *Die Gärungsvorgänge und der Zuckerumsatz*, Jena, 1913.
4 G. Embden, W. Griesbach and E. Schmitz, *Z. physiol. Chem.*, 93 (1914–1915) 1.
5 O. Meyerhof, *Z. physiol. Chem.*, 101 (1924) 165; 102 (1924) 185.
6 H. von Euler, K. Myrbäck and S. Karlsson, *Z. physiol. Chem.*, 143 (1925) 243.
7 H. Jost and H. Emde, *Z. physiol. Chem.*, 261 (1939) 225.
8 A. von Lebedev, *Ber. deut. chem. Ges.*, 47 (1914) 660, 995.
9 A. Harden and W. J. Young, *Biochem. Z.*, 40 (1912) 485.
10 H. von Euler and R. Nilsson, *Z. physiol. Chem.*, 148 (1925) 211.
11 C. Neuberg and M. Kobel, *Biochem. Z.*, 166 (1925) 488.
12 R. Nilsson, *Svensk Vet. Akad. Arkiv Kemi*, 9 (1928) 1.
13 O. Meyerhof, *Naturwiss.*, 14 (1926) 1175.
14 O. Meyerhof and K. Lohmann, *Biochem. Z.*, 185 (1927) 113.
15 O. Meyerhof, *Die chemische Vorgänge in Muskel*, Berlin, 1930.
16 D. M. Needham, *Machina Carnis. The Biochemistry of Muscular Contraction in its Historical Development*, Cambridge, 1971.
16a O. Meyerhof and W. Kiessling, *Biochem. Z.*, 283 (1936) 83.
17 R. Nilsson, *Biochem. Z.*, 258 (1933) 198.
18 O. Meyerhof, *Biochem. Z.*, 178 (1926) 395.
19 O. Meyerhof, *Biochem. Z.*, 178 (1926) 462.
20 O. Meyerhof, *Biochem. Z.*, 183 (1927) 176.
21 O. Meyerhof and K. Lohmann, *Biochem. Z.*, 168 (1926) 128.
22 F. Laquer, *Z. physiol. Chem.*, 93 (1914) 60.
23 O. Meyerhof and K. Lohmann, *Biochem. Z.*, 168 (1926) 128.
24 O. Warburg, *Biochem. Z.*, 164 (1925) 481.
25 H. von Euler and E. Adler, *Z. Physiol. Chem.*, 235 (1935) 122.
26 C. Lutwak-Mann and T. Mann, *Biochem. Z.*, 281 (1935) 140.
27 O. Meyerhof, *Naturwiss.*, 23 (1935) 850.
28 S. P. Colowick and H. M. Kalckar, *J. Biol. Chem.*, 148 (1943) 117.
29 S. P. Colowick and H. M. Kalckar, *J. Biol. Chem.*, 137 (1941) 789 (reprinted in Kalckar[55]).
30 A. Gottschalk, *Austral. J. Exptl. Biol. Med. Sci.*, 21 (1943) 13.
31 R. H. Hopkins and M. Horwood, *Biochem. J.*, 47 (1950) 95.
32 H. M. Kalckar, *J. Biol. Chem.*, 143 (1942) 299 (reprinted in Kalckar[55]).
33 H. M. Kalckar, *J. Biol. Chem.*, 153 (1944) 355.
34 J. K. Parnas, *Klin. Wochschr.*, 14 (1935) 1017.
35 J. K. Parnas and T. Baranowski, *Compt. Rend. Soc. Biol.*, 120 (1935) 307.
36 P. Ostern and J. A. Guthke, *Compt. Rend. Soc. Biol.*, 121 (1936) 282.
37 E. Lundsgaard, *Biochem. Z.*, 264 (1933) 209.
38 P. Ostern, J. A. Guthke and J. Terszakowec, *Compt. Rend. Soc. Biol.*, 121 (1936) 1133.
39 G. T. Cori and C. F. Cori, *J. Biol. Chem.*, 116 (1936) 119.
40 C. F. Cori, S. P. Colowick and G. T. Cori, *J. Biol. Chem.*, 121 (1937) 465.
41 J. K. Parnas, *Ergeb. Enzymforsch.*, 6 (1937) 57.
42 C. F. Cori and G. T. Cori, *Proc. Soc. Exptl. Biol. Med.*, 34 (1936) 702 (reprinted in Kalckar[55]).
43 G. T. Cori and C. F. Cori, *Proc. Soc. Exptl. Biol. Med.*, 36 (1937) 119.
44 V. A. Najjar, *J. Biol. Chem.*, 175 (1948) 281.
45 G. T. Cori, S. P. Colowick and C. F. Cori, *J. Biol. Chem.*, 123 (1938) 375.

46 G. T. Cori, S. P. Colowick and C. F. Cori, *J. Biol. Chem.*, 124 (1938) 543.
47 K. Lohmann, *Biochem. Z.*, 262 (1933) 137.
48 O. Meyerhof, *Biochem. Z.*, 273 (1934) 80.
49 C. S. Hanes, *Proc. Roy. Soc. (London), Ser. B*, 129 (1940) 174.
50 G. F. Somers and E. L. Cosby, *Arch. Biochem.*, 6 (1945) 295.
51 P. Ostern, J. A. Guthke and J. Terszakowec, *Z. physiol. Chem.*, 243 (1936) 9.
52 H. von Euler and E. Adler, *Z. physiol. Chem.*, 235 (1935) 122.
53 J. K. Parnas, *Enzymol.*, 5 (1938) 166.
54 H. Ohle, *Ergeb. Physiol.*, 33 (1931) 558.
55 H. M. Kalckar, *Biological Phosphorylations. Development of Concepts*, Englewood Cliffs, 1969.

Chapter 22

The Scission of Hexosediphosphate into Triosephosphates

1. Kluyver and Struyk's theory

At the end of the 1920's, it was generally accepted, as shown in Chapter 21, that an "active form" of hexosemonophosphate was an intermediate in the glycolytic pathway, the sequence of steps between this "active" hexosemonophosphate and pyruvate, also recognized as an intermediate (Chapter 17), remaining unknown. A tentative answer to the problem of the scission of the hexose molecule was formulated by Kluyver and Struyk[1]. These authors suggested that two categories of catalysts were at work in the glycolytic pathway: agents of esterification and agents of oxido-reduction. In the oxido-reduction process, as taking place in living yeast, the "active" hexosemonophosphate was considered as splitting into two parts: a molecule of glyceraldehyde and a molecule of triosephosphate. In the case of experiments with yeast juice, according to the authors, the process is upset and a variable proportion of the intermediates condenses and is stabilized in the form of hexosemonophosphate or of hexosediphosphate.

In the theory of Kluyver and Struyk, it was supposed that the introduction of one phosphate molecule into a hexosephosphate molecule renders that molecule able to split into two parts, triosephosphate and triose, the triose being the substrate of the oxido-reduction which liberates energy.

2. Phosphoglyceric acid

After 1929, Nilsson published a series of papers[2–5] mainly concerned with the first steps of glycolysis and particularly with the phenomena of phosphorylation and of the scission of the 6-carbon chain. At the time, as stated above, a growing number of biochemists believed that a phosphoryl-

References p. 115

ation took place before the scission of the carbon chain, as proposed by Meyerhof's school. A requisite was that the ester involved should be fermented at least as rapidly as glucose by yeast and neither G-6-P, F-6-P nor FDP could fulfill this requirement, due, as generally believed today, to the alteration of ATPase in yeast juice. But in the case of hexosediphosphate, as had been shown by Harden, there was a striking parallelism between its formation and CO_2 evolution which was difficult to account for by the "stabilization product" concept.

In one of the experiments performed, Nilsson incubated a dried yeast suspension with FDP and acetaldehyde in the presence of fluoride, and isolated 3-phosphoglyceric acid from the products.

$$\begin{array}{l} CH_2OH \\ | \\ CHOH \\ | \\ CH_2{-}O{-}PO_3H_2 \\ \quad \alpha \end{array} \qquad\qquad \begin{array}{l} CH_2OH \\ | \\ CH{-}O{-}PO_3H_2 \\ | \\ CH_2OH \\ \quad \beta \end{array}$$

Glycerophosphoric acid
(phosphorylated glycerol)

$$\begin{array}{l} COOH \\ | \\ CH{-}O{-}PO_3H_2 \\ | \\ CH_2OH \\ 2 \text{ or } \alpha \end{array} \qquad\qquad \begin{array}{l} COOH \\ | \\ CHOH \\ | \\ CH_2{-}O{-}PO_3H_2 \\ 3 \text{ or } \beta \end{array}$$

Phosphoglyceric acid
(phosphorylated glyceric acid)

At about the same time, in Meyerhof's laboratory, Lohmann had observed that in fluoride-inhibited rabbit muscle extracts, the phosphate added in the form of FDP turned acid stable. The product he called[6] "Ester I". With Lipmann who worked at that time in Meyerhof's laboratory, Lohmann showed that the effect could arise without addition of fluoride in winter frog muscles in which the glycolysis of FDP to lactic acid was minimal[7]. Lipmann[8] recalls that, just after this research was completed, Nilsson visited Meyerhof's laboratory at Heidelberg:

"Nilsson and I stood around for a long time discussing his baffling findings but we were completely blind to the close connection between his conversion of FDP + acetaldehyde to PGA + ethanol and our conversion of FDP to esters difficult to hydrolyze. We saw the trees, but not the woods."

3. First scheme of the oxido-reduction events following the scission of FDP

In the theoretical paper of Ohle[9] cited in the previous chapter and published after the identification of 3-phosphoglyceric acid in Nilsson's experiments, it was suggested on a chemical basis that after the sequence of steps leading to FDP

$$\text{G-6-P} \rightarrow \text{F-6-P} \rightarrow \text{FDP}$$

the molecule of hexosediphosphate is dehydrogenated into 5-keto-fructose-1,6-PP which spontaneously decomposes into one molecule of phospho-dihydroxyacetone and one molecule of oxymethyl-glyoxalphosphate. The latter accepts the hydrogen resulting from the oxidation of FDP and becomes phosphoglyceric acid. The two triosephosphates lose their phosphate residues and are transformed into methylglyoxal. Ohle's theory thereby joins the theory of Neuberg which was soon to be discarded.

The identification of 3-phosphoglyceric acid by Nilsson in 1930 became the origin of new and decisive developments. During the summer of 1930, Nilsson stayed for some time in Embden's laboratory at Frankfurt am Main where he studied the phosphorylations occurring in rabbit muscle juice in the presence of fluoride and acetaldehyde, an aspect he had already studied in Sweden using dried muscle in a mixture containing glycogen, HDP, fluoride and acetaldehyde (Nilsson[3]).

After leaving Embden's laboratory, Nilsson paid a visit to Meyerhof's laboratory in Heidelberg, as alluded to by Lipmann[8]. These visits gave Nilsson many opportunities to discuss his finding of 3-phosphoglyceric acid with Embden and his collaborators as well as with Meyerhof and with Lohmann and Lipmann. Nilsson considered 3-phosphoglyceric acid as a side-product of yeast fermentation and did not recognize it as an intermediate in glycolysis, a concept which was introduced by the Embden's school.

Embden, Deuticke and Kraft[10] published in 1933 a very important preliminary paper which was to be the last publication during his life-time to bear the signature of Embden, who died suddenly in July 1933*.

The authors isolated phosphoglyceric acid from muscle mince to which they had added FDP and fluoride. They suggested that Lohmann's acid-stable ester, obtained under similar conditions, was a mixture of two phosphorylated 3-carbon derivatives formed during the scission of FDP,

---

* A complete report written by Deuticke who was at the time in Svedberg's laboratory, appeared in 1934 together with other papers concerning the last studies performed under Embden's direction[37–43].

References p. 115

Plate 76. Ragnar Nilsson.

while it had been believed before by Meyerhof and Lohmann, as well as by Nilsson, that it was an *hexosemonophosphate* which underwent the scission, leading to one triosephosphate and one triose. The recognition of *hexosediphosphate* as an intermediate discarded the obstacle which resulted from Harden's views formulated 25 years before.

The scission of FDP is formulated by Embden, Deuticke and Kraft, as follows:

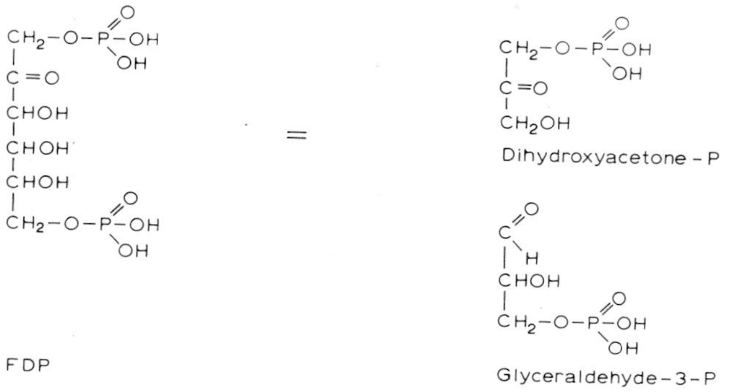

CH_2–O–P(=O)(OH)–OH
|
C=O
|
CHOH
|
CHOH
|
CHOH
|
CH_2–O–P(=O)(OH)–OH

FDP

=

CH_2–O–P(=O)(OH)–OH
|
C=O
|
CH_2OH

Dihydroxyacetone - P

C(=O)H
|
CHOH
|
CH_2–O–P(=O)(OH)–OH

Glyceraldehyde-3-P

A dismutation of the two triosephosphates takes place

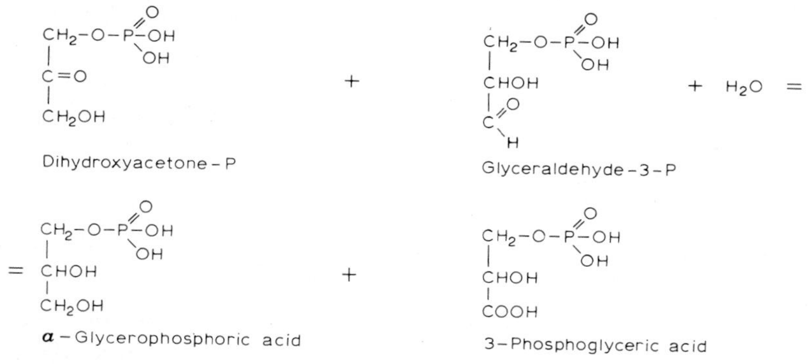

CH_2–O–P(=O)(OH)–OH
|
C=O
|
CH_2OH

Dihydroxyacetone - P

+

CH_2–O–P(=O)(OH)–OH
|
CHOH
|
C(=O)H

Glyceraldehyde-3-P

+ H_2O =

CH_2–O–P(=O)(OH)–OH
|
= CHOH
|
CH_2OH

α - Glycerophosphoric acid

+

CH_2–O–P(=O)(OH)–OH
|
CHOH
|
COOH

3-Phosphoglyceric acid

The scission of phosphoglyceric acid into phosphoric acid and pyruvic acid is formulated by the authors as follows:

CH_2–O–P(=O)(OH)–OH
|
CHOH
|
COOH

3-Phosphoglyceric acid

=

CH_3
|
C=O
|
COOH

Pyruvic acid

+ H_3PO_4

References p. 115

Glyceric acid is not produced and there is no participation of water. Pyruvate is reduced to lactic acid at the expense of the oxidative formation of triosephosphoric acid from the α-glycerophosphoric acid supposed to have been formed in a previous step.

$$\begin{matrix} CH_3 \\ | \\ CO \\ | \\ COOH \end{matrix} + \begin{matrix} CH_2O-P(=O)(OH)-OH \\ | \\ CHOH \\ | \\ CH_2OH \end{matrix} = \begin{matrix} CH_3 \\ | \\ CHOH \\ | \\ COOH \end{matrix} + \begin{matrix} CH_2O-P(=O)(OH)-OH \\ | \\ CHOH \\ | \\ C(=O)H \end{matrix}$$

Pyruvic acid | α-Glycerophosphoric acid | Lactic acid | Triosephosphoric acid

In this scheme, not only does pyruvic acid appear as an intermediate of glycolysis, but it is also the case with α-glycerophosphate, which was actually identified[40] but had to be discarded when, as seen below, it was shown that the reduction of pyruvate to lactic acid was due to triosephosphates. We also know that the accumulation of glycerophosphate was the consequence of the presence of fluoride preventing the formation of pyruvate from phosphoglyceric acid. In this first coherent scheme of the oxido-reduction events following the scission of FDP, it was visualized for the first time that FDP yields one keto- and one aldo-triosephosphates.

Pyruvic acid results from the decomposition of the phosphoglyceric acid supposed to result from a dismutation of hexosediphosphate leading to α-glycerophosphoric acid and to 3-phosphoglyceric acid.

As Lipmann[8] remarks, Embden and his colleagues

> "managed to piece together, courageously and with great vision, the first coherent scheme of the oxidoreductive events that followed the scission of FDP."

As stated in Chapter 17 there were valid arguments (mainly due to Neuberg) in favour of considering pyruvic acid as an intermediate in alcoholic fermentation. Is it also an intermediate in muscle glycolysis? In 1929, A. Hahn[11] had found pyruvic acid among the products of muscle glycolysis. But he had suggested that, in anaerobic conditions it was an oxidation product of lactic acid. Meyerhof and McEachern[12], in 1933, found with the help of bisulphite as a trapping agent, that when glycogen or FDP is added to muscle extracts, pyruvic acid is produced. What is the reducing partner in these muscle extracts? Meyerhof and Kiessling[13] identified it as phosphoglyceric acid.

4. Embden's and Meyerhof's unified theory of glycolysis

At the stage reached at that date in the study of muscle glycolysis, it remained to Embden, Deuticke and Kraft[10,37] to propose a new scheme of yeast fermentation in which, contrary to Neuberg's scheme in which a dismutation was postulated between acetaldehyde and methylglyoxal, glyceraldehyde-3-P entered into the pathway. The D-form of glyceraldehyde-3-P had been shown to be easily fermentable[14]. It was thought that it could arise from glucose:

$$\underset{\text{Glucose}}{C_6H_{12}O_6} + 2\ H_3PO_4 \rightarrow 2\ \underset{\text{Glyceraldehyde-3-P}}{CH_2O(PO_3H_2)\cdot CHOH\cdot CHO}$$

and the oxidation–reduction phase was represented as follows:

$$\underset{\text{Glyceraldehyde-3-P}}{CH_2O(PO_3H_2)CHOH\ CHO} \xrightarrow{\text{oxidation}} \underset{\text{3-Phosphoglyceric acid}}{CH_2O(PO_3H_2)CHOH\ COOH}$$

In this scheme, the phosphoglyceric acid formed is converted into pyruvate thence to acetaldehyde which is reduced to alcohol:

$$\underset{\text{Pyruvic acid}}{CH_3CHO} \xrightarrow{\text{reduction}} \underset{\text{Ethanol}}{CH_3CH_2OH}$$

As long as phosphate and sugar are present, the reactions are supposed to proceed as follows:

(*A*) Glucose → glyceraldehyde-3-P → 3-phosphoglyceric acid

(*B*) 3-phosphoglyceric acid → pyruvic acid → acetaldehyde → CO_2

(*C*) Acetaldehyde → ethanol

During phase *A*, acetaldehyde, not yet being formed, cannot play the role of dismutation partner for glyceraldehyde-3-P, a role which is played by another molecule of glyceraldehyde-3-P. From this Cannizzaro reaction, phosphoglyceric acid and glycerophosphoric acid are supposed to arise:

$$2\ \underset{\text{Glyceraldehyde-3-P}}{CH_2O(PO_3H_2)CHOH\cdot CHO} + H_2O \rightarrow \underset{\text{Phosphoglyceric acid}}{CH_2O(PO_3H_2)CHOH\cdot COOH} + \underset{\text{Glycerophosphoric acid}}{CH_2O(PO_3H_2)CHOH\cdot CH_2OH}$$

When sufficient amounts of acetaldehyde are present, one mole of glyce-

References p. 115

raldehyde-3-P is replaced by one mole of acetaldehyde. A trace of FDP is considered to be necessary at the start, as a catalyst (this phase is called "stationary phase" by Meyerhof). When the FDP is formed according to Harden and Young's equation, it is used, along with glucose and phosphoric acid, in the formation of glyceraldehyde-3-P

$$1\ \text{FDP} + 1\ \text{glucose} + 2\ \text{phosphoric acid} \rightarrow 4\ \text{glyceraldehyde-3-P}$$

In order to compare a number of aspects of lactic acid glycolytic production in muscle and alcohol production in fermentation by yeast, Meyerhof and Kiessling[15] applied to yeast maceration juice the methods (use of inhibitors, etc.) they had relied upon for the study of muscle juice and confirmed the unity of the pathway of glycolysis in both instances. They showed that glycerophosphoric acid is not an intermediary step in glycolysis. Added to a fresh maceration juice of yeast, phosphoglyceric acid is transformed into acetaldehyde and CO_2 by way of pyruvate. The effect is not increased by an addition of α-glycerophosphoric acid. This acid does not react with any other intermediary product of alcoholic fermentation and for instance, not with acetaldehyde.

In order to account for the formulation given by Harden and Young's reaction, Meyerhof and Kiessling[15] distinguished two phases of glycolysis, an "Angärungsphase", and a stationary phase. As they had observed the fermentability of phosphopyruvic acid, they revised their scheme of glycolysis and considered as intermediates on the way to acetaldehyde the sequence of triosephosphates to pyruvate (phosphodihydroxyacetone + glyceraldehyde-3-P), phosphoglyceric acid, phosphopyruvic acid and pyruvic acid[6]. The concept of the phases of glycolysis, representing a difference between yeast fermentation and muscle glycolysis, remained until 1945 when Meyerhof[16] himself showed that the occurrence of two phases was an artifact resulting from the fact that in yeast maceration juice, but not in muscle juice, adenylpyrophosphatase is destroyed in the drying and extracting processes. A complete conformity was thereby established between both forms of glycolysis.

One of the most important advances of the decade following 1926 was the realization that the formation of sugar esters is not, as Harden had proposed, a side reaction of glycolysis, but an intermediate reaction. In the course of successive catabolic steps the formation of sugar esters is followed by the production of a series of derived phosphate compounds.

5. Identification of the triosephosphates resulting from the scission of FDP

As recalled above, several authors, during the first phases of alcoholic fermentation studies assumed that in the course of the process, the hexose splits into two trioses, but no evidence was produced. When Harden and Young discovered hexosediphosphate, this compound had been wrongly considered by Ivanov to be triosephosphate. But no triosephosphate had been synthesized before 1932, when Fischer and Baer[17,18] synthesized glyceraldehyde-3-P, the D-compound of which was shown to be easily fermented by yeast maceration juice[14].

Meyerhof and Lohmann[19] discovered in 1934 that in dialyzed extracts of muscle and yeast, F-1,6-PP in the presence of an enzyme they called "zymohexase" is reversibly split into two molecules of phosphotriose which they identified as phosphodihydroxyacetone*. Two years later, Meyerhof, Lohmann and Schuster[21] corrected this conclusion and showed that the phosphotriose obtained is a mixture of phosphodihydroxyacetone and glyceraldehyde-3-P. The reversible reaction between F-1,6-PP and the two phosphotrioses has been shown to be catalyzed by the enzyme aldolase, and the two phosphotrioses were found to reach an equilibrium (about 95% of phosphodihydroacetone and 5% of phosphoglyceraldehyde[22,23]) catalyzed by the enzyme triosephosphate isomerase. This work confirmed the validity of the consideration of F-1,6-PP as an intermediate in the glycolysis pathway.

Meyerhof[24] later summarized as follows the evidence for the composite nature of what was first called "zymohexase":

"In the following years, the composite nature of this enzyme system was deduced from the following observations:

(*1*) The D-compound of synthetic 3-glyceraldehyde phosphate was rapidly transformed into dihydroxyacetone phosphate by extracts containing zymohexase, in this way showing the presence of a triosephosphate isomerase. (ref. 22)

(*2*) Other aldehydes including free D-glyceraldehyde reacted with dihydroxyacetone phosphate by aldol condensation between the aldehyde group and the free alcoholic group of the keto triose. This partial enzyme of the zymohexase system was therefore called "aldolase" (ref. 21)

(*3*) Interception of the triosephosphate by means of hydrazine during the splitting of hexosediphosphate gave about equal amounts of glyceraldehyde phosphate and dihydroxyacetone phosphate, definitely proving that the zymohexase reaction was a sequence of two steps.

* At the suggestion of Meyerhof, Kiessling[20] attempted the synthesis of phosphodihydroxyacetone, but the product obtained by biochemical methods was not related to glyceraldehyde phosphate.

As a result of the work of Fischer and Baer[28] and Kiessling and Schuster[29], dihydroxyacetone phosphate was recognized as the precursor of the biological L-α-glycerophosphate and D-glyceraldehyde phosphate as the precursor of D-3-phosphoglyceric acid."

The equilibrium is in favour of phosphodihydroxyacetone but, as observed by Kiessling and Schuster[29] the ultimate end compound produced is D-3-phosphoglyceric acid, which shows that further breakdown occurs *via* the D-glyceraldehyde-3-P. The glyceraldehyde-1-P is not fermented by yeast.

The enzyme catalyzing the cleavage of F-1,6-PP, called aldolase, has been crystallized by Warburg and Christian[30] from rat muscle and by Taylor, A. Green and Cori[31] from rabbit muscle. Its action has been studied in detail by Herbert, Gordon, Subranmayan and D. E. Green[32].

6. The liquidation of the methylglyoxal theory and of the Harden equation

It may be said that the follow-up of the finding of 3-phosphoglyceric acid by Nilsson in his experiments started the process which gave the death blow to the methylglyoxal theory. It had already been shaken when Lohmann[33], in 1932, showed that if a muscle extract is dialyzed and if to such a solution deprived of glutathione (the coenzyme of methylglyoxalase) the coenzymes of glycolysis are added, the extract transforms glycogen into lactic acid, though it had no action on methylglyoxal.

After the papers by Embden, Deuticke and Kraft, and by Meyerhof and his collaborators no place was left for the methylglyoxal theory, and it collapsed, leaving the field to what was from thereon called the Embden–Meyerhof pathway.

It is interesting to note that the reasons which led to discarding methylglyoxal as an intermediary step of glycolysis were not always confirmed afterwards. One of these reasons was that glyoxalate produced D-lactate, a substance considered at the time as unnatural. It has more recently been shown that the D-lactate is the only form metabolized by kidney and liver mitochondria of the rabbit[34,36]. This illustrates the danger of deciding that a compound is unnatural. Another example is the case of D-amino acid oxidases, considered for some time as an artifact on the basis of the assumed unnatural character of D-amino acids, since recognized as occasionally occurring in biological material.

REFERENCES

1 A. J. Kluyver and A. P. Struyk, *Koninkl. Ned. Akad. Wetenschap., Proc.*, 30 (1927) 871; 31 (1928) 882.
2 R. Nilsson, *Svensk. Kem. Tidskr.*, 41 (1929) 169.
3 R. Nilsson, *Kgl. Svenska Vetenskapsakad. Arkiv Kemi*, 10 A, No. 7 (1930).
4 R. Nilsson, *Biochem. Z.*, 258 (1933) 198.
5 R. Nilsson, *Angew. Chem.*, 46 (1933) 647.
6 K. Lohmann, *Biochem. Z.*, 222 (1930) 324.
7 F. Lipmann and K. Lohmann, *Biochem. Z.*, 222 (1930) 389.
8 F. Lipmann, *Wanderings of a Biochemist*, New York, 1971.
9 H. Ohle, *Ergeb. Physiol.*, 33 (1931) 558.
10 G. Embden, H. J. Deuticke and G. Kraft, *Klin. Wochschr.*, 12 (1933) 213 (a translation in English is found in Kalckar[44]).
11 A. Hahn, *Z. Biol.*, 88 (1929) 516.
12 O. Meyerhof and D. McEachern, *Biochem. Z.*, 260 (1933) 417.
13 O. Meyerhof and W. Kiessling, *Biochem. Z.*, 264 (1933) 40.
14 C. V. Smythe and W. Gerischer, *Biochem. Z.*, 260 (1933) 414.
15 O. Meyerhof and W. Kiessling, *Biochem. Z.*, 281 (1935) 249.
16 O. Meyerhof, *J. Biol. Chem.*, 157 (1945) 105.
17 H. O. L. Fischer and E. Baer, *Ber. deut. chem. Ges.*, 65 (1932) 337.
18 H. O. L. Fischer and E. B. Baer, *Ber. deut. chem. Ges.*, 65 (1932) 1040.
19 O. Meyerhof and K. Lohmann, *Biochem. Z.*, 271 (1934) 89.
20 W. Kiessling, *Ber. deut. chem. Ges.*, 167 (1934) 869.
21 O. Meyerhof, K. Lohmann and P. Schuster, *Biochem. Z.*, 286 (1936) 301.
22 O. Meyerhof and W. Kiessling, *Biochem. Z.*, 279 (1935) 40.
23 O. Meyerhof and R. Junowicz-Kocholaty, *J. Biol. Chem.*, 145 (1942) 443.
24 O. Meyerhof, in J. B. Sumner and K. Myrbäck (Eds.) *The Enzymes. Chemistry and Mechanism of Action*, Vol. II, Part I, New York, 1951.
25 O. Meyerhof, *Bull. Soc. Chim. Biol.*, 20 (1938) 1345.
26 O. Meyerhof, *Bull. Soc. Chim. Biol.*, 20 (1938) 1033.
27 O. Meyerhof, *Bull. Soc. Chim. Biol.*, 21 (1939) 965.
28 H. O. L. Fischer and E. Baer, *Naturwiss.*, 25 (1937) 589.
29 W. Kiessling and P. Schuster, *Ber. deut. chem. Ges.*, 71 (1938) 123.
30 O. Warburg and W. Christian, *Biochem. Z.*, 314 (1943) 149.
31 J. F. Taylor, A. A. Green and G. T. Cori, *J. Biol. Chem.*, 173 (1948) 591.
32 D. Herbert, H. Gordon, W. Subranmayan and D. E. Green, *Biochem. J.*, 34 (1940) 1108.
33 K. Lohmann, *Biochem. Z.*, 254 (1932) 332.
34 H. G. Wood, *Physiol. Rev.*, 33 (1955) 841.
35 E. H. Kaplan, J. L. Still and H. R. Mahler, *Arch. Biochem.*, 34 (1951) 16.
36 H. R. Mahler, A. Tomisek and F. M. Huennekens, *Exptl. Cell. Res.*, 4 (1955) 208.
37 G. Embden, H. J. Deuticke and G. Kraft, *Z. Physiol. Chem.*, 230 (1934) 12.
38 G. Embden and H. J. Deuticke, *Z. Physiol. Chem.*, 230 (1934) 29.
39 G. Embden and H. J. Deuticke, *Z. Physiol. Chem.*, 230 (1934) 50.
40 G. Embden and T. Ickes, *Z. Physiol. Chem.*, 230 (1934) 63.
41 G. Embden and H. Jost, *Z. Physiol. Chem.*, 230 (1934) 69.
42 E. Lehnartz, *Z. Physiol. Chem.*, 230 (1934) 90.
43 H. Jost, *Z. Physiol. Chem.*, 230 (1934) 96.
44 H. M. Kalckar, *Biological Phosphorylations. Development of Concepts*, Englewood Cliffs, 1969.

Chapter 23

The Unravelling of the Nature of Harden and Young's "Coenzyme" and the Discovery of Energy-Rich Phosphate Bonds

1. Discovery of the adenosine phosphates of muscle

Embden and Zimmermann[1], in 1927, found adenylic acid in muscle, (adenosine monophosphate, AMP) which had already been isolated from yeast. At the same time, Parnas and Mozolowski[2] were pursuing their researches on ammonia formation in muscle and observed that the very low content present in fresh muscle rapidly increased when the muscle was injured. This resulted from the presence of a specific adenylic acid deaminase which was prepared from muscle by Embden and Wassermeyer[3]. Parnas[4] identified the nature of the changes in the muscle purines due to traumatic injury. He saw that the total muscle purine, initially 82% as adenine and 18% as hypoxanthine in his material, altered within two minutes of traumatic injury to give 77% hypoxanthine and 23% adenine.

Lohmann[5,6] introduced a method of analysis which was to become of great importance in following developments. It is based on the rate of hydrolysis of phosphoric esters. He found the velocity constant of hydrolysis of hexose diphosphate to be $23 \cdot 10^{-3}$ for the first phosphoric group and $3.5 \cdot 10^{-3}$ for the second. The velocity constant for the phosphoric group of the Embden ester was $0.2 \cdot 10^{-3}$. Applying this method to frog and to rabbit muscles, Lohmann detected the presence of a more rapidly hydrolysing ester, with a velocity constant of $250 \cdot 10^{-3}$. He found approximately 0.25–0.35 mg P of this hydrolyzable ester, a pyrophosphate, per g of frog's muscle. When the muscle was stimulated to a state of fatigue or was in a state of rigor, the concentration of the pyrophosphate diminished due to the action of a heat-labile enzyme, and was replaced by orthophosphate.

References p. 132

Plate 77. Karl Lohmann.

Let us come back to the concept of lactacidogen, considered by Embden as the *immediate* precursor of lactic acid in muscle. Embden, as described in Chapter 20, first applied the term to hexosediphosphate and later when it was shown that the presence of this compound could not be recognized in untreated muscle, to hexosemonophosphate. Embden and his school, in order to determine the amount of "lactacidogen" in different conditions measured the inorganic phosphate before and after incubation in 1% bicarbonate for two hours at 37°. They assumed the difference to correspond to the amount of hexosemonophosphate. It is in relation to this special aspect of the "lactacidogen" story that Lohmann, as said above, applied his hydrolysis curves to the supposed "hexosemonophosphate" and showed that most of the phosphate liberated during the incubation in *N* HCl at 100° came from a compound which, unlike hexosemonophosphate, could be easily hydrolyzed. This illustrates the role of an erroneous concept, such as the "lactacidogen" theory in the emergence of such a great discovery as that of ATP. Lohmann, in 1928, isolated pyrophosphate from the incubation product obtained in Embden's method for the determination of "lactacidogen" but shortly afterwards he showed that his compound as well as the AMP isolated from muscle by Embden and Zimmermann[1] were artifacts resulting from the method of extracting the muscle with an alkaline solution of barium or calcium salts which results, as Lohmann showed, in the splitting of ATP into AMP and pyrophosphate.

In the issue of *Naturwissenschaften* dated August 2nd, 1929, Lohmann[7] reported on the isolation, from muscle, of this new compound which could be split into one molecule of adenylic acid (AMP) and one molecule of pyrophosphate. In order to record the importance of Lohmann's discovery of what became known as ATP (adenosine triphosphate) and which promised to have a great potential, it is necessary to quote from his paper of August 2nd, 1929 in *Naturwissenschaften* which leaves no doubt about this importance:

... "jetzt die Isolierung einer Verbinding gelungen, die bei der neutralen Hydrolyse ihres Bariumsalzes in Pyrophosphorsäure und Adenosinphosphorsäure (Adenylsäure) zerfällt." ... "Bei der Spaltung der von mir isolierten Adenylpyrophosphorsäure durch kurzes Erhitzen in heissen verdünnten Säuren werden 2 Mol. *O*-Phosphorsäure und je 1 Mol. Adenin- und Pentose-(Ribose-)phosphorsäure erhalten."
... "Die Pyrophosphatfraktion im Muskel sowie in anderen Zellen kann ... durch das Verhältnis von 2:1 für das leicht und schwer hydrolysierbare P characterisiert bezw. bestimmt worden." ... "dass das in frischen Muskeln vorhandene säurelösliche Adenin zum ganz überwiegenden Teil in der Pyrophosphatfraktion vorliegt."

References p. 132

... "Es ist bemerkenswert, dass die analoge aus Hefe isolierte Pyrophosphatverbindung, die wahrscheinlich mit der aus Frosch- und Kaninchenmuskulatur identisch ist, die Milchsäurebildung in gereinigter, komplementfreier aber hochfermenthaltiger Muskelfermentlösung ebenfalls auszulösen vermag."

At the XIIIth International Physiological Congress held at Boston (August 19–24th, 1929) Lohmann spoke on August 21st, 1929. The precirculated abstract of his report reads as follows[8]:

"*Über die Pyrophosphatfraktion in Muskel*
Die im Muskel vorhandene, autolytisch und in heisser Säure leicht zu *O*-Phosphat ausspaltbare P-Verbindung (Pyrophosphatfraktion) liegt in chemischer Bindung an einen P- und N-haltigen organischen Rest vor. Die Abspaltung des Pyrophosphats aus dieser Verbindung erfolgt spontan schon bei Zimmertemperatur aus dem in Wasser, schwer löslichen Ba-Salz bei neutraler Reaktion".

This precirculated abstract was prepared in the first months of 1929 and, on Meyerhof's advice, written in very general terms. In the corrected text which appeared after the Congress in the *American Journal of Physiology*[9] (October 1929 issue) and which gave the substance of the facts presented by Lohmann on August 21st, 1929, facts having already been stated in *Naturwissenschaften* of August 2nd 1929, the first sentence of the abstract of the communication to the Congress read:

"Die im Muskel vorhandene, autolytisch und in heisser Säure leicht zu *O*-Phosphat aufspaltbare P-Verbindung (Pyrophosphatfraktion) liegt in chemischer Bindung an Adenylsäure vor."

and a new sentence was added at the end:

"Die biologische Bedeutung dieser Adenylpyrophosphorsäure für die Kohlenhydratspaltung des Muskels und der Hefe wird besprochen."

It is puzzling to read in a text by Kalckar[45]

"In 1929 when ATP was first discovered by Fiske and Subbarow at Harvard it was the "stella nova" at the International Congress of Physiology held in Boston".

Neither Fiske, nor Subbarow, were listed among the attendants at the Congress, and no mention was made of their work at the occasion of this communication of Lohmann who only heard of Fiske and Subbarow's work two months later when reading their paper in *Science* of October 18th, 1929*.

No trace can be detected of any excitement resulting from Lohmann's communication.

The French biochemist René Fabre[9a], in a report on the Congress, states that, as Embden, Parnas, Lehnhartz, Lohmann, Knoop and Palladin were present, it was natural to expect discussions on the nature of lactacidogen, but nothing new, he concludes, was worth mentioning.

Fulton[9b] also wrote a report on the XIIIth Physiological Congress. Concerning the important work reported at the sessions, he mentions the discovery of the growth hormone by Cushing and Teel, of the sleep centre by Hess, of the liver constituent active in pernicious anemia by Cohn, McMeekin and Minot and of the relation of the sympathetic nervous system to emotions by Cannon, but he finds nothing of importance in the sessions on muscle physiology and he does not mention ATP.

In the issue of *Science* of October 18th, 1929, Fiske and Subbarow[10] reported on the isolation and crystallization of a compound

"containing, in addition to adenine and carbohydrate, three molecules of phosphoric acid."

Fiske and Subbarow's studies[10,11] were situated in the frame of the work pursued in the laboratory of Folin on the nitrogenous constituents of muscle. They were puzzled by the presence of purine in the insoluble calcium precipitate.

As observed by Lohmann[12], and confirmed by Meyerhof, Lohmann and Meyer[13], ATP is necessary for the glycolytic activity of muscle extract. ATP is found in the fresh extract but not in the autolyzed one. An addition of ATP restores the activity of the latter. Inosinic acid has no such activity. This explains why an autolyzed extract is no longer able to produce lactic acid, as adenylic acid, in such an extract, is split into ammonia and inosinic acid, which has no activity.

The importance of ATP as a cellular constituent was appreciated only in the late thirties, and mostly as a consequence of the publication of several papers by Lohmann. In a paper with Schuster, published in 1934, Lohmann[13a] showed that in the muscles of frog, rabbit and lobster the adenylic acid which is acid soluble is in the living muscle in the form of ATP

* "Von Fiske und Subbarow ist mir nur ihre Mitteilung in der *Science* von 18 Oktober 1929 bekannt." (From a letter of K. Lohmann to the author, March 13th, 1974.)

References p. 132

Plate 78. Cyrus Hartwell Fiske (1932).

(within an error of 2 or 3%), while it had been claimed in 1930 by Embden and Schmidt[13b] that adenylic acid exists in muscle mostly in the free form. Ferdmann[13c] had found that in protein-free extracts of muscle pyrophosphate was present. In 1935, Lohmann and Schuster[13d] proved that almost all of the adenosine nucleotide was in the form of ATP in calfs' heart muscle. This was in opposition with the finding by Embden[13e] of a preparation from horses' heart with a N:P ratio of 10:5 (instead of 5:3 in ATP). The same conclusion was reached by Beattie *et al.*[13f], by Ostern[13g] and by Ostern and Baranowsky[13h]. These authors concluded that a di-adenosine-pentaphosphoric acid was present in muscle.

Barrenscheen[13i] found in 1934 a ratio N:P between 10:4 and 10:3, and he, as well as Ferdmann and Galperin[13j] was in favour of the presence of an adenine-dinucleotide.

The mononucleotidic nature of the adenine-nucleotide of hearts' muscle was demonstrated by Lohmann and Schuster[13d] whose work was based on the utilization of calf's heart, a material which can be obtained at the slaughterhouse before the skinning of the animal, while in the case of a horse or a bull, the animals are only opened after the skinning, a circumstance which brings about a delay during which chemical alterations take place.

In 1937, Lohmann and Schuster[13h] isolated ATP from rabbits' muscle and from pigs' red blood cells and measured its molecular weight. This work confirmed the nature of ATP as a simple nucleotide, discarding the concept proposed by Parnas[13k] to consider ATP (molecular weight 507) as a diadenosine hexaphosphoric acid (molecular weight 994).

2. Discovery of the guanidine phosphates

Embden and other authors had determined the inorganic phosphate content of muscle. Eggleton and Eggleton[14], in 1927, discovered the existence of a very rapidly hydrolysable substance (phosphagen) which, during the course of the analysis, liberated free phosphate and which corresponded to about 70% of the so-called "inorganic phosphate" of muscle. Fiske and Subbarow[15] isolated the pure substance in a crystalline form and identified it as creatine phosphate. Eggleton and Eggleton[16,17] showed the presence of creatine phosphate in the voluntary muscles of all the vertebrate species studied by them. Creatine phosphate was recognized later as playing the role of a store of easily mobilized phosphate bond energy.

References p. 132

Arginine phosphate, the phosphagen of crustacean muscles was discovered by Meyerhof and Lohmann[18] in 1928. It is still sometimes designated as "the phosphagen of invertebrates", a statement which ignores the presence in invertebrates of a multiplicity of phosphagens: creatine phosphate, arginine phosphate, glycocyamine phosphate, taurocyamine phosphate, lombricine phosphate, hypotaurocyamine phosphate and opheline phosphate (recent literature in Chapter 5, by Van Thoai, of Vol. 6 of this Treatise).

3. Discovery of the "energy-rich bonds"

Very shortly after the discovery of creatine phosphate, Meyerhof and Lohmann[19] determined the ΔH of its hydrolysis and found a value of approximately 150 cal. per g ($\pm$14800 cal. per mole). The following year (1928), Meyerhof and Lohmann[18] determined the ΔH of the creatine phosphate and arginine phosphate by acids as 11000–12000 cal per mole. That the exothermy of the splitting of ATP to adenylic acid and two molecules of inorganic phosphate corresponded to 25000 cal, *i.e.* 12500 cal per mole was shown by Meyerhof and Lohmann[20] in 1932.

It is sometimes written that the metabolic importance of "energy rich bonds" was not recognized when their existence was detected. That such a thesis is far from true is shown by the perusal of a number of early publications of the Meyerhof school. In a paper of 1927, for instance, Meyerhof and Suranyi[21], commenting on the value of 150 cal for the splitting of creatine phosphate (per g H_3PO_4), comment on the participation of such splitting in the catabolic pathway of lactic acid formation.

"Es sei nur hingewiesen auf die Bedeutung die dieser Wärme für den im lebenden Muskel bestimmte kalorische Quotienten der Milchsäure zukommt. Bei 0.2% anaerober Milchsäurebildung zerfällt pro Gramm Muskel eine Kreatinphosphorsäure entsprechend etwa 1.5 mg H_3PO_4, auf 1 g Milchsäure kommen also ungefähr 0.75 mg zerfallende H_3PO_4 mit einer Wärmebildung von 110 bis 120 cal." (ref. 21)

Meyerhof and Lohmann[22] could, as early as 1931, concerning ATP, describe the utilization of a high-energy bond in biosynthesis.

"Die Kofermentrolle der ATP als Koferment der Milchsäurebildung ... besteht offenbar darin, dass die Veresterung des Phosphats, die der Spaltung des Kohlenhydrats in Milchsäure vorausgeht, unter gleichzeitiger Aufspaltung der ATP verläuft während im weiteren Verlauf der Spaltung, dieses wieder resynthetisiert wird. Der Kreislauf der ATP unterhält auf diese Weise die Milchsäurebildung. Die Synthese des Phosphagens wird danach im enzymhaltigen Muskelextrakts, und wahrscheinlich ebenso im lebenden Muskel, durch die

Spaltungsenergie der ATP ermöglicht, während die Energie der Milchsäurebildung (aus Phosphorsäurenestern) dazu dient dies zerfallene Pyrophosphat wieder zu aufbauen."

Lohmann[23] was the first to recognize, as early as 1934, the importance of the high-energy bonds for the accomplishment of biochemical reactions, without heat production. According, as said above, to the data available at the time, the splitting of 2 P from ATP was in the calorimeter accompanied by a liberation of heat of 12 500 cal per mole of phosphate and the splitting of creatine phosphate with a liberation of 11 000–12 000 cal.

To quote Lohmann[23]:

"Auf Grund der gemessenen Reaktionswärmen der Gleichungen (1) und (3)

Adenylpyro-Phs. = Adenylsäure + 2 Phs. + 25 000 cal. (1)

2 Kreatin-Phs. + Adenylsäure = Adenylpyro.-Phs. + 2 Kreatin (2)

---

2 Kreatin-Phs. = 2 Kreatin + 2 Phs. + 22 000 bis 24 000 cal. (3)

ergibt sich also, dass die Spaltung der Kreatin-Phs. unter gleichzeitiger Veresterung der Adenylsäure zu Adenylpyro-Phs. innerhalb der Messgenauigkeit ein thermisch neutraler Vorgang ist. Bei dem eigentlichen Zerfall der Kreatin-Phs. wird also keine Wärme frei (ob die Reaktion auch energielos verläuft, muss vorläufig wegen der bisher unbekannten freien Energie unentschieden bleiben). Die Resynthese der Adenylpyro-Phs. erfolgt also praktisch mit einem 100%igen Nutzeffekt".

Lohmann[24,25] has deciphered the chemical constitution of ADP and of ATP in 1932 and in 1935, and the formula proposed by him was confirmed by Lythgoe and Todd[26] in 1945.

The recognition of the fact that the phosphate ester bond resulting from the esterification of phosphate in glycolysis changes metabolically into a new type of bond ("energy-rich phosphate bond", in which a large part of the energy made available by the catabolic process accumulates) remains one of the landmarks of metabolic studies (Lohmann and Meyerhof[26a]).

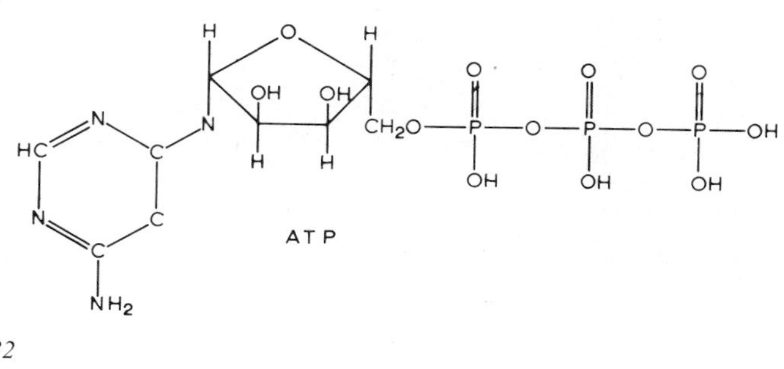

ATP

References p. 132

Plate 79. Yellapragada Subbarow.

4. Discovery of enzymatic transphosphorylation

Another major contribution of Lohmann was the recognition that there was no specific phosphatase for the phosphagen hydrolysis but that it could only be split enzymatically by a transphosphorylation to a member of the adenyl system and principally to ADP (discovered by Lohmann[25], 1935). This was the first example known of an enzymatic transphosphorylation. At this occasion the reverse aspect of the reaction was considered but recognized as not yet elucidated[23]. This aspect was further studied by Lehmann[27,28] at Meyerhof's suggestion. Using muscle extracts kept for a long time and submitted to a prolonged dialysis, Lehmann recognized that the hydrolysis of phosphagens (creatine phosphate or arginine phosphate) was of the nature of an equilibrium reaction subject to the law of mass action. The phosphagen synthesis only took place at pH values higher than 8 (Lohmann[23] had observed that at pH 9 no ATP synthesis from adenylic acid could be recognized). From the studies of Lohmann and of Lehmann emerged the knowledge of the nature of the equilibrium which, in the presence of magnesium ions and of the enzyme phosphoguanidine transphosphorylase, occurs in muscle extracts (Reaction of Lohmann)

$$\text{Creatine phosphate} + \text{AMP} \rightleftharpoons \text{creatine} + \text{ADP}$$

$$\text{Creatine phosphate} + \text{ADP} \rightleftharpoons \text{creatine} + \text{ATP}$$

and which was to provide, for the first time, the description of the mechanism for the utilization of phosphate bond energy.

One of the services rendered by Lehmann's studies was the description of an elegant synthesis of creatine phosphate from the easily available phosphoglyceric acid, which spared the experimenters from the time-consuming process of isolation from large amounts of frog or rabbit muscle.

5. Mg^{2+} as a constituent of coenzyme

Lohmann[12] recognized Mg^{2+} as a part of the enzyme of Harden and Young when he observed that if Mg^{2+} is precipitated by NH_3 (as $MgNH_4PO_4$) the muscle juice becomes inactive and its activity is restored by the addition of Mg^{2+}.

At this time, he writes[12]:

"Vorläufig nenne ich nun das System Adenylpyrophosphat + anorganisches Phosphat + Mg das "Kofermentsystem der Milchsäurebildung".

References p. 132

Plate 80. Hans von Euler-Chelpin.

As he observed that dried yeast washed with an acid extracting solution did not accomplish fermentation, except if complemented by a salt of magnesium, he concluded:

"Milchsäurebildung und alcoholische Gärung finden also nur in Gegenwart von Magnesium statt."

An activating action of Mg^{2+} had been observed by Meyerhof[29] in the case of the splitting of hexosediphosphoric acid.

6. Nature of the adenylic compounds present in the "coenzyme"

Von Euler[30] was convinced that the enzymatic dismutation of aldehydes, a reaction which he considered to take place at the origin of alcoholic fermentation, was impossible without the presence of a co-factor which he identified, in his theory, with Harden's "coenzyme of alcoholic fermentation" (cozymase in von Euler's language).

It is to the influence of this concept that the determination of Von Euler[31,32] to unravel the nature of cozymase is due. He prepared from yeast an active product he considered as a mononucleotide of the nature of adenylic acid, to which he attributed a mol. wt. of 486 ± 6 and which he considered as the "coenzyme of alcoholic fermentation".

In 1918 and in 1921, Meyerhof[33–35] showed that muscle also contains Harden's coenzyme.

In our present theory, the coenzyme is considered as composed of ATP, NAD(P), cocarboxylase and Mg^{2+}. When von Euler, as stated above, did propose to identify coenzyme with adenylic acid, this theory was in contradiction with the views of Lohmann who, after his discovery of ATP had recognized it as a constituent of Harden's coenzyme.

In his first studies on ATP, Lohmann[8] had found that, for the formation of ethyl alcohol in alcoholic fermentation, and of lactic acid in muscle glycolysis, not only was cozymase necessary, but also ATP, defined as adenylpyrophosphate. In subsequent studies, ATP alone was considered as required (Lohmann[36]), while von Euler considered cozymase alone as necessary (at the time, cocarboxylase had not yet been discovered). This was the origin of a polemic between Lohmann and von Euler concerning which adenylic acid derivative was essential. In 1932, von Euler and Nilsson[37] stated that they were able to confirm the conclusions of Lohmann qualitatively but not quantitatively and they concluded as follows:

References p. 132

"Für die von Lohmann und für die in diesem Institut gefundene Tatsachen haben wir bis jetzt keine befriedigende und einheitliche Deutung gefunden".

We now know that the difficulty of obtaining each of their compounds in the pure state was at the basis of the discrepancies between Lohmann's and von Euler's results. Both schools were partly right, ATP being required for the phosphorylation and NAD (von Euler's cozymase) for the oxido-reduction. As we shall see, in 1934, Warburg and Christian (see Chapter 36) studying biological oxidations, isolated from red blood cells a preparation which released adenine on acid hydrolysis (as was the case with von Euler's cozymase). They realized a system of hexose phosphate* oxidation by associating this preparation with the "yellow ferment" they had isolated from yeast, and with a colorless fraction of yeast which they called *Zwischenferment*. Warburg and Christian[38] showed that their coferment contained nicotinamide. That the pyridine component of this coferment was the active group, the catalytic action depending on the alternation of the oxidation states of its pyridine component was shown by Warburg, Christian and Griese[39]. These authors concluded that their own coferment was composed of adenine, nicotinamide, pentose and phosphate in the ratio 1:1:2:3 (three phosphates for one adenine). From this coferment, considered as active in biological oxidations, Warburg and Christian[40] distinguished the cozymase of von Euler, which they designated as "fermentation coferment" and recognized as containing two phosphates for one adenine, instead of three.

They called the two compounds "Triphospho-Pyridinnucleotid" and "Diphospho-Pyridinnucleotid".

Von Euler[41], in 1936, stated that, as he recognized them as the general cofactors in dehydrogenations, his cozymase (diphospho-compound) should be called codehydrogenase I and Warburg's coferment (triphospho-compound), codehydrogenase II. The terms cozymase I (old cozymase) and cozymase II (new cozymase) were also used and the abbreviations DPN and TPN later replaced by NAD and NADP (see Chapter 36).

7. Reasons for the delayed elucidation of the nature of Harden's coenzyme

A long delay separates the discovery of the "coenzyme of fermentation" by Harden and Young in 1906 (see Chapter 21) and the unravelling of its

* By this time hexosephosphates had become available in the pure state.

composition (ATP, NAD(P), cocarboxylase, Mg^{2+}). It took five years after the preparation of muscle juice by Meyerhof in 1926, for the recognition by Lohmann, in 1931, of Mg^{2+} as a component of the coenzyme. Auhagen[42] discovered cocarboxylase only in 1932 (see Chapter 24).

The long-lasting controversy between Lohmann and von Euler in favor either of ATP or of NAD as the adenylic system involved resulted from the difficulty of obtaining either of these compounds deprived of traces of the other, even at the price of prolonged dialysis or repeated washings. By the way, it is, as stated above, at the occasion of variations of such washings that Lohmann had recognized Mg^{2+} as a component of the coenzyme. Auhagen[42] in his report on the recognition of cocarboxylase, has commented on the difficulty of obtaining preparations free from cozymase and Lohmann[43] has also commented on this point in his paper on the identification of cocarboxylase as thiamine diphosphate.

Another obstacle came from the view of von Euler and his school[31,32,41] who first identified cozymase as adenylic acid and determined, by diffusion method, its molecular weight as 486 ± 6, as stated above, while, as we know, the molecular weight of AMP is equal to 347, of ADP to 427, of ATP to 507, and of NAD to 663. The history of the development of our knowledge of the nature, configuration and properties of NAD has been reviewed in Vol. 14 of this Treatise by Colowick, Van Eys and Park, and the reader is referred to this excellent presentation.

References p. 132

REFERENCES

1 G. Embden and N. Zimmermann, *Z. physiol. Chem.*, 167 (1927) 114.
2 J. K. Parnas and W. Mozolowski, *Biochem. Z.*, 184 (1927) 399.
3 G. Embden and H. Wassermeyer, *Z. physiol. Chem.*, 179 (1928) 226.
4 J. K. Parnas, *Biochem. Z.*, 206 (1929) 16.
5 K. Lohmann, *Biochem. Z.*, 202 (1928) 466.
6 K. Lohmann, *Biochem. Z.*, 203 (1928) 164.
7 K. Lohmann, *Naturwissenschaften*, 17 (1929) 624.
8 K. Lohmann, in *Abstracts of Communications, XIIIth International Congress held at Boston, Aug. 19–24, 1929* (Preprint).
9 K. Lohmann, *Am. J. Physiol.*, 90 (1929) 434.
9a R. Fabre, *Bull. Soc. Chim. Biol.*, 11 (1929) 999.
9b J. F. Fulton, *Bull. Harvard Med. Sch. Alumni Assoc.*, 4 (1929) 4.
10 C. H. Fiske and Y. Subbarow, *Science*, 70 (1929) 381.
11 C. H. Fiske and Y. Subbarow, *J. Biol. Chem.*, 81 (1929) 629.
12 K. Lohmann, *Biochem. Z.*, 237 (1931) 445.
13 O. Meyerhof, K. Lohmann and K. Meyer, *Biochem. Z.*, 237 (1931) 437.
13a K. Lohmann and Ph. Schuster, *Biochem. Z.*, 272 (1943) 24.
13b G. Embden and G. Schmidt, *Z. physiol. Chem.*, 186 (1930) 205.
13c D. Ferdmann, *Z. physiol. Chem.*, 216 (1933) 205.
13d K. Lohmann and Ph. Schuster, *Biochem. Z.*, 282 (1935) 104.
13e G. Embden, cited after H. J. Deuticke, *Arch. ges. Physiol.*, 230 (1932) 537.
13f F. Beattie, Th. H. Milroy and R. W. M. Strain, *Biochem. J.*, 28 (1934) 84.
13g P. Ostern, *Biochem. Z.*, 270 (1934) 1.
13h P. Ostern and T. Baranowski, *Biochem. Z.*, 281 (1935) 157.
13i H. K. Barrenscheen, *Handbuch der Biochemie, Ergänz. Werk*, 2 (1934) 238.
13j D. Ferdmann and W. Galperin, *Biochem. Z.*, 277 (1935) 191.
13k T. K. Parnas, cited after Lohmann and Schuster[13l].
13l K. Lohmann and Ph. Schuster, *Biochem. Z.*, 294 (1937) 183.
14 P. Eggleton and G. P. Eggleton, *Biochem. J.*, 21 (1927) 190.
15 C. H. Fiske and Y. Subbarow, *Science*, 65 (1927) 401. (reprinted in Kalckar[45]).
16 P. Eggleton and G. P. Eggleton, *J. Physiol. (London)*, 63 (1927) 155.
17 P. Eggleton and G. P. Eggleton, *J. Physiol. (London)*, 65 (1928) 15.
18 O. Meyerhof and K. Lohmann, *Biochem. Z.*, 196 (1928) 49.
19 O. Meyerhof and K. Lohmann, *Naturwissenschaften*, 15 (1927) 670.
20 O. Meyerhof and K. Lohmann, *Biochem. Z.*, 253 (1932) 431.
21 O. Meyerhof and J. Suranyi, *Biochem. Z.*, 191 (1927) 106.
22 O. Meyerhof and K. Lohmann, *Naturwissenschaften*, 19 (1931) 575.
23 K. Lohmann, *Biochem. Z.*, 271 (1934) 264.
24 K. Lohmann, *Biochem. Z.*, 254 (1932) 381.
25 K. Lohmann, *Biochem. Z.*, 282 (1935) 120.
26 B. Lythgoe and A. R. Todd, *Nature*, 155 (1945) 695.
26a K. Lohmann and O. Meyerhof, *Biochem. Z.*, 273 (1934) 60.
27 H. Lehmann, *Biochem. Z.*, 281 (1935) 271.
28 H. Lehmann, *Biochem. Z.*, 286 (1936) 336.
29 O. Meyerhof, *Biochem. Z.*, 178 (1926) 462.
30 H. von Euler, K. Myrbäck and R. Nilsson, *Erg. Physiol.*, 26 (1928) 531.
31 H. von Euler, K. Myrbäck and R. Nilsson, *Z. physiol Chem.*, 168 (1927) 177.

32 R. Nilsson, in H. von Euler (Ed.), *Chemie der Enzyme*, 2er Teil, 3. Abschnitt, Munich, 1934.
33 O. Meyerhof, *Z. physiol. Chem.*, 101 (1918) 165.
34 O. Meyerhof, *Z. physiol. Chem.*, 102 (1918) 185.
35 O. Meyerhof, *Arch. ges. Physiol.*, 188 (1921) 114.
36 K. Lohmann, *Biochem. Z.*, 141 (1931) 67.
37 H. von Euler and R. Nilsson, *Z. physiol. Chem.*, 208 (1932) 173.
38 O. Warburg and W. Christian, *Biochem. Z.*, 275 (1935) 464.
39 O. Warburg, W. Christian and A. Griese, *Biochem. Z.*, 279 (1935) 143.
40 O. Warburg and W. Christian, *Biochem. Z.*, 287 (1936) 291.
41 H. von Euler, *Erg. Physiol.*, 38 (1936) 1.
42 E. Auhagen, *Z. Physiol. Chem.*, 204 (1932) 149.
43 K. Lohmann, *Biochem. Z.*, 294 (1937) 188.
44 H. von Euler, *Biokatalysatoren*, Stuttgart, 1930.
45 H. M. Kalckar, *Biological Phosphorylations, Development of Concepts*, Englewood Cliffs, N.J., 1969.

Chapter 24

The Complete Glycolysis Pathway

1. Phosphate esterification connected with the second (oxidative) phase of glycolysis

It has been repeatedly hinted, on a chemical basis, that glycolysis probably ended in a dismutation process, one half of the hexose molecule oxidizing the other half, whatever form was suggested for these halves.

Parnas[1], as early as 1910, had insisted on the importance of oxido-reduction in glycolysis. He reported that an "aldehyde mutase" present in animal tissues was catalyzing the conversion of an aldehyde, with the participation of the elements of water, into alcohol and acid (Cannizzaro reaction)

$$2\ RCHO + H_2O \rightarrow RCH_2OH + RCOOH$$

this "dismutation" involving the participation of one molecule of aldehyde as the hydrogen acceptor, forming alcohol, and of another as hydrogen donor, forming the acid.

Von Euler, as stated above, considered cozymase (later known as NAD) as a cofactor for aldehyde enzymatic dismutation, and Kluyver had also (see Chapter 22) suggested oxido-reduction as essential in fermentation.

Since 1934, the reactions from a hexose or from a polysaccharide to the phosphotriose glyceraldehyde-3-P were elucidated but the existence of a second occurrence of phosphorylation during the oxido-reductive phase of glycolysis still remained unsuspected. We find this phosphorylation mentioned for the first time by Nilsson[2] when he noted the phosphate uptake resulting from the addition of pyruvate to a glucose fermentation held up by an inhibitor.

In experiments accomplished by Dische[3], when AMP was added to washed, haemolysed red blood cells glycolysing FDP, in concentrations comparable to that of FDP, the glycolysis rate increased, phosphates

References p. 155

were taken up and ATP increased. Meyerhof and Kiessling[4], adding phosphopyruvate to fermenting glucose by yeast poisoned by fluoride, observed reactions they interpreted as follows:

$$2\ \text{phosphopyruvate} + 2\ \text{hexose} + 2\ H_3PO_4 \rightarrow 2\ CO_2 + 2\ \text{alcohol} + 1\ \text{FDP} + 2\ \text{phosphoglyceric acid}$$

According to their interpretation, this meant an oxido-reduction (between pyruvate and triosephosphate) accompanied by an esterification, but they considered it as an esterification of glucose by inorganic phosphate as it was then believed.

2. Warburg's "Gärtest"

In the same way as Buchner had provided by the discovery of cell-free fermentation a simplified soluble system to study the pathway of glycolysis, Warburg, in 1936, endeavoured to prepare an even simpler glycolysing system. The "Gärtest" method was the first manifestation of a methodological trend which would lead to the isolation of pure enzymes as a way of establishing the existence of a link in a metabolic chain. Warburg and Christian[5] isolated two protein fractions from a yeast maceration juice. The A fraction is precipitated at pH 5 by addition of acid, soluble in $\frac{1}{3}$ saturated ammonium sulphate, insoluble in $\frac{1}{2}$ saturated ammonium sulphate. Protein B was obtained, after the elimination of protein A, and neutralisation of the juice, by precipitation with 75% acetone. In order to determine the role of NAD, which had been recently discovered (see Chapter 23), Warburg and Christian devised what they called a "Gärtest". They put together the two proteins, ATP, NAD, Mg, Mn and phosphate and to this reconstituted system they added hexosemonophosphate and acetaldehyde. They expressed the reactions taking place by the following equations:

$$2\ \text{HMP} + P_i + 2\ \text{acetaldehyde} = 1\ \text{HDP} + 1\ \text{pyruvate} + 1\ \text{phosphoglycerate} + 2\ \text{ethanol}$$

Warburg and Christian believed that the system assembled could not accomplish a dephosphorylation of phosphoglyceric acid. They believed that the addition of acetaldehyde resulted in the oxidation of NADH (as in its absence, NADH was formed). The opinion of Warburg and Christian was that NAD oxidized a hexosemonophosphate which was split into pyruvate and phosphoglycerate.

Warburg's "Gärtest" was reinvestigated by Meyerhof *et al.*[6] (for an extensive discussion, see Meyerhof[7]). They concluded that in this system the breakdown of HMP involved HDP as follows

1 HDP (→ 2 triose P) + 1 HMP + P_i + 2 acetaldehyde → 2 phosphoglyceric acid + 1 HMP + ethanol

followed by

1 HMP + 1 phosphoglyceric acid (→ phosphopyruvate) → pyruvate + HDP

Even if it did not contribute to the elucidation of the oxidoreductive part of the glycolysis pathway, Warburg's "Gärtest" had a seminal value in the line of thought which led Warburg to study the biological activity of NAD, no longer by relying on crude protein preparations isolated from yeast juice but by the use of a crystalline enzyme preparation, and thereby to precisely identify a mechanism for the transfer of the oxidative energy of metabolism into a high-energy phosphate bond (see Chapter 25).

3. Phosphate transfer from phosphopyruvate to ADP (second phosphorylation step of the oxidoreductive phase of the pathway)

When studying ammonia liberation in muscle, Parnas *et al.*[8] had, in 1934, observed that, if ground with phosphate solution the liberation of ammonia observed when grinding muscle with water did not take place. Their interpretation was that the source of the ammonia observed, known as being adenylic acid (AMP) was maintained for some time in the phosphorylated state (ATP) in conditions (such as phosphate addition) favouring lactic acid formation. As the delay in ammonia formation was prevented by poisoning the ground muscle with fluoride or with fluoroacetate, known to inhibit some stages of glycolysis, they decided to try several carbohydrates to test their capacity, in the presence of fluoride or iodoacetate, to ward off ammonia liberation. This acute analysis showed that phosphoglyceric acid as well as pyruvic acid were effective in the presence of fluoride while in the presence of iodoacetate, only phosphoglyceric acid was effective. The authors suggested that phosphate was transferred to creatine from a phosphorylated substance, situated in the pathway between phosphoglyceric acid and pyruvate, the phosphate passing to ATP by the reaction of Lohmann. This theory was based on the data on ammonia formation in muscle entering rigor after iodoacetate poisoning, obtained by Mozolowski *et al.*[9].

In fact, Parnas *et al.* found a synthesis of creatine phosphate[10]. The effectiveness of pyruvate in the presence of fluoride and phosphate, led the authors to believe in a formation of phosphopyruvate from pyruvate and

References p. 155

inorganic phosphate, an interpretation which was not confirmed. As noted by D. M. Needham[11], the suppression of ammonia formation is the result of ATP formation in the process, not yet discovered at the time, of coupled oxidoreduction and phosphorylation. The fact that phosphoglycerate (in presence of fluoride and phosphate) inhibits ammonia formation in muscle *brei* was difficult to account for, as it was known that fluoride inhibits the formation of phosphopyruvate as well as its dephosphorylation. Mann[12] solved the problem by showing that the inhibitions of ammonia formation depended on the relative concentrations of fluoride and phosphoglycerate. To quote D. M. Needham[11]:

"Almost simultaneously, in 1935, Ostern, Baranowski and Reis[13,14] and Needham and van Heyningen[15,16] (in Hopkins' laboratory) showed that phosphoglyceric acid was not dephosphorylated in absence of adenylic acid and that its phosphate was transferred (presumably *via* phosphopyruvate) to adenylic acid, then from the ATP formed to creatine. This point was also realised in Meyerhof's laboratory[17,18] at almost the same time. Needham and van Heyningen clearly stated the view that the co-enzyme function of the adenylic compounds in glycolysis lay in their ability to act not only as phosphate donors but also as phosphate acceptors. They emphasized that this co-enzyme had been found unnecessary at stages involving no phosphate transfer:

hexosediphosphate → 2 triosephosphate[19]

phosphoglyceric acid → phosphopyruvic acid[20]

pyruvic acid + glycerophosphate → lactic acid + phosphoglyceric acid[21]

Parnas and Baranowski also spoke of "this circulation of phosphate" and the "donor-acceptor aspect of the participation of the adenylic compounds".

It must be recalled that, as stressed in Chapter 21, Parnas considered ADP as a P donor, rather than a P acceptor. Nevertheless Parnas suggested the importance of a definite *specific* process, a transfer of phosphate from molecule to molecule, a concept which was in opposition with current views of "energy coupling" involved in glycolysis as a whole and which was of a great seminal value (see Chapter 25).

The "Parnas reaction"

$$\text{phosphoenolpyruvate} + \text{ADP(AMP)} \rightleftharpoons \text{pyruvate} + \text{ATP(ADP)}$$

was the first example of the formation of ATP at the expense of energy derived from glycolysis.

Phosphate is transferred from phosphoenolpyruvate to ADP and regenerates ATP with the formation of pyruvic acid in presence of the enzyme phosphoenol transphosphorylase (our pyruvate kinase), first studied in muscle extracts by Parnas and his collaborators[8, 23]. The enzyme was also found in yeast extracts by Lutwak-Mann and Mann[24]. That phosphate is transferred from phosphoenolpyruvate and not from phosphoglycerate as Parnas had believed, was demonstrated by Lehmann[18].

Meyerhof *et al.*[25] first considered this reaction as irreversible, an opinion based on the failure to observe any incorporation of ^{32}P from inorganic phosphate into phosphopyruvate in a muscle extract in which ^{32}P was continuously incorporated into ATP by the coupling action of glycolysis.

4. The oxidation of triosephosphate

The oxidation of triosephosphate into phosphoglyceric acid was studied in muscle extracts, by Needham and Pillai[26] who found (1937) that inorganic phosphate is used and that a formation of ATP takes place, as the authors believed, from AMP.

$$\text{2 triosephosphate} + \text{2 pyruvate} + 2\ P_i + \text{AMP} \rightarrow \text{2 phosphoglyceric acid} + \text{2 lactate} + \text{ATP}$$

This provided a very important advance and the first definite model of coupling between phosphorylation and oxido-reduction. The reaction was found reversible by Green *et al.*[27] who observed that the reduction of phosphoglycerate triosephosphate was accompanied by a release of phosphate from ATP. This focussed attention on the relation between an esterification of inorganic phosphate into ATP and a NAD-dependent dehydrogenation of glyceraldehyde-3-P. A study of the reaction by Meyerhof and his collaborators[25, 28, 29], using yeast and muscle extracts, led them to conclude that the overall equation of the transformation of glyceraldehyde-3-P into phosphoglyceric acid could be formulated as follows:

$$\text{glyceraldehyde-3-P} + P_i + \text{NAD}^+ + \text{ADP} \rightleftharpoons \text{3 phosphoglyceric acid} + \text{NADH} + \text{ATP}$$

i.e. that the first step of the oxido-reduction phase was a reduction of NAD^+ by glyceraldehyde-3-P completed, in a way which remained unexplained, with a synthesis of ATP from ADP.

5. From 3-phosphoglycerate to phosphoenolpyruvate (2nd step of the oxido-reduction phase of glycolysis pathway)

The enquiries just mentioned left open the structure of phosphopyruvate, as at the time, uncertainty reigned between two formulas.

$$H_2C{=}C{-}C\begin{matrix}\diagup OH_2PO_3 \\ \diagdown OH\end{matrix} \text{ (with O bridge between the two C) } \quad \text{and} \quad H_2C{=}C(OH_2PO_3){-}COOH$$

References p. 155

Plate 81. Joseph Needham and Dorothy Moyle Needham.

The optical activity observed being very small was in favour of the second formula (phosphoenolpyruvate). Meyerhof and Kiessling[30] found that the natural phosphoglyceric acid, from which phosphopyruvate was derived was an equilibrium mixture of the 3- and the 2-acids, in muscle glycolysis as well as in alcoholic fermentation. To obtain the purified (−) 3-acid, they submitted the racemic 2-acid to the action of muscle extract in the presence of fluoride and isolated it in the form of its barium salt. To obtain the purified (+) 2-acid they treated the (−) 3-acid again by muscle extract in the presence of fluoride and separated the barium salt. The recognition of 2-phosphoglyceric acid as intermediate was in favour of the phosphoenol formula of phosphopyruvic acid.

It may be recalled that Embden *et al.* (1933) had found that muscle extract can act on phosphoglyceric acid to yield pyruvate and phosphoric acid. The following year (1934) the reaction had been studied further by Lohmann and Meyerhof[20] who isolated phosphopyruvic acid as an intermediary product. The phosphate group being displaced from position three to position two, the existence of another intermediate was required. and was provided by the recognition of 2-phosphoglyceric acid as intermediate, the conversion of 3-phosphoglyceric acid to pyruvic acid proceeding as follows:

$$\text{3-phosphoglyceric acid} \xrightleftharpoons{\text{phosphoglyceromutase}} \text{2-phosphoglyceric acid} \xrightleftharpoons{\pm H_2O} \text{phosphoenolpyruvic acid} \rightleftharpoons \text{pyruvic acid} + H_3PO_4$$

OH / C=O, HC–OH, $CH_2OPO_3H_2$	⇌	OH / C=O, $HCOPO_3H_2$, CH_2OH	$\xrightleftharpoons{\pm H_2O}$	OH / C=O, $COPO_3H_2$, ‖ CH_2	⇌	OH / C=O, C=O, CH_3	+ H_3PO_4
3-phosphoglyceric acid		2-phosphoglyceric acid		phosphoenolpyruvic acid		pyruvic acid	

The enzyme catalyzing the transfer from position 3 to position 2 was called phosphoglyceromutase (our phosphoglycerate phosphomutase) by Meyerhof and Kiessling[30] who showed that the subsequent loss of water was catalysed by a specific enzyme to which they gave the name of enolase (our phosphopyruvate hydratase). Sodium fluoride in the presence of Mg^{2+} and phosphate had been used for years as an inhibitor of alcoholic fermentation. It became possible now to locate its action, as under this action 2-phosphoglyceric acid accumulates, showing that the fluoride anions specifically act on enolase[20,30,31].

References p. 155

Plate 82. Otto Heinrich Warburg

6. Phosphate transfer from 1,3-diphosphoglycerate to ATP (first phosphorylation step)

As stated above, Meyerhof[25, 28, 29] had observed, in the transformation of glyceraldehyde-3-P into phosphoglyceric acid the formation of ATP from ADP, a phosphorylation which remained unexplained. In a deeper study of the oxidation of glyceraldehyde-3-P to phosphoglyceric acid, it was shown by Negelein and Brömel[32, 33] that this oxidation takes place in two steps. In the first one, glyceraldehyde-3-P is oxidised to 1,3-diphosphoglyceric acid in the presence of inorganic phosphate and NAD^+

$$\underset{\text{glyceraldehyde-3-P}}{\begin{array}{l}\mathrm{CHO}\\ |\\ \mathrm{HCOH}\\ |\\ \mathrm{CH_2OPO_3H_2}\end{array}} + \mathrm{P_i} + \mathrm{NAD^+} \rightleftharpoons \underset{\text{1,3-diphosphoglyceric acid}}{\begin{array}{l}\mathrm{C(=O)-OPO_3H_2}\\ |\\ \mathrm{HCOH}\\ |\\ \mathrm{CH_2-OPO_3H_2}\end{array}} + \mathrm{NADH}$$

This pointed to the important fact that in the oxidation of glyceraldehyde-3-P it is phosphate and not water which is involved, with a formation of acylphosphate. It was first thought that diphosphoglyceric acid was formed through 1,3-diphosphoglyceraldehyde, but this intermediate step could not be identified by Drabkin and Meyerhof[34], nor by Meyerhof and Oesper[35].

It was therefore considered that the reaction proceeds as stated above (1st step) from glyceraldehyde-3-P to 1,3-diphosphoglyceric acid and is catalyzed by triosephosphate dehydrogenase*. This enzyme was crystallized from yeast by Warburg and Christian[36, 37] and later, from rabbit muscle by Cori *et al.*[38, 39] and by Caputto and Dixon[40]. When they had prepared crystalline triosephosphate dehydrogenase (formerly 1,3-diphosphoglyceraldehyde dehydrogenase), Warburg and Christian were able to demonstrate that glyceraldehyde-3-phosphate is the substrate in a coupled oxido-reduction–phosphorylation, and on the other hand, Negelein and Brömel[32, 33] found the product of the reaction to be 1,3-diphosphoglyceric acid. The second step of the oxidation of 1,3-glyceraldehyde-3-P consists

$$\underset{\text{1,3-diphosphoglyceric acid}}{\begin{array}{l}\mathrm{C(=O)-OPO_3H_2}\\ |\\ \mathrm{HCOH}\\ |\\ \mathrm{CH_2OPO_3H_2}\end{array}} + \mathrm{ADP} \rightleftharpoons \underset{\text{3-phosphoglyceric acid}}{\begin{array}{l}\mathrm{C(=O)-OH}\\ |\\ \mathrm{HCOH}\\ |\\ \mathrm{CH_2OPO_3H_2}\end{array}} + \mathrm{ATP}$$

* When 1,3-diphosphoglyceraldehyde was suggested as an intermediate, this enzyme had received the name of 1,3-diphosphoglyceraldehyde dehydrogenase.

References p. 155

Plate 83. Theodor Bücher.

of the reaction of 1,3-diphosphoglyceric acid with ADP to give 3-D-phosphoglyceric acid and ATP[32,33,36,37].

The conversion is catalyzed by an enzyme obtained in the pure state from Lebedev's maceration juice by Bücher[41,42] who named it phosphoglycerate kinase. The reaction provides a second source (the first in the order of reactions in the pathway) of ATP in glycolysis. As rightly stated by Colowick *et al.*[43]

"The discovery of these reactions was one of the important milestones in biochemistry, since it established a mechanism whereby the oxidative energy of metabolism could be harnessed as "phosphate bond energy" in ATP".

The full significance of this discovery became generally recognized only after the writings of Lipmann[44] and of Kalckar[45] brought it, as we shall see, into proper perspective.

7. Conversion of 3-phosphoglycerate into pyruvate

It had been demonstrated that the conversion of 3-phosphoglyceric acid to pyruvic acid goes through 2-phosphoenolpyruvic acid but the fact that the phosphate group changed from position 3 to position 2 remained unexplained. A more detailed study by Meyerhof and Kiessling[30] showed that another intermediate compound, 2-phosphoglyceric acid exists and that an enzyme, phosphoglycerate phosphomutase (phosphoglyceromutase), catalyses the transport of phosphate from position 3 to position 2.

$$\underset{\text{3-Phosphoglyceric acid}}{\begin{matrix} C(=O)OH \\ | \\ HC-OH \\ | \\ CH_2OPO_3H_2 \end{matrix}} \rightleftharpoons \underset{\text{2-Phosphoglyceric acid}}{\begin{matrix} C(=O)OH \\ | \\ HCOPO_3H_2 \\ | \\ CH_2OH \end{matrix}} \underset{}{\overset{\pm H_2O}{\rightleftharpoons}} \underset{\text{2-Phosphoenolpyruvic acid}}{\begin{matrix} OH \\ | \\ C=O \\ | \\ C-OPO_3H_2 \\ \| \\ CH_2 \end{matrix}} \rightleftharpoons \underset{\text{Pyruvic acid}}{\begin{matrix} C(=O)OH \\ | \\ C=O \\ | \\ CH_3 \end{matrix}} + H_3PO_4$$

As stated above, the identification of enolase had permitted the location at the level of this enzyme of the inhibiting action of fluoride ions. This effect was clarified further by the crystallization of enolase by Warburg and Christian[46]. These authors, along with others[47–49] confirmed that Mg^{2+} was required for the action of enolase as had already been shown by Lohmann and Meyerhof[20]. In the presence of fluoride and P_i, magnesium ions are precipitated in the form of a fluorophosphate complex. Manganese ions can replace magnesium ions[47,48] and in this condition there is no fluoride inhibition.

Meyerhof had considered the second oxido-reduction reaction of glycolysis (from 2-phosphoglyceric acid to phosphoenolpyruvic acid), as irreversible. Boyer *et al.*[50,51] and Lardy and Ziegler[52] later showed it to be reversible in the presence of potassium and magnesium ions. On the other hand, the enzyme involved (phosphoenol transphosphorylase) has been crystallized from human muscle by Kubowitz and Ott[53].

In 1949, Cardini *et al.*[54] discovered that in fermenting extracts of yeast, G-1, 6-PP accumulates along with F-1, 6-PP. Paladini *et al.*[55] found that there is in muscle as well as in yeast, an enzyme that they called phosphoglucokinase, catalyzing the reaction

$$\text{ATP} + \text{G-1-P} \rightarrow \text{G-1, 6-PP} + \text{ADP}$$

8. The epilogues of glycolysis: formation of alcohol or of lactic acid

As previously mentioned, it has been admitted for a long time that in alcoholic fermentation, pyruvic acid yielded acetaldehyde by irreversible decarboxylation, acetaldehyde being reduced to ethanol. The change of pyruvic acid into acetaldehyde takes place in yeast in the presence of the enzyme carboxylase.

The discovery of carboxylase is retraced in Harden's book[56]. Neuberg and Hildesheimer[57] showed in 1911 that yeast has the power of decomposing, with an evolution of CO_2, a number of hydroxyacids and keto-acids, and among them pyruvic acid. This property also belongs to active yeast preparations and extracts (Neuberg and Karczag[58]). Neuberg has pointed to the universal presence of the enzymes in yeasts capable of producing alcohol fermentation (Neuberg and Kerb[59]). Neuberg and Rosenthal[60] have emphasized the character of carboxylase as an independent enzyme, and for instance the fact that carboxylase decomposes pyruvate in the absence of Harden's coenzyme. This is demonstrated experimentally by allowing maceration extracts to autolyse or dialyse until they are free from coenzyme.

That the decomposition of pyruvate by carboxylase constitutes a stage in the process of alcoholic fermentation was demonstrated by Neuberg and Reinfurth[61] who showed that when sodium sulphite Na_2SO_3 is added to fermenting mixtures of sugar and yeast considerable amounts of glycerol and acetaldehyde are formed, the acetaldehyde being trapped in the form of the bisulphite compound. This trapping of acetaldehyde was also demonstrated by Connstein and Lüdeke[62] who adapted the process to the industrial production of glycerol which was largely utilized in Germany during World War I.

The reaction was extensively studied by Neuberg and his school (literature in Harden[56]).

Auhagen[63, 64], in von Euler's laboratory, showed the necessary presence of Mg^{2+} and of a coenzyme, which received the name of cocarboxylase. This coenzyme was prepared from yeast by Lohmann and Schuster[65] who identified it as thiamine diphosphate, suggested by the authors to be synthesized from ATP and thiamine

$$\text{thiamine} + 2\ \text{ATP} \rightarrow \text{thiamine diphosphate} + 2\ \text{ADP}$$

a view which was later confirmed with the added precision that phosphopyruvic acid could also act as the phosphate donor[66,67].

It is recognized that lactate dehydrogenase catalyzes the final reduction of muscle glycolysis according to the reaction

$$\text{L-lactic acid} + \text{NAD}^+ \rightleftharpoons \text{pyruvic acid} + \text{reduced NAD}$$

Franke[68] has written that it is impossible to date the discovery of lactate dehydrogenase. Earlier work concerned the oxidative part of the reaction and the enzyme complex studied by von Fürth and Lieben[69,70] and by Meyerhof[71,72] obviously contained lactate dehydrogenase. After rather confused studies on the coenzyme involved, more precise data were provided by Szent-Györgyi[73], and the reasons for the identification of the coenzyme with NAD were provided by Andersson[74,75] and other authors, while conclusive evidence was afforded by spectrophotometric measurements of von Euler, Adler, Günther and Hellström[76] and of Meyerhof and Ohlmeyer[77].

The end-reaction of alcoholic fermentation has a very long story of confused concepts about the reductive action involved in the reaction leading from acetaldehyde to ethanol. The reverse reaction was familiar enough, as the appearance of acetaldehyde from alcohol was made familiar by the vinegar industry. Buchner and Meisenheimer[78] tried vainly to isolate the enzyme, which was first prepared from animal tissues by Battelli and Stern[79,80]. Ever since, a large number of studies were made with this enzyme (acetaldehyde reductase, alcohol dehydrogenase), of very wide distribution in nature[81]. The participation of coenzyme (NAD) in alcohol dehydrogenation was demonstrated by Andersson[74,75].

The theory of alcoholic fermentation and of muscle glycolysis now

References p. 155

accepted is in agreement with the fact that D-glucose in which C-1 is labelled with ^{14}C is fermented by yeast to alcohol with practically the whole of the radioactivity in the methyl group[82,83].

9. The complete pathway. Epistemological aspects of its unravelling

Fig. 7 shows the present theory of the unified pathway of glycolysis, the enzymes being mentioned with reference to the stereochemistry of the reactions[84–90]. The pathway has resulted from a convergence of studies of two phenomena (alcoholic fermentation and muscle glycolysis) which were not suspected at the start to be based on the same metabolic scheme.

The process of unravelling the intermediary steps of glycolysis has been a long one. At the time, the isotopic method was not available and, in order to identify intermediary steps (chronology of this identification, see Table V) the experimenters had recourse mainly to the methods of trapping intermediates or preventing their further transformation by changing the medium or inhibiting enzyme actions.

TABLE V

Tentative chronology of the recognition of the intermediary steps of the glycolytic pathway[a]

1911	Pyruvic acid	Neuberg and Wastenson
1918	Acetaldehyde	Neuberg and Reinfurth
1933	D-3-Phosphoglyceric acid	Embden, Deuticke and Kraft
1933	F-1,6-PP	Embden, Deuticke and Kraft
1934	G-6-P, F-6-P	Meyerhof
1934	2-Phosphoenolpyruvic acid	Lohmann and Meyerhof
1934	Phosphodihydroxyacetone	Meyerhof and Lohmann
1935	D-2-Phosphoglyceric acid	Meyerhof and Kiessling
1936	D-Glyceraldehyde-3-P	Meyerhof, Lohmann and Schuster
1936	G-1-P	Cori and Cori
1939	D-1,3-Diphosphoglyceric acid	Negelein and Brömel

[a] See Chapters 19, 21, 22 and 24. The table refers to the first presentation of experimentally founded arguments (in the writer's opinion) for the acceptance of the compound mentioned as an intermediate in the glycolysis pathway accepted at the present time. It does not refer to the first identification and isolation of the compound mentioned, nor to their implication in the so-called "reaction of Harden and Young" or in other forlorn metabolic schemes.

That pyruvic acid is an intermediate in glycolysis was demonstrated by trapping it with β-naphthylamine[91]. In the case of acetaldehyde, it

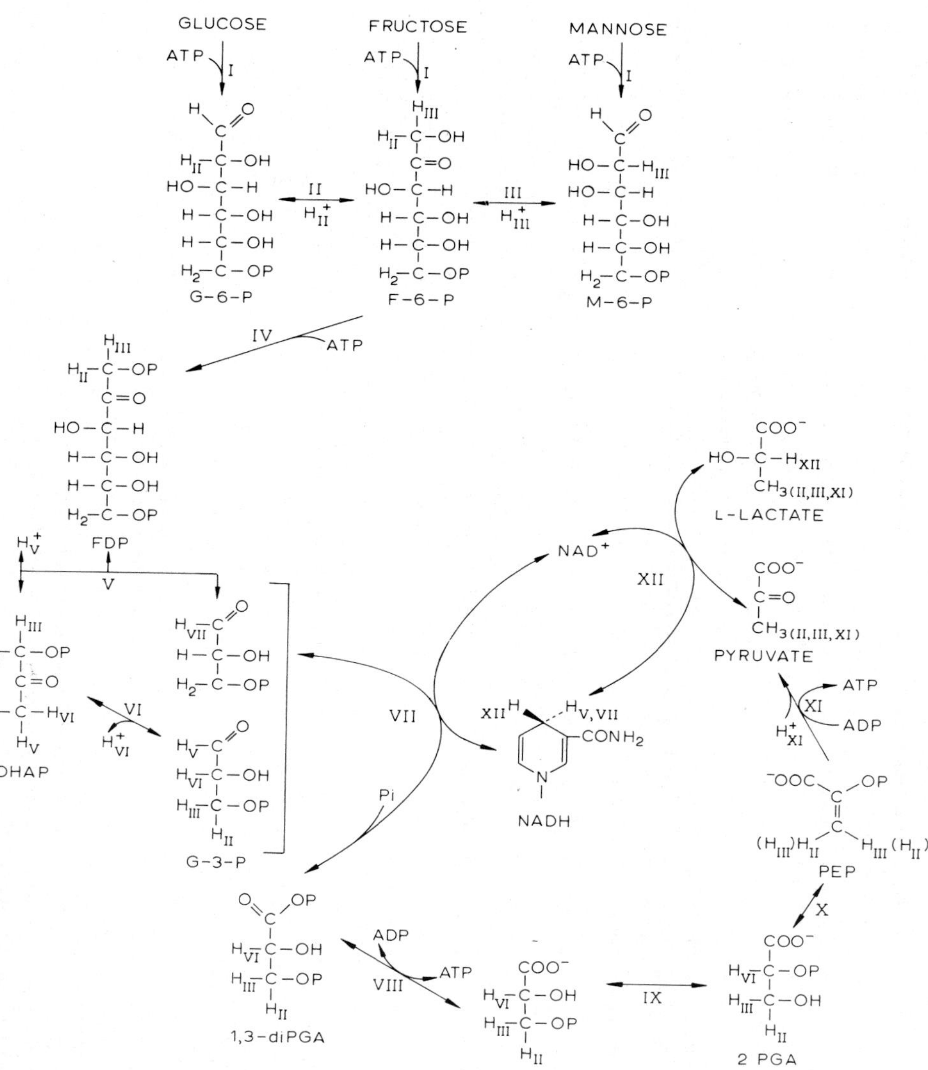

Fig. 7. Reactions of the glycolytic pathway with stereochemical detail (Rose and Rose[84]). The appropriate enzymes with references (85–90) to the stereochemistry of the reactions are as follows: I. Hexokinase (EN 2.7.7.1); II. Glucosephosphate isomerase[85] (EN 5.3.1.9); III. Mannosephosphate isomerase[85] (EN 5.3.1.8); IV. Phosphofructokinase (EN 2.7.1.11); V. Aldolase[86] (EN 4.1.2.7; 4.1.2.13); VI. Triosephosphate isomerase[86] (EN 5.3.1.1); VII. Triosephosphate dehydrogenase[87,88] (EN 1.2.1.9, 1.2.1.12); VIII. Phosphoglycerate kinase (EN 2.7.2.3); IX. Phosphoglycerate phosphomutase (EN 5.4.2.1); X. Phosphopyruvate kinase (enolase) (EN 4.2.1.11); stereochemistry unknown: XI. Pyruvate kinase (EN 2.7.1.40); XII. Lactate dehydrogenase[88–90] (EN 1.1.1.27; 1.1.2.3).

References p. 155

TABLE VI Chronology of the identification of the enzymes of glycolysis

Date	Recommended trivial name[a]	Other (non recommended) names[a]	Systematic name[a]	Number[a]	Authors
1909	Alcohol dehydrogenase	Aldehyde reductase	Alcohol:NAD$^+$ oxidoreductase	1.1.1.1	Battelli and Stern
1911	Pyruvate decarboxylase	α-Carboxylase	2-Oxoacid carboxy-lyase	4.1.1.1	Neuberg and Hildesheimer
1927	Hexokinase		ATP:D-hexose-6-phosphotransferase	2.7.1.1	Meyerhof[b]
1933	Lactate dehydrogenase		L-Lactate:NAD$^+$ oxidoreductase	1.1.1.27	Andersson[c]
1933	Glucosephosphate isomerase	Phosphohexose isomerase	D-Glucose-6-phosphate-ketol-isomerase	5.3.1.9	Lohmann
1934	Pyruvate kinase	Phosphoenolpyruvate kinase	ATP:pyruvate-2 *O*-phosphotransferase	2.7.1.40	Parnas
1935	Phosphoglycerate phosphomutase	Phosphoglyceromutase	D-Phosphoglycerate 2,3-phosphomutase	5.4.2.1.	Meyerhof and Kiessling
1935	Enolase	Phosphopyruvate hydratase	2-Phospho-D-glycerate hydro-lyase	4.2.1.11	Meyerhof and Kiessling
1936	6-Phosphofructokinase	Phosphohexokinase	ATP:D-fructose-6-phosphate 1-phosphotransferase	2.7.1.11	Ostern, Guthke and Terzakowec
1936	Fructose-biphosphate aldolase	Zymohexase, aldolase	D-Fructose-1,6-biphosphate D-glyceraldehyde-3-phosphate-lyase	4.1.2.13	Meyerhof, Lohmann and Schuster
1936	Triosephosphate isomerase		D-Glyceraldehyde-3-phosphate ketol-isomerase	5.3.1.1	Meyerhof, Lohmann and Schuster
1936	Glycogen phosphorylase		α-1,4-Glucan:orthophosphate α-Glucosyltransferase	2.4.1.1	Cori and Cori
1936	Phosphoglucomutase, glucose-phosphomutase		α-D-Glucose-1,6-biphosphate: α-D-glucose-1-phosphate phosphotransferase	2.7.5.1	Cori and Cori
1939	Triosephosphate dehydrogenase, glyceraldehyde-phosphate dehydrogenase (NADP$^+$)		D-Glyceraldehyde-3-phosphate: NADP$^+$ oxidoreductase	1.2.1.9	Warburg and Christian
1942	Phosphoglycerate kinase		ATP:3-phospho-D-glycerate 1-phosphotransferase	2.7.2.3	Bücher[d]

[a] *Enzyme Nomenclature* (EN), vol. 13 of this Treatise, third ed. 1972.
[b] Discovery reported by Meyerhof (*Biochem. Z.*, 183 (1927) 176), name given by Meyerhof in his book on muscle.
[c] The discovery of the enzyme cannot be dated. The date given here corresponds to the first identification of NAD as the coenzyme.
[d] Preliminary announcement by Warburg and Christian, 1939.

accumulated by being preserved from reduction by sodium sulphite as shown by Neuberg and Hirsch[92]. The accumulation of 2-phosphoglyceric acid in the presence of fluoride inhibiting enolase in fermenting yeast extracts proved its role as an intermediary step as well as the accumulation of FDP and triosephosphates in the presence of iodoacetate.

Another method was to introduce one of the supposed intermediates and determine whether it would promote glycolysis. Such a method has of course to be submitted to very close criticism. The intermediates tried were suggested by hypotheses derived from possible schemes based on the previsions derived from the structural formulas and properties of compounds considered on a chemical basis as possible intermediary steps between sugar on the one hand and ethanol or lactic acid on the other hand. None of the pathways proposed by the chemists proved useful, and the only one which fitted into temporary theories of biochemistry, the methylglyoxal theory, was, on biochemical arguments rejected after having reigned a long time.

Glycolysis has been the first domain of expansion and demonstration of the enzymatic theory of cell metabolism. It developed from the rupture with previous beliefs, a rupture introduced by the discovery, accomplished by E. Buchner in 1897, of the alcoholic fermentation by yeast juice, by the isolation and characterization of all the enzymes involved (see Table VI) and by the repetition *in vitro* of all the steps making up the pathways.

It must also be stressed here that the trail of research followed in the unravelling of the glycolytic pathway has involved destructive methods such as the use of yeast juice, muscle juice and minced muscle, yeast maceration juice or crystallized enzymes as well as experiments on whole yeast cells or on muscle and that occasionally the experiments were led astray, sometimes for a long time, by the unrecognized perturbation introduced by the destructive character of the methods adopted. The history of glycolysis is a demonstration of the way in which the wrong interpretations based on these artifacts have been successively discarded by a close criticism of experimental data in such a way that, when isotopic carbon became available, it allowed the demonstration that the scheme of glycolysis resulting from these studies obtains in living cells.

The process of unravelling the metabolic pathway of glycolysis lasted almost forty years. One of the reasons for it was the long-lasting persistence of the concept introduced by Harden, who did not consider the phosphorylated compounds as integrated members of the pathway. As it was repeatedly

References p. 155

stated above, the concept according to which any intermediate should be fermented at least as fast as glucose by yeast juice or glycogen by muscle juice persisted for a long time. This pitfall of common sense, later contradicted by the accumulation of experimental evidence, had a great deal of responsibility in the rejection of phosphorylated compounds, which lasted until Embden *et al.*, in 1933, introduced into the pathway the ester of Harden and Young, shown to be a hexosediphosphate by Young (1909, 1911), and identified as F-1,6-PP by Levene and Raymond, only in 1928.

It should not be forgotten that, if among the intermediates enumerated in Table I, and in Fig. 7, G-6-P and D-3-phosphoglyceric acid can easily be identified in living tissues, and if in red blood cells D-1,3-diphosphoglyceric acid has also been identified for a long time, most of the intermediates could only be detected and their concentration determined in very recent times, after the introduction of the very sensitive enzymatic methods of determination. As in the living state the intermediates are metabolized as rapidly as they are formed, direct evidence has long been lacking due to methods of analysis unable to detect the very small concentrations present.

We have already commented on the reasons which retarded the knowledge of the composition of the "coenzyme" of Harden and Young. With regard to the "zymase" of Buchner, the process of identifying its enzymatic components known to us, lasted 30 years. It started with the discovery of alcohol dehydrogenase by Battelli and Stern in 1909 and ended with the identification of phosphoglycerate kinase by Bücher (this important discovery was first mentioned in 1939 in a paper of the same laboratory, by Warburg and Christian[37]. Bücher published a first report[41] in 1942, and a more detailed one[42] in 1947). The factor responsible for such delayed progress may be found in the relative lack of available chemical knowledge. It is sometimes stated that biochemistry was able to take wing on the basis of knowledge accumulated by the organic chemists. This is far from true. When the process of the elucidation of the pathway of glycolysis did begin, from the compounds which were to be recognized as participants in the pathway, only pyruvic acid and acetaldehyde had been synthesized by organic chemists. They were rapidly recognized as intermediates. The indifference of organic chemists to the repeated suggestions of the biochemists (for instance von Euler[95]; in 1925) who experimentally detected the presence of compounds yet unsynthesized was stupefying. Fischer and Baer[96,97] who synthesized 3-phosphoglyceraldehyde in 1932 were exceptions to this statement.

The synthesis of glyceraldehyde-3-P was suggested by Warburg to his friend Hermann Fischer, the son of his teacher Emil Fischer. Fischer has told the story as follows[93]:

"This was a time when, according to the fermentation theories of Neuberg, methylglyoxal was the key substance in the breakdown of sugar in fermentation and glycolysis. I could not believe that a substance, which is not fermentable itself, could be the central compound of fermentation; therefore, remembering the classical work of Harden and Young, we set out to prepare glyceraldehyde-3-phosphate (racemic), hoping that with three carbon atoms it would be the long-sought intermediate of fermentation. The synthesis was not easy, but we finally succeeded, particularly by using the benzyl residue as a blocking group. It can be easily introduced and, under very mild conditions, removed by catalytic hydrogenation. The method had been discovered by Otto Wolfes in studies on the chemistry of morphine and then applied by Karl Freudenberg to sugar problems. For our purposes, it was just the right thing, and, in the spring of 1932, I could bring to my friend, Otto Warburg, the first one and a half grams of the crystalline calcium salt of the D,L-glyceraldehyde-3-phosphoric acid. Indeed, Dr. C. V. Smythe and Miss Gerischer found, in Warburg's laboratory, that the substance was fermentable by undiluted Lebedev juice to 5%, which is perfectly understandable since, at this time, I could produce only the racemic form of the compound. Otto Warburg later carried out some of his classical work on this substance and even has done me the honor to call it "Fischer ester". At the time, however, Gustaf Embden and later, Otto Meyerhof, utilized the glyceraldehyde-3-phosphate for the development of their now famous scheme of alcoholic fermentation and glycolysis."

That a production of triosephosphate results from the addition of hexose-diphosphate to muscle extract deprived of coferment was demonstrated by Meyerhof and Lohmann[94] in 1934 (it must be remembered that the authors considered the triosephosphate as an intermediate on the path to methylglyoxal).

While the Harden and Young ester was discovered in 1905, it was only in the twenties that organic chemists became interested in sugar phosphates.

One of the main acquisitions towards unravelling the nature and meaning of the pathway was due to the great skill displayed by Negelein and Brömel[32] in the isolation of the labile compound which they showed to be 1,3-diphosphoglyceric acid. This discovery became of great importance as Warburg and Christian[36,37] isolated in the crystalline state the enzyme triosephosphate dehydrogenase (formerly erroneously called 1,3-diphospho-glyceraldehyde dehydrogenase). The recourse to a crystalline preparation, relatively free from other catalysts, allowed Warburg and Christian to work specifically with glyceraldehyde-3-P, to obtain a compound which Negelein and Brömel recognized as 1,3-diphosphoglyceric acid. Such methodology offered a sharp contrast with the uncertainties involved, as repeatedly

References p. 155

stated above, in experiments performed with juices or extracts of muscle or yeast, or with crude enzymatic preparations.

The scheme of Fig. 7 shows that the final theory of glycolysis is at variance with the current views which had reigned for a long period, and according to which glycolysis should be considered as a dismutation process between two C_3 compounds, one oxidizing the other. What happens is that the six carbon chains, with phosphates attached at both ends, goes into two molecules of 3-phosphoglyceraldehyde.

Each of these molecules gives an H atom to NAD^+ (the oxido-reduction coenzyme) and after unloading of the phosphate bond energy, which is the function of the pathway, is transformed into pyruvate (or acetaldehyde) accepting an H atom from the NADH produced in a previous step, and going into the waste product lactic acid (or alcohol). As stated by Lipmann[98]

"... a hydrogen donor, after unloading of phosphate bond energy, is transformed into a hydrogen acceptor".

REFERENCES

1 J. K. Parnas, *Biochem. Z.*, 28 (1910) 274.
2 R. Nilsson, *Biochem. Z.*, 258 (1933) 198.
3 Z. Dische, *Naturwissenschaften*, 22 (1934) 776; 24 (1936) 462.
4 O. Meyerhof and W. Kiessling, *Biochem. Z.*, 281 (1935) 249.
5 O. Warburg and W. Christian, *Biochem. Z.*, 287 (1936) 291.
6 O. Meyerhof, W. Kiessling and W. Schulz, *Biochem. Z.*, 292 (1937) 25.
7 O. Meyerhof, *Erg. Physiol.*, 39 (1937) 10.
8 J. K. Parnas, P. Ostern and T. Mann, *Biochem. Z.*, 272 (1934) 64.
9 W. Mozolowski, T. Mann and C. Lutwak, *Biochem. Z.*, 231 (1931) 290.
10 J. K. Parnas, P. Ostern and T. Mann, *Biochem. Z.*, 275 (1934) 74.
11 D. M. Needham, *Machina Carnis. The Biochemistry of Muscular Contraction and its Historical Development*, Cambridge, 1971.
12 T. Mann, *Biochem. Z.*, 279 (1935) 82.
13 P. Ostern, T. Baranowski and J. Reis, *C.R. Soc. Biol. (Paris)*, 118 (1935) 1414.
14 P. Ostern, T. Baranowski and J. Reis, *Biochem. Z.*, 279 (1935) 85.
15 D. M. Needham and W. E. van Heyningen, *Nature*, 135 (1935) 585.
16 D. M. Needham and W. E. van Heyningen, *Biochem. J.*, 29 (1935) 2040.
17 O. Meyerhof and H. Lehmann, *Naturwissenschaften*, 23 (1935) 337.
18 H. Lehmann, *Biochem. Z.*, 281 (1935) 271.
19 O. Meyerhof and K. Lohmann, *Biochem. Z.*, 273 (1934) 73.
20 K. Lohmann and O. Meyerhof, *Biochem. Z.*, 273 (1934) 60.
21 O. Meyerhof and W. Kiessling, *Biochem. Z.*, 264 (1933) 40.
22 J. K. Parnas and T. Baranowski, *C.R. Soc. Biol. (Paris)*, 120 (1935) 307.
23 J. K. Parnas, *Bull. Soc. Chim. Biol.*, 18 (1936) 53.
24 C. Lutwak-Mann and T. Mann, *Biochem. Z.*, 251 (1935) 140.
25 O. Meyerhof, P. Ohlmeyer, W. Gentner and H. Maier-Leibnitz, *Biochem. Z.*, 298 (1938) 396.
26 D. M. Needham and R. K. Pillai, *Biochem. J.*, 31 (1937) 1837.
27 D. E. Green, D. M. Needham and J. G. Dewan, *Biochem. J.*, 31 (1937) 2327.
28 O. Meyerhof, P. Ohlmeyer and W. Möhle, *Biochem. Z.*, 297 (1938) 90.
29 O. Meyerhof, W. Schulz and P. Schuster, *Biochem. Z.*, 293 (1937) 309.
30 O. Meyerhof and W. Kiessling, *Biochem. Z.*, 276 (1935) 239.
31 O. Meyerhof and W. Schulz, *Biochem. Z.*, 297 (1938) 60.
32 E. Negelein and H. Brömel, *Biochem. Z.*, 303 (1939) 132.
33 E. Negelein and H. Brömel, *Biochem. Z.*, 301 (1939) 135.
34 D. L. Drabkin and O. Meyerhof, *J. Biol. Chem.*, 157 (1945) 571.
35 O. Meyerhof and P. Oesper, *J. Biol. Chem.*, 170 (1947) 1.
36 O. Warburg and W. Christian, *Biochem. Z.*, 301 (1939) 221.
37 O. Warburg and W. Christian, *Biochem. Z.*, 303 (1939) 40.
38 G. T. Cori, M. W. Slein and C. F. Cori, *J. Biol. Chem.*, 159 (1945) 56.
39 G. T. Cori, M. W. Slein and C. F. Cori, *J. Biol. Chem.*, 173 (1948) 60.
40 R. Caputto and M. Dixon, *Nature*, 156 (1945) 630.
41 T. Bücher, *Nature*, 30 (1942) 756.
42 T. Bücher, *Biochim. Biophys. Acta*, 1 (1947) 292.
43 S. P. Colowick, J. Van Eys and J. H. Park, in M. Florkin and E. H. Stotz (Eds.), *Comprehensive Biochemistry*, vol. 14, Amsterdam, 1966.
44 F. Lipmann, *Advan. Enzymol.*, 1 (1941) 71.
45 H. M. Kalckar, *Chem. Rev.*, 28 (1941) 71.
46 O. Warburg and W. Christian, *Biochem. Z.*, 310 (1941–42) 384.

47 M. F. Utter and C. H. Werkman, *J. Biol. Chem.*, 146 (1942) 289.
48 M. F. Utter and C. H. Werkman, *Biochem. J.*, 36 (1942) 485.
49 L. Massart and R. Dufait, *Z. physiol. Chem.*, 272 (1942) 157.
50 P. D. Boyer, H. A. Lardy and P. H. Philips, *J. Biol. Chem.*, 146 (1942) 673.
51 P. D. Boyer, H. A. Lardy and P. H. Phillips, *J. Biol. Chem.*, 149 (1943) 529.
52 H. A. Lardy and J. A. Ziegler, *J. Biol. Chem.*, 159 (1945) 343.
53 F. Kubowitz and P. Ott, *Biochem. Z.*, 317 (1944) 193.
54 C. E. Cardini, A. C. Paladini, R. Caputto, L. F. Leloir and R. E. Trucco, *Arch. Biochem.*, 22 (1949) 87.
55 A. C. Paladini, R. Caputto, L. F. Leloir, R. E. Trucco and C. E. Cardini, *Arch. Biochem.*, 23 (1949) 55.
56 A. Harden, *Alcoholic Fermentation*, London, 1911; 4th ed., 1932.
57 C. Neuberg and A. Hildesheimer, *Biochem. Z.*, 31 (1911) 170.
58 C. Neuberg and L. Karczag, *Biochem. Z.*, 36 (1911) 76.
59 C. Neuberg and J. Kerb, *Biochem. Z.*, 53 (1913) 406.
60 C. Neuberg and O. Rosenthal, *Biochem. Z.*, 51 (1913) 128.
61 C. Neuberg and E. Reinfurth, *Biochem. Z.*, 89 (1918) 365.
62 W. Connstein and W. Lüdeke, *Ber. deut. Chem. Ges.*, 52 B (1919) 1385.
63 E. Auhagen, *Z. physiol. Chem.*, 204 (1932) 149.
64 E. Auhagen, *Z. physiol. Chem.*, 204 (1932) 20.
65 K. Lohmann and P. Schuster, *Biochem. Z.*, 294 (1937) 188.
66 H. Weil-Malherbe, *Biochem. J.*, 33 (1939) 1997.
67 M. A. Lipton and C. A. Elvehjem, *J. Biol. Chem.*, 136 (1940) 637.
68 W. Franke, in H. von Euler (Ed.), *Chemie der Enzyme*, vol. II, Part 2, Munich, 1934.
69 O. von Fürth and F. Lieben, *Biochem. Z.*, 128 (1922) 144.
70 O. von Fürth and F. Lieben, *Biochem. Z.*, 132 (1922) 165.
71 O. Meyerhof, *Arch. f.d. Ges. Physiol.*, 175 (1919) 20.
72 O. Meyerhof, *Arch. f.d. Ges. Physiol.*, 175 (1919) 88.
73 A. von Szent-Györgyi, *Biochem. Z.*, 157 (1925) 50.
74 B. Andersson, *Z. physiol. Chem.*, 217 (1933) 186.
75 B. Andersson, *Z. physiol. Chem.*, 225 (1934) 57.
76 H. von Euler, E. Adler, G. Günther and H. Hellström, *Z. physiol. Chem.*, 245 (1937) 217.
77 O. Meyerhof and P. Ohlmeyer, *Biochem. Z.*, 290 (1937) 334.
78 E. Buchner and J. Meisenheimer, *Ber. deut. Chem. Ges.*, 36 (1903) 634.
79 F. Battelli and L. Stern, *C.R. Soc. Biol. (Paris)*, 67 (1909) 419.
80 F. Battelli and L. Stern, *C.R. Soc. Biol. (Paris)*, 68 (1910) 5.
81 T. Thunberg, *Erg. Enzymol.*, 7 (1938) 163.
82 D. Koshland Jr. and F. H. Westheimer, *J. Am. Chem. Soc.*, 71 (1949) 1139.
83 D. Koshland Jr. and F. H. Westheimer, *J. Am. Chem. Soc.*, 72 (1950) 3383.
84 I. A. Rose and Z. B. Rose, in M. Florkin and E. Stotz (Eds.), *Comprehensive Biochemistry*, vol. 17, p. 93, Amsterdam, 1969.
85 I. A. Rose and E. L. O'Connell, *Biochim. Biophys. Acta*, 42 (1960) 159.
86 I. A. Rose, *J. Am. Chem. Soc.*, 80 (1958) 5835.
87 F. A. Loewus, H. R. Levy and B. Vennesland, *J. Biol. Chem.*, 223 (1956) 589.
88 J. W. Cornforth, G. Ryback, G. Popjack, C. Donninger and G. Schroepfer Jr., *Biochem. Biophys. Res. Commun.*, 9 (1962) 371.
89 F. A. Loewus and H. A. Stafford, *J. Biol. Chem.*, 235 (1960) 3317.
90 F. A. Loewus, P. Ofner, H. F. Fischer, F. H. Westheimer and B. Vennesland, *J. Biol. Chem.*, 202 (1953) 699.
91 M. von Grab, *Biochem. Z.*, 123 (1921) 69.

92 C. Neuberg and J. Hirsch, *Biochem. Z.*, 96 (1919) 175.
93 H. O. L. Fischer, *Ann. Rev. Biochem.*, 29 (1960) 1.
94 O. Meyerhof and K. Lohmann, *Naturwissenschaften*, 22 (1934) 134.
95 H. von Euler, *Chemie der Enzyme*, 1st Part, Munich, 1925.
96 H. O. L. Fischer and E. Baer, *Ber. deut. chem Ges.*, 65 (1932) 337.
97 H. O. L. Fischer and E. Baer, *Ber. deut. chem. Ges.*, 65 (1932) 1040.
98 F. Lipmann, *Wanderings of a Biochemist*, New York, 1971.

Chapter 25

Glycolysis as a Source of Free Energy

1. From calorimetry to free-energy studies

The first developments of bioenergetics are found in the field of the principle of energy conservation, when Mayer[1] formulated the concept of the transformation of chemical force into mechanical force by muscle. In his well-known lecture of 1847, Helmholtz[2] recognized that the heat of combustion of the nutrients is equivalent to the heat given off by an animal. This was expressed more quantitatively by Rubner as shown in Chapter 11*. But, as we read in a letter of Mayer to his friend the physiologist Griesinger, dated 14 june 1844,

"For some time a logical instinct has led physiologists to the axiomatic proposition: No action without metabolism. From the physical side, this thesis is already definitely expressed in my theory. However, in physiology, it is a question of "How and what and when and where?" (ref. 3, translation by Rosen[4]).

The work of A. V. Hill on the thermochemistry of muscle started in 1912. At that time Meyerhof had embarked on solving the query of Mayer, "How and what and when and where?" by thermochemical studies of intermediary metabolism. Meyerhof's main interest was to find whether the energy liberated by the catabolism of nutrients would account for the heat evolved by the animal body, and on the other hand, he wondered how the potential energy of food could be made available for the accomplishment of work by the cells. In calorimetric studies, Rubner had, in his studies on the nutrients of animal organisms, determined the heat values of a number of biological compounds present in food.

* That, in organisms, the various forms of power are interconvertible without loss is a concept which is sometimes attributed to Liebig, as a consequence of misreading his *Animal Chemistry* (except if we are ready to recognize vital force as one of the forms of energy).

References p. 182

According to present views, which have developed since Lipmann[29] formulated the ADP–ATP phosphate cycle in 1941, free energy released by metabolic reactions is either stored in ATP or dissipated as heat. The free energy stored is available either for work or for biosynthesis. This concept of the phosphate cycle has been confirmed by a large number of data to be dealt with in the subsequent chapters of this History.

The system ADP–ATP dominates all forms of phosphate transfer.

Phosphorylation by ATP is required before monosaccharides can be oxidized. As we shall later see it is also required for the activation of fatty acids and of amino acids. By these activations, the autocatalytic nature of catabolic pathways is insured. In a uniform medium, this cycle would take place smoothly with the maintenance of a steady state of ATP, ADP and P_i concentrations. But a sudden change of exigence would upset the whole system if a buffer (phosphagen) was not involved in muscle, with a transfer potential similar to that of ATP. Just as in biosynthesis, the free energy of energy-rich phosphate bonds is involved in mechanical work, in active transport, *etc.* This conceptual system has got its historical base from studies on glycolysis, and the thermochemistry of this pathway was first approached by the bias of muscle contraction. Muscle was long considered as a thermal machine, though by the turn of the century evidence pointed to a direct utilization of chemical energy, and not of heat, in muscle contraction. The name of "inogen" had been given by Hermann[5] to what he conceived as a store of nitrogen containing substances splitting with the liberation of energy. (On the history of inogen, see D. M. Needham[6]).

When it appeared that glycogen was the source of energy in muscle contraction, Meyerhof, taking into account the calories produced or removed during anaerobic contraction and during oxidative recovery, established, to use the appropriate phrase of D. M. Needham, the first balancing of the thermochemical books.

Hill's work had greatly influenced Meyerhof and the Hill-Meyerhof theory, current in the twenties, includes the contributions of both researchers to the concept that the provision of energy necessary for muscle contraction was derived from the whole process of lactic acid formation (muscle glycolysis). From a proportionality between the work performed and the lactic acid produced, Meyerhof concluded (a pitfall of common sense) that this exothermic process provided the free energy for muscle contraction. It should be noted, nevertheless, that Meyerhof never did, as he is sometimes accused of, claim that contraction resulted from a direct action of lactic acid.

In a paper of 1926 (sent in November 1925) Meyerhof and Lohmann[7] write:

"Nun ist es ja niemals in Vorschlag gebracht worden, den chemischen Vorgang der Milchsäure-bildung *direkt* die mechanische Arbeit leisten zu lassen."

How did Meyerhof, who played such a determinant role in its development, become involved in bioenergetics?

Otto Meyerhof (1884–1951), the pioneer of the thermochemical study of intermediary metabolism, graduated at Heidelberg in 1909 as a doctor of medicine, with a thesis on a psychiatric subject. During his studies he had already displayed the universality of interests which was a trait of his personality, and he had received an initiation to chemistry and to physics as well as to philosophy. His interest in the latter was at the time as active as to lead him to become the editor of the *Abhandlungen der Friesschen Schule*, publishing the works of the philosopher Leonard Nelson, of Göttingen, and of his group of disciples of Fries.

Fries, before being professor at Jena, had taught in Heidelberg. He was classified as neokantian and he was the leader of a theory recognizing in psychology the basis of philosophy (psychologism). If the fact that his thesis was devoted to a psychiatric subject may be interpreted as an influence of his philosophical preoccupations, this aspect should not be overrated and we should remember that in his book on muscle Meyerhof[8] states, quoting Bertrand Russell, his belief that science will not return to this gross form of causality to which the natives of the Fiji Islands, *and the philosophers*, still believe.

Meyerhof had become interested in the field of the accounting, for the heat evolved in the animal body, of the combustion of nutrients, a field opened by Rubner. He wondered about the mechanisms by which the potential energy of food could be made available as free energy to the cells. After graduating in 1909, at the age of twenty-five, he went to the Klinik of Krehl. Krehl was the author of important contributions to the study of basal metabolism, and he had developed in Heidelberg an active center of reseach.

In Krehl's laboratory, Otto Warburg (1883–1970) worked as assistant. Warburg had obtained his doctor's degree in chemistry in 1906 with a thesis prepared under the direction of E. Fischer, on the preparation

References p. 182

of optically active peptides and on their enzymatic hydrolysis, and while contributing to the activity of Krehl's laboratory he studied medicine until 1911, following a shortened curriculum (without "Approbation").

It is in Krehl's department that Warburg and Meyerhof met and became friends.

Lipmann[9] writes:

> "Meyerhof was really a pupil of Warburg. He had begun with a strong tendency to philosophical generalizations, but a short period of collaboration with Warburg was a turning point for him. It seemed to have convinced him that the best hope of coming near to explain the life phenomena was in physicochemical exploration".

Meyerhof[8] himself, in the preface of his book on muscle, states that he owed to Warburg his personal initiation to the problems and methods of cell physiology, and Warburg considered Meyerhof as one of his disciples[16]. But, as far as the author knows, there is no reason to believe that Warburg inclined Meyerhof towards thermochemical researches. These, Meyerhof had since 1911 initiated by several studies[10–13], when he published in collaboration with Warburg[14], in 1912, a paper on the respiration of killed cells and cell fragments.

Calorimetric measurements had occupied Meyerhof since the start of his research (he had perfected the methods of Rubner) while Warburg had so far devoted his experimental work to measurements of oxygen consumption and had never been active in the field of thermochemical studies. The paper Meyerhof published in collaboration with Warburg[14] was in the line of the latter and cannot be considered as having influenced Meyerhof in the choice of his own line of research. As has been stated by Weber[15] it appears that it is in the orientation of Meyerhof's career towards physiology that the influence of Warburg has been active. In 1913, Meyerhof joined, as Privatdozent, the University of Kiel, where he became professor of physiology in 1918. His friend Warburg was appointed to the Kaiser Wilhelm Institut für Biologie, in Berlin-Dahlem, in 1914. Owing to the outbreak of World War I, he did not start working there until 1919. In 1930, with the help of the Rockefeller Foundation, he got a building to himself, the Kaiser Wilhelm Institut für Zellphysiologie, of which he was the director and in which he worked until he died in 1970 (on Warburg's biography, see Krebs[16]).

Shortly after moving to Kiel in 1913, Meyerhof delivered a lecture on cellular energetics ("Zur Energetik der Zellvorgänge"[17]).

He starts from the law of energy conservation as formulated by Mayer, Joule and Helmholtz. At this time, he states that a transformation of energy is necessary for the maintenance of life, *i.e.*:

"Zur Verhütung oder Rückgängigmachung der sich von selbst abspielenden Vorgänge."

He recognizes the imperative of dealing with free energy (maximal work) and he states that

"erst im jüngster Zeit" the heat theorem of Nernst[18] has made possible the calculation of free energy,

and he proceeds by writing:

"Danach ist die Arbeitsfähigkeit [the free energy] der Oxydation des Kohlenstoffs und von Kohlenwasserstoffverbindungen bei Körpertemperatur tatsächlich der Wärmetonung ungefähr gleich".

As he was unable to calculate free energy, Meyerhof (as well as Warburg[19] in a review paper published in 1914) admitted that for the oxidation of hydrogen or carbon it is permissible to set, with some degree of approximation, the heat liberated as equal to the decrease of free energy.

For a long time it had been believed by biologists that the heat liberated in a reaction was equal to the work accomplished in the organism, following this reaction.

Willard Gibbs, in 1878, and Helmholtz, in 1882, introduced the distinction between the "total energy" (ΔH) resulting of a chemical process in the form of heat and the "free energy" made available for work (ΔF, later designated as ΔG, in recognition of the importance of the contribution of Gibbs).

Nernst's heat theorem was taken advantage of by Báron and Polanyi[20], in the laboratory of Tangl in Budapest, to determine the free energy change in the catabolism of glucose. They found a value about 13% higher than the heat value.

In 1912, Hill[21] had underlined that besides the application of the heat theorem of Nernst, there was another possible approach to free energy measurement:

"In many cases it is already possible to calculate the equilibrium constant K of a reaction and hence also... its free energy. The development of ferment chemistry, especially in the case of reversible changes carried out by ferments, may make it possible to calculate directly the equilibrium constants of many breakdowns of organic material. If *e.g.* the biochemist can decide what chemical reactions go on in order in the process of carbohydrate

References p. 182

breakdown, and if we can determine directly the values of K_1, K_2, K_3,...for the several reactions, then it will be possible not only to give the total free energy... but the free energy at every stage."

The fact that Meyerhof, in a lecture published in 1913, is already concerned with free energy is worth consideration. In the U.S.A., biochemists were little aware of the distinction of ΔH and ΔG until the publication, in 1923, ten years later, of the book of Lewis and Randall[22] while in European circles Ostwald had popularized much earlier the concepts introduced by Gibbs.

As foreseen by Hill, the acquisition of the knowledge of the sources of free energy in glycolysis has been conditioned by the unravelling of its metabolic pathway and of its intermediary reactions.

When we know the equilibrium constant K of a reversible reaction, ΔG^0 can be calculated by the equation:

$$\Delta G^0 = -RT \ln K$$

In the case of the reaction

$$\text{G-1-P} \rightleftharpoons \text{G-6-P}$$

for instance, K^0 being equal to 19, ΔG^0 is calculated as -1800 cal.[23,24]. This method has been extensively used.

The heat of hydrolysis of other energy-rich compounds (such as phosphoglycerate and phosphopyruvate), besides that of creatine phosphate and ATP has been determined by Meyerhof and Lohmann[25] in 1934 and by Meyerhof and Schulz[26] in 1935.

It was made clear by these studies that the phosphate bond of hexosephosphates was in the course of glycolysis transformed into a new form of bond, more exothermic when hydrolyzed and the energy of which can be transferred without being liberated, to another phosphate bond in a transphosphorylation reaction.

Of course heat values are not an accurate indication of the free-energy values of these compounds, but they can be used to arrive at the free-energy value if the entropy can be calculated and deduced. It is also possible to determine ΔG^0 by making use of the heat released by a reaction and of the heat capacities of reactants and products (see Pardee and Ingraham[27], or Kaplan[28]).

Though they were inclined to accept that free energy values were approximately equivalent to heat values, it was clear to biochemists that *to gain this work* (corresponding to the work they believed to be approximately equivalent to the heat released in the calorimeter) devices must be available to permit catabolism to take place in a reversible way. This is already understood by Meyerhof in his review lecture of 1913 when he considers the existence of reaction cycles. The Kiel lecture also indicates the origins of Meyerhof's interest in muscle physiology, as a domain in which work could be the object of measurement and about which biophysical data were at the time contributed by A. V. Hill. Meyerhof has also stated[8] that the researches of Fletcher and Hopkins as well as those of Parnas, played a seminal role in the development of his own interest in muscle bioenergetics.

Returning to the device for obtaining and utilizing free energy in cells, a device which since Lipmann[29] formulated the theory in 1941, we consider as constituted by the phosphate metabolic cycle*, it is worth remembering that in a monograph on oxido-reduction potentials published in English in 1930 about the same time as it appeared in German, Michaelis[30], one year after the discovery of ATP by Lohmann, wrote these prophetic lines:

"Thus physiology is faced with the following problems. First in metabolism it must seek reversible chemical processes, obscured perhaps by the many irreversible processes of combustion. Second, a cell contrivance must be sought similar to an accumulator in which the free energy of these reversible processes can be stored. Third, devices are to be looked for, capable of transforming the stored energy into new forms of energy." (from the translation of L. B. Flexner)

At the time the studies on biological oxidations (see Chapters 26–37) which had been devoted to the concept of oxygen and of hydrogen activation had already led to the comparison of "combustion" in organisms, to oxido-reduction phenomena. The thermodynamic theory of oxido-reductions had been formulated, and by the turn of the century the relation was known between the equilibrium constants of reversible oxidation–reduction reactions, and the free energy changes of these reactions at constant temperature. Oxido-reductions were since the beginning of the present century recognized as electron exchanges and when the electronic theory of valency was formulated, the organic chemists began to figure out the organic reactions as involving electronic mechanisms. Oxidation–

* This cycle consists of three phases: the incorporation of inorganic phosphate into energy-rich bonds in a coupling with oxidation, a transfer of $\sim$P to the adenylic acid system; and the utilization of the free energy of the pyrophosphate high energy bond of ATP with a liberation of inorganic phosphate.

References p. 182

reduction potentials referred to the normal hydrogen electrode (platinized platinum electrode in one atmosphere of hydrogen, in a solution normal with respect to the hydrogen ion) were determined for many dyes and other organic compounds (see Clark[31]).

The device for obtaining ATP was unravelled by pathway studies as the coupling of oxido-reduction and phosphorylation.

2. Coupled chemical reactions

The concept was introduced by Ostwald in 1900 after he noted the difference between the transfer of heat or electrical energy, and the transfer of chemical energy (coupled reactions).

To quote Ostwald[32]:

"A direct reciprocal transformation of chemical energies is only possible to the extent that chemical energies can be set in connection with each other, that is, within such processes that are represented by a stoichiometric equation. Coupled reactions of this kind may be distinguished from those that proceed independently of each other; their characteristic lies in the fact that they can be represented by a single chemical equation with definite integral coefficients. On the other hand, chemical processes that are not coupled cannot transfer energy to each other." (translation by Fruton[33])

The concept has often been loosely applied, an aspect which would well deserve a detailed searching of the literature.

We have quoted in Chapter 23 a text of Meyerhof and Lohmann[34], dating of 1931 in which these authors comment on the utilization of the energy-rich bonds of ATP and creatine phosphate. In this text they clearly realize the participation of these high energy bonds in the biosynthesis of biochemical compounds but what they call the cycle of ATP remains by force, due to lack of knowledge undefined in physicochemical terms. The authors describe ATP as being restored at the end.

"Der Kreislauf der ATP unterhält auf diese Weise die Milchsäurebildung."[34]

The synthesis of creatine phosphate is recognized as made possible by the splitting of ATP, while the energy of the whole glycolysis permits the synthesis of ATP.

"Die Synthese des Phosphagens wird danach ... durch die Spaltungsenergie der ATP ermöglicht, während die Energie der Milchsäurebildung (aus Phosphorsäuren-estern) dazu dient dies zerfallene Pyrophosphat wieder zu aufbauen".

We find here, as well as in the "Meyerhof cycle" (of glycogen catabolism and oxidative resynthesis at the expense of a portion of the lactate) different aspects of what D. M. Needham[6] appropriately designates as a balancing of thermochemical books, but no trace of mechanism for the energy transfer which could only take place by "reaction coupling" as was underlined by A. Hahn[35] who pointed to the vicious use of the expression "reaction coupling" by Meyerhof[36]:

"Denn wir haben es mit einer gekoppelten Reaktion zu tun, bei der die Energie der Oxydation zur Resynthese der Milchsäure verwandt wird."

Die Bedeutung der gekoppelten Reaktion, die in der Erholungsperiode des Muskels abläuft, ist darin zu suchen, dass vermittels der Oxydationsenergie ... die Milchsäure endotherm in Glykogen resynthetisiert wird".

As remarked by Hahn[35], "coupled reactions" can only involve free energy, which does not appear in Meyerhof's argumentation.

"Stets findet man die Angabe dass die Energie die zur Bildung von Glykogen aus Milchsäure benötigt wird, geliefert werde durch die gleichzeitige Oxydation eines Teiles der Milchsäure. Energie kann aber nach dem thermodynamischen Sprachgebrauch nur die Änderung der Gesamtenergie, d.h. die Wärmetönung der Reaktion bedeuten: und sie bedeutet dies auch tatsächlich in den Ausführungen von Meyerhof ... Auf diese kommt es aber bei gekoppelten Reaktionen gar nicht an, sondern nur auf die freie Energie."

In his book on muscle, Meyerhof[8], commenting on the so-called Meyerhof-cycle, introduced the concept of "energy coupling" (as distinct of "reaction-coupling").

To quote Meyerhof[8]:

"Schon das Schwanken des Oxydationsquotienten beweist, dass Oxydation und Resynthese nicht einen chemisch gekoppelten Vorgang darstellen, für den eine stöchiometrische Gleichung anzugeben ist, sondern einen energetisch gekoppelten."

Commenting on this sentence, Hahn[35] writes:

"Hier taucht der Begriff der *energetischen Koppelung* auf. Es muss festgestellt werden, dass er nicht aus der Thermodynamik stammt. Man kennt zwar gekoppelte Reaktionen, ausserdem noch eine energetische Koppelung anzunehmen, ist der Thermochemie völlig fremd. Somit dürfte es sich bei Meyerhof um die Aufstellung eines *biologischen* Begriffes handeln..."

Meyerhof[37] did not oppose this statement and in 1931, he explained that the expression "energetic coupling" was intentionaly vague, and used when the common reagents taking part in "coupled reactions" remained unknown. The "principle of energetic coupling" as expressing the balancing of the thermochemical books ("bilanzmässige Betrachtungsweise") was developed in Meyerhof's laboratory, by von Muralt[38] who, taking for granted the existence of the gluconeogenesis was supposed to operate for the reparation of muscle glycogen in the so called "Meyerhof cycle" stated that:

References p. 182

"Der Wirkungsgrad der Muskelmaschine, berechnet aus gemessener Wärmebildung und Arbeitsleitung ist um 5% grösser als der Wirkungsgrad, berechnet auf die Änderung der freien Energie. Praktisch folgt daraus, dass die Muskelmaschine eine chemodynamische Maschine ist, die die Änderung der freien Energie der Kohlehydratverbrennung, messbar an der Wärmetömung—ΔH_n dieses Vorganges, mit einem ganz bestimmten Wirkungsgrad in Arbeit umsetzen kann. Solange die Erholung des Muskels vollständig ist, wird an dieser Feststellung durch den Umstand, dass in den Muskelchemismus Zwischenglieder eingeschaltet sind, bei denen Wärmetönung und Änderung der freien Energie ganz verschiedene Werte haben, nichts geändert. Diese Tatsache ist A. Hahn bei seiner Kritik der energetischen Vorstellung von Meyerhof entgangen."

That the concept of "energetic coupling" was a biological concept, as reproached by A. Hahn to Meyerhof, is of course true and is in the favour of the tendency of Meyerhof to follow the teaching of experimentation rather than current ideas. He was evidently quite aware of the lack of any stoichiometric formulation for what he designated as "energetic coupling" but he expected that the formulation could come. To get an idea of the conceptual background of the expression "energetic coupling", an extract of a paper of Meyerhof and Lohmann[39] may be quoted (see also the quotation, in Section 3 of Chapter 23 of a paper of 1931 by the same authors):

"... the present experiments lay the foundation of the thesis that the endothermic synthesis of creatine phosphate can take place through a coupling of this process with the exothermic and spontaneous breakdown of ATP, whilst the resynthesis of ATP out of adenylic acid and inorganic phosphate is made possible through the energy of lactic acid formation. One may also assume here a coupling of the synthesis with the metabolism of the intermediate hexose esters and so see in the phosphate groups contained in all these compounds, the unique carriers of the chemical coupling process. This would at once make understandable how the ATP acts as coenzyme of the lactic acid formation, in that the taking up of phosphate by the hexose stands in connection with the giving up of phosphate on the side of the ester is concerned with the rebuilding of the pyrophosphate groups on the adenylic acid. This conception would also make possible a special meaning for the fact that for the anaerobic splitting of hexoses intermediate esterification with phosphoric acid is above all necessary." (translation by D. M. Needham[6]).

That the resynthesis of creatine phosphate and of ATP could be conceived as energetically linked to glycolysis as a whole (what Meyerhof designates as "energetic coupling") was opposed by Parnas, Ostern and Mann[40] in 1934 on a different basis of that of Hahn. Parnas recognized that "definite partial processes" were involved, and not glycolysis as a whole, a concept which has received confirmation subsequently.

3. The discovery of the alactacid muscle contraction

Lundsgaard[41,42] showed in 1930 that a complete blocking of glycolysis does not prevent anaerobic muscle contraction, accompanied by a breakdown of creatine phosphate, the exhaustion of which coincided with the end of the contraction power. From this Lundsgaard concluded that the breakdown of creatine phosphate was more closely related to the contraction mechanism than as accepted in the Hill–Meyerhof lactic acid theory. It must be stated that in 1926, Embden, Hirsch-Kauffmann, Lehnartz and Deuticke[43] had observed the formation of lactic acid after the termination of a tetanic stimulation under anaerobic conditions. This observation led them to believe that contraction and lactic acid formation were not linked, a view which was also proposed in other subsequent publications of the Embden school. As Meyerhof, in experiments accomplished by indirect stimulation, could not observe the same effect he repudiated Embden's conclusions. When Lundsgaard, in the Physiological Laboratory of the Copenhagen Medical Faculty, started a research on the specific dynamic action of amino acids, he studied the effect of iodinated amino acids. He used monoiodo-acetate as an analogue to iodinated glycine and he tried to study its effects on the metabolism of the rabbit. He found that even small amounts of iodoacetate were very toxic for rabbits. Shortly before death, the animals entered into a state of generalized muscle contracture and Lundsgaard recognized that this contracture was of "alactacid" nature. Two french authors, Schwartz and Oschmann[44,45] of Strasbourg, had already observed that the contractures due to bromacetate were not accompanied by the production of lactic acid nor of phosphoric acid (the theory of lactacidogen was widely spread at the time).

Lundsgaard, experimenting with frogs, found that the performance of work and the amounts of creatine phosphate hydrolyzed were proportional. He made the same mistake as Meyerhof did with lactic acid when he concluded from this proportionality that this exergonic process directly provided the contraction energy. To quote Lundsgaard[42]:

"... hier die Phosphagenspaltung der energielieferende Prozess ist" (p. 78).
"... dass man deshalb im Moment damit rechnen muss, dass die Phosphagenspaltung für sich unmittelbar die Kontraktionsenergie liefert." (p. 81).

As stated above, Meyerhof himself had never attributed such direct action to lactic acid.

But if this pitfall of proportionality induced Meyerhof as well as Lundsgaard in error, the recognition of its dangers acted as an epistemological obstacle in the opposite way when Lohmann[46], in 1934, showed that

Plate 84. Einar Lundsgaard.

ATP or AMP is required for enzymatic breakdown of creatine phosphate in muscle extracts. This was in favour of attributing to ATP rather than to creatine phosphate, the role of immediate energy source for the contraction. But from this conclusion. Lohmann carefully guarded himself and he underrated his conclusions by designating the ATP breakdown as

"die zeitlich erst bis jetzt nachgewiesene exotherme Reaktion bei der Muskelzuckung",

and leaving to Engelhardt and Lyubimova, as will be shown in a later chapter, the merit of showing the direct relation of ATP to the contraction of muscle fibers.

As stated by Lipmann[9], concerning Lundsgaard's discovery

"The turmoil that this news created in Meyerhof's laboratory is difficult to realize today. The lactic acid concept had been deep-rooted but seemed to have to be abandoned for a functioning of Cr$\sim$P breakdown ...

"... The explanation for the appearance of Cr$\sim$P breakdown with blocked glycolysis came when one began to realize that glycolysis is a mechanism for the production of energy-rich phosphate bonds delivered into the $\sim$P pool of ATP, its $\sim$P turned out to be the source of contraction energy. Although not realized immediately, it soon became clear that the Cr$\sim$P in muscle represented a reservoir of $\sim$P which, in an emergency, could feed into ATP."

But the observations of Lundsgaard came at a time when Meyerhof was inclined to renounce the concept that creatine phosphate is involved in muscle energetics (see for instance a paper by Meyerhof, Mc Cullagh and Schulz[47]).

On the other hand, Lundsgaard demonstrated the existence of a relation between glycolysis and the formation of energy-rich phosphate bonds.

To quote Lipmann[29]:

"Using the heat data of Meyerhof, Lundsgaard[42] calculated the tension/heat quotient of Hill (Tl/H).

"He found practical agreement with the quotient calculated earlier from normal muscle where heat of glycogen breakdown was compared with tension[49]. In other words, equal amounts of ultimate heat energy, irrespective of its origin from either creatine phosphate or glycogen, did the same amount of mechanical work. By this finding of Lundsgaard the applicability of phosphate bond energy for the driving of the muscle machine was established".

To quote Kalckar[50]:

"The iodoacetate experiments furnished the first demonstration of the biological role of high-energy phosphates. This role became apparent ten years later from Engelhardt's and Szent-Györgyi's experiments on myosin and ATP."

These studies could not have been accomplished if Fiske and Subbarow on the one hand and Lohmann on the other had not opened the field of study of the energy rich phosphate bonds.

Though the implication of creatine phosphate in the energy metabolism of muscle represents, as it appeared later, a specific adaptation feature representing a reservoir which, in emergency, feeds into ATP, its consideration was determinant in defining the energetic relations in muscle glycolysis.

In a paper published in 1934, Lundsgaard[51] himself presented arguments for not considering creatine phosphate any more as the immediate source of muscular free energy.

He observed that, in poisoned muscle, at low temperatures, in anaerobic conditions, an important part of the creatine phosphate breakdown takes place after a short tetanus, showing that creatine phosphate breakdown, as well as glycolysis, is involved in the restitution processes.

That creatine phosphate may be resynthesized anaerobically at the expense of glycolysis* had been demonstrated in 1928 by Nachmansohn[53], in Meyerhof's laboratory. Lundsgaard[54], in 1931, determined the efficiency of glycolysis in that respect and found it high. Two moles of creatine phosphate were reformed after one half mole of glucose was broken down to lactic acid. The two reactions were estimated thermochemically as each about 24000 cal. and of opposite signs which, as stated above, led to the conclusion that the total free energy released in glycolysis was converted into energy-rich phosphate bonds, a conclusion which was not confirmed by subsequent work.

As resulted from Lundsgaard's experiments, energy-rich phosphate bonds were available in the absence of glycolysis as well as of respiratory oxidations. On the other hand, these energy-rich phosphate bonds were easily supplied by glycolysis as well as by respiratory oxidations. Therefore Lundsgaard[54] suggested that the energy used by muscle was derived from energy-rich phosphate bonds constantly supplied by glycolysis and by respiratory oxidations.

This may be considered as the essential contribution of Lundsgaard to what was to become the "phosphate cycle" concept, a contribution more

* That creatine phosphate, when exhausted by a series of contradictions, is rapidly reconstituted during recovery in oxygen had been shown by Eggleton and Eggleton[86] and by Fiske and Subbarow[87].

important than his experiments on "alactacid contracture" (which had raised no interest when first discovered by the French authors).

Lundsgaard was one of the pioneers of the introduction of the concept of the phosphate cycle. It appeared from his work that creatine phosphate splitting was compensated not by glycolysis but at the expense of ATP generated from glycolysis. This concept, which was to win universal recognition after the formulation of the phosphate cycle concept by Lipmann[29] in 1941, was not accepted by all, and we may recall such attacks as those launched in 1940 by Sacks[55,56]. This author maintained that creatine phosphate and ATP were mere buffers and had no other functions in cells.

On the other hand, at that time, Korzybski and Parnas[57] had already studied the turnover, *in vivo*, of the phosphorus atoms of ATP using $Na_2H^{32}PO_4$ provided by Hevesy. This work showed that the turnover of the phosphorus atom of adenylic acid is very slow, whereas the turnover of the other two atoms of ATP is very rapid.

4. Phosphate bond energy and its generation in glycolysis

In 1941, Lipmann[29] published the review paper in which he formulated the general "metabolic principle" which, as we shall see, became of such essential importance in bioenergetics.

Lipmann, in 1940, had recognized, as will be developed in Chapter 33, the enzymatic acetylphosphate synthesis in *Lactobacillus*, this compound acting as a donor of phosphoryl as well as of acetyl. This led him, in association with the advances which, in the thirties, had increased the knowledge of glycolysis to acutely conceive a widespread utilization of energy-rich phosphate (and other) bonds as energy carriers, not only in such a specialized feature as muscle contraction but in the general frame of metabolic events including biosynthesis.

Lipmann introduced the concept of "high group potential" to designate energy-rich linkages, in order to emphasize, rather than the escape and dissipation of energy through cleavage, the positive aspect of

"the largeness of the energy present in the linkage before cleavage", the "escaping tendency of the group".

Lipmann recognized that a variety of energy-rich groups could be derived from the energy of certain metabolic processes and transferred to certain

References p. 182

metabolic situations. He did differentiate two classes of phosphate esters. In one of these classes, the phosphate was linked to an alcohol (free-energy change on hydrolysis corresponding to −2000 to −3000 cal/mole). As the equilibrium constant of the action of phosphatase on glycerophosphate was known (Kay[58]), Lipmann calculated a value of −2280 cal by using the equation

$$\Delta G^0 = -RT \ln K$$

in which G^0 represents the free energy available under standard conditions (the first calculation of the ΔG^0 of a phosphate ester). A second group of esters was recognized by Lipmann: those containing P-O-P (as in the pyrophosphate bond of ATP), N-P (guanidine phosphates), carboxyl-phosphate (acyl-phosphate, O:C.O∼P) or enol-phosphate (C:C.O∼P) bonds. He introduced the squiggle (∼) to represent an energy-rich bond.

Taking as an example the conversion of 2-phosphoglycerate to pyruvate, and taking into consideration the ΔG^0 of intermediate stages and the heat of combustion of glyceric acid and pyruvic acid, Lipmann calculated an approximate value of −11250 cal. for the energy-rich phosphate bonds.

Lipmann has suggested that "high group potential" was also obtaining at the level of bonds other than phosphate bonds, a concept which was confirmed by the metabolic implications of the methyl group of choline, for example.

That the potentially labile phosphate bond is protected against non-enzymatic hydrolysis was underlined by Lipmann[59].

Returning to the coupled reactions of glycolysis (Chapter 24) we may, using the squiggle (∼) introduced by Lipmann, represent the first one as follows:

```
CHO                                     C(=O)O~PO3H2   + NADH
|                                       |
HCOH  +  Pi  +  NAD+      <==>          HCOH
|                                       |
CH2OPO3H2                               CH2OPO3H2
glyceraldehyde-3-P                      1,3-diphosphoglyceric acid

C(=O)O~PO3H2                            C(=O)OH
|                                       |
HCOH          +  ADP      <==>          HCOH      +  ATP
|                                       |
CH2OPO3H2                               CH2OPO3H2
1,3-diphosphoglyceric acid              3-phosphoglyceric acid
```

The energy of the oxidation of the aldehyde group of glyceraldehyde-3-P by NAD^+ is stored in the pyrophosphate bond of ATP, through a labile intermediate containing an energy-rich acylphosphate bond, 1,3-diphosphoglyceric acid.

The second coupled reaction of glycolysis consists (Chapter 24) in the dehydration of 2-phosphoglyceric acid, accompanied by the formation of the enol-phosphate bond of phosphoenol-pyruvic acid, from which the phosphate is transferred to ATP.

$$\underset{\text{2-phosphoglyceric acid}}{\begin{matrix} C\overset{OH}{\lessgtr}O \\ | \\ HCOPO_3H_2 \\ | \\ CH_2OH \end{matrix}} \underset{+H_2O}{\overset{-H_2O}{\rightleftharpoons}} \underset{\text{phosphoenolpyruvic acid}}{\begin{matrix} C\overset{OH}{\lessgtr}O \\ | \\ CO\sim PO_3H_2 \\ \| \\ CH_2 \end{matrix}}$$

$$\underset{\text{phosphoenolpyruvic acid}}{\begin{matrix} C\overset{OH}{\lessgtr}O \\ | \\ CO\sim PO_3H_2 \\ \| \\ CH_2 \end{matrix}} + ADP \underset{+H_2O}{\rightleftharpoons} \underset{\text{pyruvic acid}}{\begin{matrix} \overset{OH}{C=O} \\ | \\ C=O \\ | \\ CH_3 \end{matrix}} + ATP$$

To quote Lipmann[29], commenting on this second coupled reaction

"Although the total of energy over the whole compound is equal on both sides of the equilibrium, the intramolecular energy distribution is changed by dehydration in such a way that a much larger part is concentrated now in the organo-phosphoric linkage."

"... The dehydration of free glycerate to pyruvate must be a largely exergonic★ reaction (occurring with loss of free energy) in contrast to the dehydration of ph~glycerate to ph~pyruvate, *where through attachment of the phosphate residue this energy, dissipated otherwise, is retained in the enol ph~bond.*"

From thermal data, Lipmann calculated a value of -15900 cal for the ΔG^0 of the enolphosphate bond. Phosphoglycrate and phosphopyruvate were thus identified as the reagents at the metabolic origin of the pyrophosphate bond of ATP which was soon to be recognized as the store from which free energy can be doled out to energy-requiring reactions.

As it was noted by Huennekens and Whiteley[61] the phosphate group transferred from phosphoenolpyruvate to ADP was introduced earlier *via* an ATP-depending reaction

$$\text{glucose} + \text{ATP} \rightarrow \text{G-6-P} + \text{ADP}$$

★ ... Coryell[60] introduced recently the terms "exergonic" and "endergonic" specifically for reactions occurring with negative and positive change in free energy, to contrast with exo- and endothermic designating now exclusively heat change."

In this case the "energy-rich bond" has been conserved rather than synthesized as is the case for the phosphate group transferred from 1,3-diphosphoglycerate to ATP (net synthesis of ATP from P_i, as the acylphosphate arose from the phosphate-dependent dehydrogenation of glyceraldehyde-3-P).

In a rview article on energy-rich phosphate bonds, and in other writings, Kalckar[62–64] drew attention to the resonance stabilization between the energy-rich compound and its products. The same concept has inspired Oesper[65,66] who, besides the resonance stabilization of the hydrolysis products, has also stressed that the total free energy of hydrolysis of a compound was apportioned between the inherent energy of hydrolysis and the energy of ionization if the reactants and products had very different p*K* values. T. L. Hill and Morales[67] state that the ΔG of energy-rich bond hydrolysis is not a reflection of a localized bond energy but rather, the difference in free energies between products and reactants and that contributions to the net difference in free energy may originate in resonance stabilization, ionization effects or electrostatic repulsions (on the chemical basis for energy rich phosphate bonds, see Huennekens and Whiteley[61] and George and Rutman[61a]).

After ATP was discovered by Lohmann in 1929 and was shown to contain an energy-rich phosphate bond by Meyerhof and Lohmann in 1934, it was demonstrated in several papers of the Meyerhof school that ATP was produced by glycolysis which, after this work, appeared as a device for the conversion of the low-energy bonds of hexosephosphates into energy-rich bonds.

Meyerhof and Lohmann, as we have stated before had given for the ΔH of hydrolysis of ATP a value of 12500 cal/mole. Meyerhof[68] in 1944 estimated the ΔG^0 of the pyrophosphate linkage of ATP from the double coupled reactions leading from glyceraldehyde-3-P to 3 phosphoglyceric acid, at the time still unsolved with respect to its stage through 1,3-diphosphoglyceric acid, and he obtained a value of -12000 cal. When instead of muscle juice as ATPase, purified myosin ATPase was used by Podolsky and Morales[69], who also took into consideration the heat resulting of the neutralization, by the buffer, of the H^+ produced, a value of $\Delta G^0 = -4700$ was obtained.

Levintow and Meister[70] calculated a value of $\Delta G^0 = -8000$ for the transphosphorylation of ATP in the reaction

$$\text{glutamate} + NH_4^+ + \text{ATP} \rightleftharpoons \text{glutamine} + \text{ADP} + P_i + H^+$$

Robbins and Boyer[71] used ^{14}C-labeled substrates in order to determine, in both directions, the equilibrium of the reaction

$$\text{ATP} + \text{glucose} \rightleftharpoons \text{G-6-P} + \text{ADP}$$

and found the value of ΔG^0 to be -4.7 kcal/mole for the reaction at 30°, pH 7.0.

From data of Meyerhof and H. Green[72] the ΔG^0 of the hydrolysis of G-6-P was found to be -3.1 kcal/mole. Summing the two reactions the authors find for the ΔG^0 (in the presence of high Mg^{2+} concentrations) a value of -7.8 kcal/mole (a lower value -7.6 is obtained in the absence of Mg^{2+}). For the concentration in animal tissues, a value of 13–16 kcal was calculated for the hydrolysis of ATP to ADP. In these calculations, Burton and Krebs assumed a concentration of 10 m*M* for all reactants

Minakami and Yoshikawa[74], taking advantage of the characteristics of red cells, neither tricarboxylic cycle, nor respiration being present, determined the concentrations of the reactants in the cells and calculated the values of changes of the free energies at the individual steps in erythrocytes glycolysis.

Their conclusions read as follows:

"The free energy change of ATP hydrolysis in red cells is calculated to be -13.3 kcal from the standard free energy change of -8.9 kcal (at pH 7.5, Burton[75]) and the concentrations of ATP, ADP and P_i. The free-energy change for the erythrocyte glycolysis may be calculated directly from the standard free energy (-47.4 kcal, Burton[75]) and the concentrations of glucose and lactate as -51 kcal. It can also be calculated to be -49 kcal from the sum of free energy changes and ΔF for ATP hydrolysis. The efficiency of ATP formation by the steady-state glycolysis of erythrocyte is thus calculated to be 26.6/50 = 53%".

To quote D. M. Needham[6]:

"In order to assess the free energy released under conditions *in vivo* the ΔG^0 values given above for standard conditions must be corrected by a term involving the actual concentration of the reactants and product, since

$$\Delta G = \Delta G^0 - RT \ln (\text{ADP})(\text{H}_3\text{PO}_4)/(\text{ATP})$$

It is difficult to guess how much effect this will have, since the concentrations of these substances will vary with the metabolic state; also they may be compartmentalised or may be subject to extensive binding. Burton[76], discussing the question gives a value of -11000 to -12000 cal based on the concentrations in the whole tissue but emphasises that this is only a very rough guide to the proportion of metabolic energy in actual fact transferred to ATP or made available on its hydrolysis in the living cell."

From the whole of the data it may be concluded that the values of approximately 8.0 in standard conditions and approximately 11–12.0 *in vivo* can be accepted today as the values of the ΔG of the pyrophosphate bond of ATP.

As stated above, the ΔG^0 of the enolphosphate bond was calculated from thermal data by Lipmann[29] who found it equal to approximately 16000 cal.

The equilibrium constant for the transphosphorylation reaction

$$\text{phosphoenolpyruvate} + \text{ADP} + H_2O \rightleftharpoons \text{pyruvate} + \text{ATP}$$

was determined by Meyerhof and Oesper[77] to be 1900, from which a ΔG^0 of -4300 can be calculated. If the ATP value of -12000, a value of 16300 cal. for the enolphosphate bond is obtained, a value which agrees with the one obtained through a different method by Lipmann, and which is of a much higher magnitude than the pyrophosphate bond of ATP.

Bücher[78] found K for the reaction

$$\text{1,3-diphosphoglyceric acid} + \text{ADP} \rightleftharpoons \text{3-phosphoglyceric acid} + \text{ATP}$$

to be 3×10^3, from which the free energy of the reaction can be calculated.

$$\Delta G^0 = -RT \log K$$

$$K = 3000$$

$$\Delta G^0 = -4300 \text{ cal.}$$

Therefore the ΔG^0 of the acylphosphate bond $= -16300$ cal $[(-12000)+(-4300)]$.

The ΔG^0 of the energy-rich phosphate bonds of phosphoenolpyruvate and of 1–3 diphosphoglycerate appear as much higher than the ΔG^0 of ATP, a feature which accounts for the easy transfer of these energy-rich bonds to the adenylic system.

It was once concluded that an almost stoichiometric conversion into phosphate bond energy occurred, while Lipmann[78a] arrived at a 70% efficiency, his calculation being based on a value of 10 kcal as the average bond energy level. As we have stated, the accurate measurements of Minakami and Yoshikawa[74] lead to the conclusion that the efficiency of ATP formation by the steady-state glycolysis of mammalian erythrocytes corresponds to 53%.

In the anaerobic breakdown of glucose to ethyl alcohol or lactic acid, four ATP are formed but two ATP are required for the phosphorylation of glucose and for the phosphorylation of F-6-P into F-1,6-PP. A net gain of two ~ ph thus takes place. Burk[79] did, in 1929 calculate that the transformation of glucose into lactic acid was accompanied by a free energy change of −58000 cal/mole of glucose.

The acquisition of two pyrophosphate bonds of ATP represents the net acquisition in the course of the process. As the ATP obtained from glyceraldehyde-3-P corresponds to a conservation of the two ATP introduced we may conclude that from one mole of glucose undergoing the process of glycolysis the generation of phosphate bond energy takes place at the level of the coupled reaction starting with glyceraldehyde-3-P. We have, in Chapter 24, recalled the history of the unravelling of these reactions by Warburg and his colleagues. The first (and so far the only one to be elucidated in its mechanism) of the sources, in metabolism, of the free energy in the utilizable form of the pyrophosphate bond of ATP was elucidated in this work.

With respect to the mechanism of the reaction, Warburg[80,81] first proposed that the aldehyde group of phosphoglyceraldehyde was non-enzymatically phosphorylated and then enzymatically oxidized to produce the acylphosphate bond.

This proposal did not win recognition and a large body of evidence has been obtained in favour of oxidation preceding phosphorylation. (For literature the reader is referred to Chapter 1 by Colowick, Van Eys, and Park in Volume 14 of this Treatise, where the history of mechanism unravelling is retraced.)

In short, the enzyme glyceraldehyde-3-phosphate dehydrogenase is composed of four identical subunits each of which contain a bound NAD^+ and a catalytic site of which an SH group is essential. In a first step NAD^+ binds to a subunit, masking the SH group which in the next step binds to the aldehyde group of glyceraldehyde-3-phosphate in a thiohemiacetal linkage

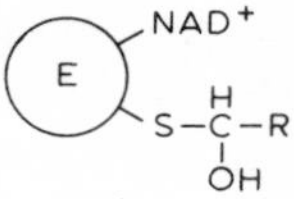

E (with bound NAD^+ and S–CH(OH)–R)

Hydrogen is then transferred from the bound glyderaldehyde-3-P to the bound NAD^+: the result is the formation of a thioester between the enzyme SH and the carboxyl of the substrate. In this stage NADH (the reduced NAD) does not leave the enzyme. It transfers a hydrogen atom to NAD^+

References p. 182

(E)–NADH, –S–C(=O)–R

in the medium, leaving the bound NAD^+ in the oxidized form of the acyl enzyme, the acyl group of which, in the presence of inorganic phosphate and of NAD, is transferred from the SH group to the inorganic phosphate,

$$\text{(E)}\begin{matrix}\text{NAD}^+\\ \text{S–C(=O)–R}\end{matrix} + P_i \longrightarrow \text{(E)}\begin{matrix}\text{NAD}\\ |\\ \text{S}\end{matrix} + R–C–O–PO_3^{2-}$$

acyl enzyme — 1,3-diphosphoglyceric acid

to form the oxidation product, 1,3-diphosphoglyceric acid. This new cleavage of the thioester is suggested to take place *via* an acetyl-imidazole intermediate. It can be seen from this short statement of the mechanism that the SH group is essential because the transferase reactions proceed *via* a thioester acetyl-enzyme intermediate.

The progress of the unravelling of the mechanism is historically based on the classical paper in which Warburg and Christian[82], in 1936, showed that the reduction of NAD, as well as that of NADP, is accompanied by the appearance of a new absorption band at 340 mμ, an optical test which became of great value and allowed to observe the kinetics of oxidation and reduction of NAD. They also showed that the reduction consisted in the conversion of a quaternary hydrogen of a pyridinium structure to a tertiary nitrogen of a dihydropyridine. The acid lability of the reduced coenzyme was also described in this paper.

It was in the same paper that by the application of the "Gärtest" the role of NAD in alcoholic fermentation was explained by two oxidation-reduction reactions.

It was demonstrated in Meyerhof's laboratory that the first of these oxido-reductions involved glyceraldehyde-3-phosphate and was, in an unexplained way, linked to a synthesis of ATP.

This work of Meyerhof was the prelude to the establishment of the mechanism by which the free energy of anaerobic oxidation was stored in ATP, when Negelein and Brömel[83] and Warburg and Christian[84] recognized the consecutive reactions leading from glyceraldehyde-3-phosphate to ATP. We shall have occasions, in subsequent chapters to relate the discovery of many dehydrogenations involving pyridine nucleotides.

The whole process in the increasing of knowledge was a consequence of Warburg's endeavour to reproduce intermediate reactions by using purified enzymes.

Lipmann, who worked in Meyerhof's laboratory between 1927 and 1932, when Meyerhof was still in Dahlem (until 1930) had been in contact with Warburg who dominated the Institute from the 4th floor writes[9]:

"Warburg's puritanical, pragmatic approach which rejected compromises and speculation filtered through to us from the fourth floor. He became our hero; his stern insistence on letting experiments speak and keeping interpretation to a minimum has dominated our generation".

For many chemists, biochemistry is a part of unified chemistry, an accurate study of molecules or systems which can be derived from organisms. It may be maintained that if we had waited for the availability of isotopes, the complicated pathway of glycolysis would have been unravelled in a short time but with the proviso that we should have known what queries to answer.

Besides the chemical and physical studies and concepts on which it leans, biochemistry has to solve, as amply emphasized in Part I and II of this History, its own problem, the super-chemistry of the cells and of the organisms. This is the domain of its exploration which has been brilliantly exemplified in the career of Warburg. To quote him[85]:

"A scientist must have the courage to tackle the great unsolved problems of his time, and solutions have to be forced by carrying out numerous experiments without much critical hesitation".

This bold kind of endeavour remains the lot of leaders in biochemistry, even today, and the lack of such spirit cannot be compensated by any amount of chemical and physical knowledge. On the other hand it remains a compulsory duty to insure the greatest amount of such knowledge to prospective biochemists while preparing them for a kind of exploratory experimental work of which chemists and physicists for a long period of time did not have the slightest idea. The history of an enterprise such as the unravelling of the pathway of glycolysis, will remain an exemplary narration whenever it is said that in order to solve the problems of biochemistry, it is enough to know chemistry.

References p. 182

REFERENCES

1 R. Mayer, *Die organische Bewegung in ihrem Zusammenhänge mit dem Stoffwechsel*, Heilbronn, 1845.
2 H. von Helmholtz, *Ueber die Erhaltung der Kraft*, Berlin, 1847.
3 W. Preyer, *Robert von Mayer und die Erhaltung der Energie*, Berlin, 1889.
4 G. Rosen, in C. McC.Brooks and P. F. Cranefield (Eds.), *The Historical Development of Physiological Thought*, New York, 1959.
5 D. L. Hermann, *Physiologie des Menschen* (5th ed.), Berlin, 1874.
6 D. M. Needham, *Machina Carnis. The Biochemistry of Muscular Contraction in its Historical Development*, Cambridge, 1971.
7 O. Meyerhof and K. Lohmann, *Biochem. Z.*, 168 (1926) 128.
8 O. Meyerhof, *Die chemischen Vorgänge im Muskel und ihre Zusammenhang mit Arbeitsleistung und Wärmebildung*, Berlin, 1930.
9 F. Lipmann, *Wanderings of a Biochemist*, New York, 1971.
10 O. Meyerhof, *Biochem. Z.*, 33 (1911) 219.
11 O. Meyerhof, *Biochem. Z.*, 35 (1911) 246.
12 O. Meyerhof, *Sitzungsber. Heidelberger Akad. d. Wiss., Mathem. Naturw. Klasse, Abt. B*, 1 (1912) 1.
13 O. Meyerhof, *Arch. f.d. ges. Physiol.*, 146 (1912) 159.
14 O. Warburg and O. Meyerhof, *Arch. f.d. ges. Physiol.*, 148 (1912) 295.
15 H. Weber, *Naturwissenschaften*, 39 (1952) 217.
16 H. A. Krebs, *Biogr. Mem. of Fellows of the Royal Soc.*, 18 (1972) 629.
17 O. Meyerhof, *Abhandl. der Fries'schen Schule*, N.F. 4 (1913) 3.
18 H. W. Nernst, *Theoretische Chemie von Standpunkte der Avogadroschen Regel und der Thermodynamik*, Stuttgart, 1893.
19 O. Warburg, *Erg. Physiol.*, 14 (1914) 258.
20 J. Báron and M. Polanyi, *Biochem. Z.*, 53 (1913) 1.
21 A. V. Hill, *J. Physiol.*, 44 (1912) 466.
22 G. N. Lewis and M. Randall, *Thermodynamics and the Free Energy of Chemical Substances*, New York, 1923.
23 O. Meyerhof and H. Green, *J. Biol. Chem.*, 178 (1949) 655.
24 A. Kornberg, *J. Biol. Chem.*, 182 (1950) 779.
25 O. Meyerhof and K. Lohmann, *Biochem. Z.*, 273 (1934) 73.
26 O. Meyerhof and W. Schulz, *Biochem. Z.*, 281 (1935) 292.
27 A. B. Pardee and L. L. Ingraham, in D. M. Greenberg (Ed.), *Metabolic Pathways* (2nd ed. of *Chemical Pathways of Metabolism*) Vol. 1, New York, 1960.
28 N. O. Kaplan, in J. B. Sumner and K. Myrbäck, (Eds.), *The Enzymes, Chemistry and Mechanism of Action*, Vol. II, Part I, New York, 1951.
29 F. Lipmann, *Advan. Enzymol.*, 1 (1941) 99.
30 L. Michaelis, *Oxidation-Reduction Potentials* (translation by L. B. Flexner), Philadelphia, 1930.
31 W. M. Clark, *Public Health Reports (U.S.)*, 38 (1923) 443.
32 W. Ostwald, *Z. physik. Chem.*, 34 (1900) 248.
33 J. S. Fruton, *Molecules and Life. Historical Essays on the Interplay of Chemistry and Biology*, New York, 1972.
34 O. Meyerhof and K. Lohmann, *Naturwissenschaften*, 19 (1931) 575.
35 A. Hahn, *Z. Biol.*, 91 (1931) 444.
36 O. Meyerhof, *Erg. Physiol.*, 22 (1923)
37 O. Meyerhof, *Naturwissenschaften*, 19 (1931) 923.
38 A. von Muralt, *Erg. Physiol.*, 37 (1935) 406.

39 O. Meyerhof and K. Lohmann, *Biochem. Z.*, 253 (1932) 431.
40 J. K. Parnas, P. Ostern and T. Mann, *Biochem. Z.*, 272 (1934) 64.
41 E. Lundsgaard, *Biochem. Z.*, 217 (1930) 162.
42 E. Lundsgaard, *Biochem. Z.*, 227 (1930) 51.
43 G. Embden, E. Hirsch-Kauffmann, E. Lehnartz and H. J. Deuticke, *Z. physiol. Chem.*, 151 (1926) 207.
44 A. Schwartz and A. Oschmann, *Compt. Rend.*, 91 (1924) 275.
45 A. Schwartz and A. Oschmann, *Compt. Rend.*, 92 (1925) 169.
46 K. Lohmann, *Biochem. Z.*, 271 (1934) 264.
47 O. Meyerhof, R. McCullagh and W. Schulz, *Arch. f. d. ges. Physiol.*, 224 (1930) 230.
48 A. V. Hill, *Physiol. Rev.*, 2 (1922) 339.
49 O. Meyerhof, E. Lundsgaard and H. Blashko, *Naturwissenschaften*, 18 (1930) 787.
50 H. M. Kalckar, *Biological Phosphorylations. Development of Concepts*, Englewood Cliffs, 1969.
51 E. Lundsgaard, *Festschrift zu Henriques*, cited after Kruhoffer and Crone[52].
52 P. Kruhoffer and C. Crone, *Erg. Physiol.*, 65 (1972) 1.
53 D. Nachmansohn, *Biochem. Z.*, 196 (1928) 73.
54 E. Lundsgaard, *Biochem. Z.*, 233 (1931) 322.
55 J. Sacks, *Am. J. Physiol.*, 129 (1940) 227.
56 J. Sacks, *Am. J. Physiol.*, 140 (1943–1944) 316.
57 T. Korzybski and J. K. Parnas, *Bull. Soc. Chim. Biol.*, 21 (1939) 713.
58 H. D. Kay, *Biochem. J.*, 22 (1928) 855.
59 F. Lipmann, in W. D. McElroy and B. Glass, (Eds.), *Phosphorus Metabolism*, Baltimore, 1951.
60 C. D. Coryell, *Science*, 92 (1940) 380.
61 F. M. Huennekens and H. R. Whiteley, in M. Florkin and H. S. Mason (Eds.), *Comparative Biochemistry*, Vol. I, New York, 1960.
61a P. George and R. J. Rutman, *Progr. Biophys. Biophys. Chem.*, 10 (1960) 1.
62 H. M. Kalckar, *Chem. Rev.*, 28 (1941) 71.
63 H. M. Kalckar, *Nature*, 160 (1947) 143.
64 H. M. Kalckar, in D. E. Green (Ed.), *Currents in Biochemical Research*, New York, 1946.
65 P. Oesper, in W. D. McElroy and B. Glass (Eds.), *Phosphorus Metabolism*, Vol. I, Baltimore, 1951.
66 P. Oesper, *Arch. Biochem.*, 27 (1950) 255.
67 T. L. Hill and M. F. Morales, *J. Am. Chem Soc.*, 73 (1951) 1656.
68 O. Meyerhof, *Ann. N.Y. Acad. Sci.*, 45 (1944) 377.
69 R. J. Podolsky and M. F. Morales, *J. Biol. Chem.*, 218 (1956) 945.
70 L. Levintow and A. Meister, *J. Biol. Chem.*, 209 (1954) 265.
71 E. A. Robbins and P. D. Boyer, *J. Biol. Chem.*, 224 (1957) 121.
72 O. Meyerhof and H. Green, *J. Biol. Chem.*, 178 (1949) 655.
73 K. Burton and H. Krebs, *Biochem. J.*, 54 (1953) 94.
74 S. Minakami and H. Yoshikawa, *Biochem. Biophys. Res. Commun.*, 18 (1965) 345.
75 K. Burton, *Erg. Physiol.*, 49 (1957) 275.
76 K. Burton, *Nature*, 181 (1958) 1594.
77 O. Meyerhof and P. Oesper, *J. Biol. Chem.*, 179 (1949) 1371.
78 T. Bücher, *Biochim. Biophys. Acta*, 1 (1947) 292.
78a F. Lipmann, *Ann. Rev. Biochem.*, 12 (1943) 13.
79 D. Burk, *Proc. Roy. Soc. London, Ser. B*, 104 (1929) 153.
80 O. Warburg, *Naturforsch.*, 126 (1957) 47.
81 O. Warburg, *Wasserstoffübertragende Fermente*, Freiburg i. Br., 1949.
82 O. Warburg and W. Christian, *Biochem. Z.*, 287 (1936) 291.

83 P. Negelein and H. Brömel, *Biochem. Z.*, 303 (1939) 132.
84 O. Warburg and W. Christian, *Biochem. Z.*, 303 (1939) 40.
85 O. Warburg, *Ann. Rev. Biochem.*, 33 (1964) 1.
86 P. Eggleton and G. P. Eggleton, *Biochem. J.*, 21 (1927) 190.
87 C. H. Fiske and Y. Subbarow, *J. Biol. Chem.*, 81 (1929) 629.

Section II

Aerobic Phosphorylation

Chapter 26

Early Theories of the "Biological Oxidations" of Intracellular Respiration

1. Introduction

According to our present views, the "biological oxidations" of intracellular respiration consist of a process of transfer of four electrons to oxygen, with the formation of water. This process takes place at the level of the base pieces of the inner membranes of mitochondria, the electrons involved being liberated largely in the mitochondrial matrix in the operation of the Krebs cycle. The operation of the "respiratory chain" in the base pieces leads to a generation of ATP (or of other unknown high-energy intermediates involved in active translocation).

The history of the emergence of this theory is a long process starting with Lavoisier's formulation of the notion of slow combustion, according to which oxygen was bound to the carbon of the foodstuffs and produced CO_2 and heat. We have seen in Chapter 8 how the concept of intracellular respiration became current. The oxygen of the air does not itself produce any combustion of the foodstuffs; hence activation of oxygen was considered a preliminary requisite for intracellular respiration.

2. Theories of oxygen activation

(a) The ozone theory

It had been known since the beginning of the 19th century that an old drug, guaiacum, turns blue in the presence of oxygen and light[1,2].

It was observed by Planche[3] that guaiacum is turned blue by fresh horse radish, as well as by other parts of plants and by milk, an effect which does not obtain if the material of biological origin is previously boiled.

References p. 202

In 1840, Schönbein discovered ozone[4] and five years later[5], he observed that guaiacum turns blue under the effect of ozone. Some mushrooms turn blue when bruised. From one of them, *Boletus luridus*, Schönbein[6,7] extracted by alcohol a substance which does not oxidize in the presence of air unless it is placed in contact with the mushroom tissues. He concluded from a number of experiments that living tissues contain a substance which is altered by heat and which could ozonize oxygen and form with ozone a compound able to give up active oxygen to chromogens (oxygen-exciter and oxygen-carrier).

As a consequence of the many years of study he had devoted to it, oxygen was highly valorized by Schönbein. To him, oxygen was not an element like any other, but, to use his own expression, it was the king of elements, the Jupiter of the scientific Olympus. He spoke of oxygen as his hero, which he regarded as omnipotent ("Allgewalt"). Any evidence presented to deprive oxygen of

"sein früher genossenes königliches Ansehen unter den Elementen"

grieved him, and he hoped that it would be possible

"den Sauerstoff in seine Würden und Rechte einzusetzen, ihn wieder in seiner Macht zu betrachten, die ihres gleichen auf dem ganzen Gebiet der Chemie nicht kennt."[8].

For Schönbein, biological oxidative processes are due to "active oxygen", by which he means ozone, either in the free or in the combined form. If there were no substance capable of changing oxygen into ozone (which is actually a toxic substance), the animal would, according to Schönbein, die of asphyxiation in the midst of the plentiful oxygen of the atmosphere[8]. He was the first upholder of the theory of the activation of oxygen ("Sauerstoffaktivierung der biologischen Verbrennung").

(b) Peroxide theories

(i) The theory of Traube

Hoppe-Seyler[9] had proposed a theory according to which in intracellular respiration, oxidation was accomplished by atomic oxygen derived from the oxygen molecules by reduction:

$$2\,H + O_2 = H_2O + O$$

A discussion with Traube followed the presentation of Hoppe-Seyler's theory, and Traube put forward a new theory[10] according to which the water molecule is decomposed, giving rise to atomic hydrogen, which reduces the resulting oxygen molecule to H_2O_2, the form of activated oxygen according to his theory.

(ii) The peroxide theory of Engler[11] *and of Bach*[12]

Bach postulates that one of the valences of O=O is easier to split than two valences, and that the substrates take the –O–O– group as such, and form peroxides, the type of which is H_2O_2.

$$H{-}O{-}O{-}H$$

Engler[11] had figured out the process as follows. Take for instance an easily oxidizable substance which he calls "auto-oxydator" (for instance turpentine oil) in the presence of a not easily oxidizable one such as indigo blue, which he calls "acceptor". In the presence of molecular oxygen, A takes up a molecule of oxygen with the formation of a peroxide:

$$A + O_2 \rightarrow AO_2$$

The peroxide reacts with the "acceptor" and half of the oxygen taken by A is transferred on to B

$$AO_2 + B \rightarrow AO + BO$$

In this way during the slow combustion of nascent hydrogen, H_2O_2 is formed and it reacts with the hydrogen not yet oxidized

$$H_2 + O_2 \rightarrow H_2O_2$$

$$H_2 + H_2O_2 \rightarrow 2\ H_2O$$

Most peroxides are hydrolyzed by water and converted into oxides and hydroperoxides. The reaction takes place in two steps.

1st step: formation of a peroxyhydrate

A(–O–O–) + H–OH → A(OH)(O–OH)

peroxyhydrate

2nd step: splitting into oxide and hydroperoxide

A(OH)(O–OH) → AO + HOOH

oxide hydroperoxide

References p. 202

In short, the peroxide theory of Bach and of Engler postulates a series of steps (see Bach[13]): (*1*) Incomplete splitting of O_2 by the free energy of the substrate. Formation of –O–O–; (*2*) The substrate takes up –O–O– with a formation of peroxides; (*3*) The peroxides contain half of the oxygen in an active form which is utilized in the oxidation processes.

3. Biological oxidations according to the theories of Pflüger, of Ehrlich and of Bernard

Pflüger, who published a number of papers on the subject between 1875 and 1893 epitomizes as follows the theory he maintains through this series of publications:

"... the life process is the intramolecular heat of highly unstable protein molecules which, on dissociation, are decomposed into carbonic acid, water and amide-like substances and which undergo continual regeneration and growth through polymerization.

"... According to my theory of vital processes, carbonic acid arises continuously within very large molecules. The ensuing explosion which accompanies its formation brings all the atoms in the particular molecule into a powerful vibration ... The moment the explosion has torn a gap in the molecule, free chemical activities are available within it ... Hence the assimilation of oxygen and combustible material takes place to saturate these affinities ... As all the power of life originates in the attraction of oxygen to carbon and hydrogen, I attribute all strong intramolecular movement in the living protein molecule to this source and to be especially dependent on the explosions. ... As long as the explosions tear gaps, atoms exist in *statu nascendi*, assimilation of oxygen and nutritive molecules takes place and the living substance vibrates and produces heat." (ref. 14, translation by Keilin[15]).

It is not necessary to comment on such unsubstantiated speculations, a judgement which also applies to the theory of Ehrlich[16], who injected into rabbits several dyes simultaneously and observed that they were taken up by different tissues, in which they appeared in the oxidized or in the reduced state. This illustrates the impossibility of providing an interpretation of certain correct observations when the necessary concepts are yet to be developed, and at the time of Ehrlich's experiments no knowledge was available concerning the oxidation–reduction potential of dyes nor of the intracellular systems. This did not prevent Ehrlich[16] from formulating a theory of biological oxidations according to which the cells are alternatively bathed by acid and by alkaline fluids and that the oxygen uptake can be attributed to the alkaline phase and its consumption to the less alkaline or acid phase.

While Claude Bernard, who had adhered to Pflüger's concept of the intracellular respiration, never conceived a definite role for oxygen, to which he vaguely attributed an excitatory role in the tissues, he did not consider

CO_2 to be a product of combustion. Though not an orthodox member of the positivist school, Bernard had definite links with the adepts of Auguste Comte and he was probably influenced by their tendency to disagree with Lavoisier's views. Without attributing to him any premonition of the decarboxylating steps of the Krebs cycle, it is remarkable that he wrote:

"The formation of carbon dioxide ... is the result of real organic destruction, a decomposition analogous to those which are produced in fermentation." (ref. 17)

and stated, in a prophetic way, that what Lavoisier really established was not that respiration is a combustion, but that it is equivalent to a combustion.

4. Enzymatic theories of oxygen activation

The rise of the enzymatic theory of metabolism, retraced in Chapter 13, stimulated the search for oxygen-activating enzymes.

When it was recognized that oxygen reaching the tissues of an organism was the origin of phenomena leading to a production of energy, it became clear that the components of tissues, when put in contact with molecular oxygen, did not undergo the transformation observed in the process of biological oxidations. In order to understand these transformations, Traube[18] postulated the existence of a hypothetical oxygen transmittor, carrying oxygen to acceptors and he admitted the participation of an enzyme activating the oxygen.

In the experiments on the oxidation of benzilic acid and salicylic aldehyde in the presence of blood or of animal tissues, accomplished by Schmiedeberg[19,20], Jaquet[21], Salkowsky[22], Abelous and Biarnès[23], the amount of oxidized substances was so small that it did not carry any conviction in favor of the implication of enzymatic action. The first demonstration of the existence of a new category of enzymes, catalyzing a biological action of oxygen, was afforded by Bertrand[24] who studied the mechanism of the darkening and hardening of the latex of lacquer trees. At the time the phenomenon was already recognized as being due to the oxidation of urishiol, a phenol present in the latex of Japanese lacquer trees (Yoshida[25]) or of laccol, another phenol present in the latex of Indo-Chinese lacquer trees.

Bertrand studied what he called the "oxidase" of laccase. He found an

References p. 202

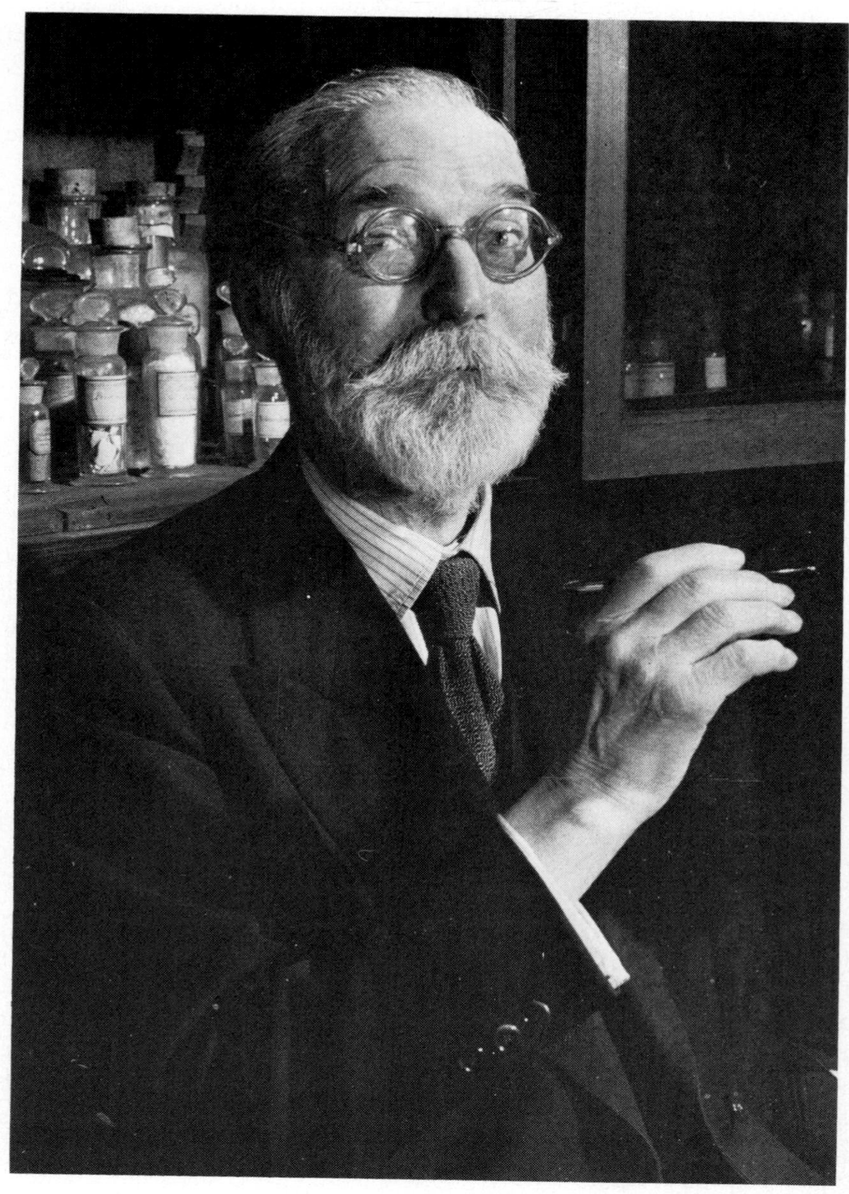

Plate 85. Gabriel Bertrand.

appreciable concentration of manganese in the ash of his preparation and observed that the activity was enhanced by the addition of manganese salts. Though the metal active in laccase is not manganese but copper (Keilin and Mann[26]), Bertrand was the first to introduce the concept according to which a metal may constitute an essential part of an enzyme. Bertrand[27,28] considered oxidases as manganese derivatives, easily split into an organic substance acting as an acid and which can be decomposed, liberating MnO

$$R''Mn + H_2O = R''H_2 + MnO$$

The inert oxygen molecule is split by the action of MnO with production of MnO_2 and atomic oxygen

$$MnO + O_2 = MnO_2 + O^{2-}$$

The MnO_2 is further attacked by the organic part of the enzyme, regenerating the oxidase which liberates the second oxygen atom

$$RH_2 + MnO_2 = RMn + H_2O + O^{2-}$$

The active part of the oxidase is considered by Bertrand as being the Mn, and the organic part is the coenzyme. Though he introduced the term, Bertrand's views were the opposite of our own conceptions.

Accepting the views of Hoppe-Seyler, Bertrand considered that the oxidizable substance combines with the inert molecular oxygen, using one atom of oxygen for its oxidation and liberating the second atom of oxygen in an activated state, able to perform the oxidation of organic substrates.

Bertrand[29] has also described tyrosinase, catalysing the conversion of tyrosine to melanin.

Bertrand introduced the term oxidase[30] to designate tyrosinase and laccase.

In our present terminology, the denomination is reserved for the cases in which oxygen is the acceptor, which is the case with tyrosinase and laccase.

Although Bertrand did not deal with the system involved in intracellular respiration, he was important in introducing the concepts of the participation of metals in the constitution of enzyme molecules, and the concept of the coenzyme.

References p. 202

Plate 86. Aleksei Nikolaevich Bach.

5. The oxygenase-peroxidase theory

Thenard[31,32], the discoverer of H_2O_2, had observed that it is decomposed into H_2O and O_2 by plant and animal material (due to the presence of an enzyme later called catalase by Loew[33]) and brought to oxidize guaiacum in the presence of the same material (due to the enzyme later called peroxidase by Linossier[34]).

In 1903, Chodat and Bach[35] revised the peroxide theory by giving an enzymatic flavor to it. They admitted that the system of oxygen activation in intracellular respiration was separable into two enzymatic components: an oxygenase introducing oxygen into an organic peroxide, and a peroxidase. Bach's oxygenases (later called oxidases by him) are simply easily oxidisable substances taking up molecular oxygen with a formation of peroxide, followed in the presence of a peroxidase by a transfer of half of the oxygen to the substrate. In the views proposed by Bach in many papers (see Bach[13]) molecular oxygen is supposed to be activated as such. Bach admits also that the oxygen of water can be involved in biological oxidations. In this issue he accepts the hypothesis of Traube, according to which labile complexes of water are formed in which oxygen has a valency of 4.

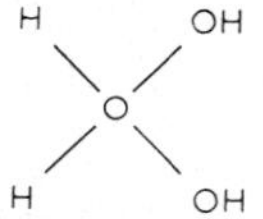

hydroperoxide hydrate

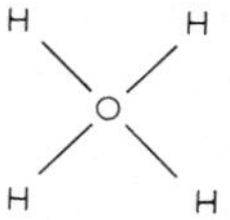

oxyperhydride

These two compounds he considers as being in equilibrium between themselves and also with water. The first gives up active oxygen and the second active hydrogen. The theory proposes a coupled oxido-reduction. In this scheme active oxygen results either from the activation of free molecular oxygen by "oxygenases" (a term which in our present terminology is reserved for the cases in which oxygen is directly incorporated into the substrate) or oxygen is derived from water and the peroxide formed is split by peroxidase, providing active oxygen. As he postulates a coupled oxido-reduction, Bach, to account for the supply of hydrogen, postulates the existence of enzymes, which he calls perhydridases, acting on the oxyperhydride which he considered to be formed besides the peroxide. The type of Bach's perhydridase was the Schardinger enzyme of milk[36]. In the Schardinger reaction, a mixture of methylene blue and of formaldehyde

References p. 202

is bleached by milk, at a temperature of 70° C. We now know that the Schardinger enzyme catalyses the oxidation of the aldehyde by methylene blue, an oxidizing dye. Schardinger considered aldehyde the catalyst enabling the enzyme to reduce methylene blue. Bach's interpretation of the reaction was that the aldehyde is oxidized into acid by the oxygen of the water, the hydrogen liberated acting, in the presence of the enzyme involved, on the methylene blue which was reduced into the leucobase. The "Atmungsfermente" of Bach (oxygenase, peroxidase, perhydridase) he considered, in the context of the oxidation of glucose in intracellular respiration, as providing the H_2O_2 which in the presence of peroxidase acted on intermediaries of glycolysis, as shown by Kostychev (see below) and resulted in the formation of CO_2 and H_2O. With his theory which was irrelevant and was based on experiments with systems which have, as we now know, no relation with intracellular respiration, Bach while rejecting the concept of direct oxidation accepted since Lavoisier, stressed the notion that the liberation of energy in metabolism goes through numerous intermediary steps and that these are catalyzed by different enzymes[13]. The oxygenase-peroxidase theory, an enzymatic variant of the peroxide theory of Engler and of Bach, was accepted with several variants for more than twenty years and it only collapsed when it became known that Bach's "oxygenase" was in fact catechol oxidase which produces a quinone acting as oxidizing agent. Furthermore, catechol oxidase is not found in animal tissues.

Later, peroxidase was shown to be a hematin derivative[37–39].

6. Respiration as a follow-up of glycolysis

It is sometimes said that the information concerning higher plant respiration has in a large part lagged behind the results of studies on animal cells or on micro-organisms.

There is nevertheless a concept, one of the leading ones in the field, which was introduced in biological studies by the Russian school of plant physiologists: it is the recognition of the respiratory pathway as grafted onto one of the intermediary products of glycolysis. We know now that this intermediate is pyruvate but at the time nothing was known of the glycolytic pathway and it was no mean achievement to recognize respiratory metabolism as derived, in the presence of oxygen, from an intermediate product of glycolysis. In 1878, Pfeffer[40], in a general review, suggested that the primary act in respiration is glycolysis (intramolekulare Atmung). In 1899, Devaux[41] found alcohol in the deep tissues of wood stems of plants, which he considered as being due to a relative anaerobiosis in these tissues. Berthelot[42] (1899), after the communication to the French Academy of the results obtained by Devaux recalled that he had also found alcohol in plant leaves without giving an interpretation of the fact but stressing that it did afford a proof against the theory according to which certain micro-organisms act by their vital process to produce alcohol. At the time the concept of the production of alcohol in anaerobic conditions by vegetable cells (including those of yeast and higher plants) had not yet been formulated. During the same year Mazé[43–45] proposed the theory according to which the presence of alcohol in plant tissues in aerobic conditions was not to be considered as the result of local conditions of asphyxia in plant tissues but as the indication that alcohol was an intermediary product in the respiratory pathway, a theory developed by him in several publications.

Godlewski[46] had already pointed to the fact that intermediary products are directed in the respiratory path before reaching the stage of alcohol. The experimental demonstration resulted from observations made by Kostychev[47] and Palladin[48]. They observed that the glycolytic system of plants producing alcohol is at work only in the absence of oxygen and that it is not the case when the plant tissues have been frozen[49]. Alcohol is produced in this case even in the presence of oxygen. This led Kostychev[50–52] to observe that when a fermenting juice (prepared with zymin) was added to wheat seedlings, their respiration was stimulated. Kostychev was able to show that additions of alcohol did not increase the respiration. He tried to isolate the intermediate product of glycolysis which acted as a starting point for the respiratory process and he was led to discard lactic acid, inorganic phosphate and organic phosphate as the relevant intermediaries.

7. Palladin's theory

After the agreement on Kostychev's views on the succession of glycolysis and respiration metabolic pathways was reached, the theories of biological oxidations in respiration amalgamated this concept with a number of forms of the oxygenase-peroxidase theory. One of these schemes was proposed in Palladin's chromogen theory[53,54]. Chromogens, widely found

References p. 202

Plate 87. Vladimir Ivanovich Palladin.

in plants, are easily oxidized molecules which, in the presence of molecular oxygen, are converted into pigments. Palladin compared this aspect to the role of hemoglobin, taking up oxygen and delivering it to the tissues. But it is not Palladin's opinion that the pigments and specially "phytohematins" take up oxygen and transfer it to substrates. According to him they take up hydrogen from the substrates and transfer it to oxygen. Water is formed from the molecular oxygen of the air (aerobic phase) and the CO_2 is produced anaerobically. In this theory the role of the phytohematin is to transfer hydrogen from substrate to molecular oxygen with formation of H_2O_2 acting on intermediary products of glycolysis to liberate CO_2 and H_2O.

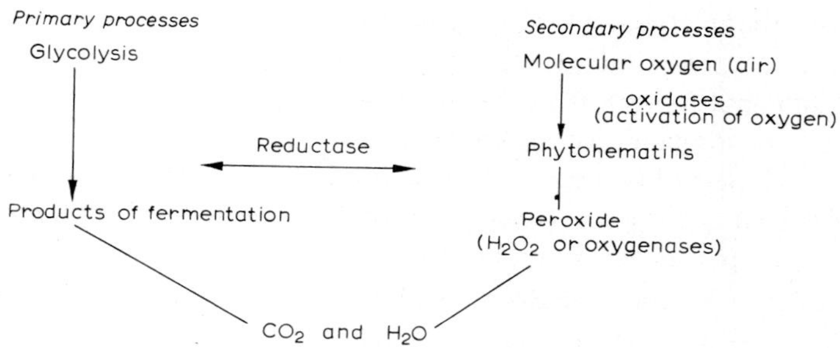

Fig. 8. Scheme of the biological oxidations of respiration according to Palladin.

Though not even mentioned by Wieland in his book on oxidation[55] nor by Keilin in his book on the history of cell respiration[15], Palladin's extensive work has played an important part in further developments and Szent-Györgyi has for instance recognized its seminal influence on his own work on muscle respiration.

8. Distinction of the specific biological oxidations of intracellular respiration from other metabolic actions of oxygen

In Bach's theory, which was accepted for a long period, the enzymes known at the time to be involved in the metabolic actions of oxygen (whatever these may be) are incorporated in the system which still accepts the old theory of H_2O_2 as the producer of the atomic active oxygen by the action of peroxidase.

References p. 202

In 1897, Portier[56], in his thesis on oxidases, already formulated the view that such enzymes as known at the time were involved, not in the respiratory process but in other metabolic processes.

In our present terminology the term oxidation which for Lavoisier meant a chemical oxygenation, means a "removal of electrons, removal of electrons and protons, removal of hydrogen atoms or incorporation of oxygen" (Mason[57]). All enzymes catalyzing oxido-reductions are called oxidoreductases. Oxidase is used in the cases in which O_2 is an acceptor and oxygenase for the cases in which oxygen is directly incorporated into the substrate.

Before 1927, the distinction between oxygenation and oxidation had not been clearly grasped. The distinction was established by Conant[58,59] and by Hill and Holden[60,61]. It was the latter authors, for instance, who established that methemoglobin, the oxidized hemoglobin, does not contain oxygen, in contradistinction to oxyhemoglobin (the oxygenated hemoglobin), but that its iron is in the Fe^{3+} (ferric) state.

Oxygen metabolism, as we know it now, appears under a number of types besides that obtaining in cellular respiration in which four electrons are transferred to molecular oxygen, forming two molecules of water. To identify this particular type in the presence of cytochrome oxidase (Atmungsferment, indophenoloxidase) as the one involved in cellular respiration as well as the chain of reactions leading to it, was like looking for a needle in a haystack. As retraced by Mason[57] in a review to which the reader is referred, the forms of oxygen metabolism are many. There are reactions in which the two atoms of an oxygen molecule are consumed per molecule of substrate, the two atoms appearing in the product, in the presence of a number of enzymes (pyrocatechase, homogentisate oxidase, *etc.*). One molecule of oxygen may be consumed by one molecule of substrate, one of its atoms appearing in the product, the other being reduced, in the presence of a "mixed-function oxidase" (phenolase, steroid hydroxylases, *etc.*). Even when electron transfer oxidases are involved, other forms of metabolism have been described besides the one involved in cell respiration and involving a transfer of four electrons to oxygen in the presence of cytochrome oxidase (Atmungsferment, indophenoloxidases). One of these other forms is the oxygen-obligative two-electron transfer in the presence of uricase, in the aerobic transformation of uric acid to allantoin, which is one of the stages of purine catabolism. There is also an oxygen-facultative two-electron transfer in the presence of glucose oxidase, the β-glucopyranose being oxidized to δ-gluconolactone.

When it was recognized that in cell respiration, oxygen does not combine with carbon, as understood in the classical combustion theory, but reacts with hydrogen to form water, *i.e.* in our present theory, that the four electrons transferred to oxygen come from hydrogen atoms of the substrate, the many dehydrogenations which take place in metabolism were considered as being involved. Another needle was finally discovered in another haystack, by the identification of the Krebs cycle as the specific pathway involved in the supply of hydrogen atoms, carriers of electrons, to the respiratory chain.

The general trend of the unravelling of biological oxidations thus shows an aspect of the identification of relevant mechanisms in a mass of possible candidates experimentally unveiled, the majority of which were irrelevant to the biological reality concerned.

References p. 202

REFERENCES

1 W. H. Wollaston, *J. Nat. Philos.*, 8 (1804) 293.
2 W. T. Brande, *Phil. Trans.*, (1806) 89.
3 L. A. Planche, *J. Pharm. chim. (Paris)*, 6 (1820) 16.
4 C. F. Schönbein, *Ann. Phys.*, 50 (1840) 616.
5 C. F. Schönbein, *Ann. Phys.*, 67 (1845) 99.
6 C. F. Schönbein, *Verhandl. Naturf. Ges.*, 1 (1856) 339.
7 C. F. Schönbein, *J. Prakt. Chem.*, 67 (1856) 497.
8 T. Thunberg, *Kungl. Fysiografiska Sällskapets i Lund Förhandlingar*, 2 (1950) 1.
9 F. Hoppe-Seyler, *Ber. deut. Chem. Ges.*, 12 (1879) 1551.
10 M. Traube, *Ber. deut. Chem. Ges.*, 15 (1882) 659.
11 C. Engler and A. Wild, *Chem. Ber.*, 30 (1897) 1669.
12 A. Bach, *Compt. Rend.*, 124 (1897) 951.
13 A. Bach, Oxidationsprozesse in der lebenden Substanz, in C. Oppenheimer (Ed.), *Handbuch der Biochemie des Menschen und der Tiere*, Ergänzungsband, Jena, 1913, p. 133.
14 E. Pflüger, *Arch. Ges. Physiol.*, 10 (1875) 251.
15 D. Keilin, *The History of Cell Respiration and Cytochrome*, Cambridge, 1966.
16 P. Ehrlich, *Das Sauerstoff-Bedürfnis des Organismus. Eine farbanalytische Studie*, Berlin, 1885.
17 C. Bernard, *Leçons sur les Phénomènes de la Vie Communs aux Animaux et aux Végétaux*, vol. I, Paris, 1878.
18 M. Traube, *Theorie der Fermentwirkungen*, Berlin, 1858.
19 O. Schmiedeberg, *Arch. Exptl. Pathol. Pharmakol.*, 6 (1876) 233.
20 O. Schmiedeberg, *Arch. Exptl. Pathol. Pharmakol.*, 14 (1881) 288, 379.
21 A. Jaquet, *Arch. Exptl. Pathol. Pharmakol.*, 29 (1892) 386.
22 E. Salkowsky, *Z. physiol. Chem.*, 7 (1882) 115.
23 J. E. Abelous et G. Biarnès, *Arch. Physiol.*, 5th Ser., 6 (1894) 591.
24 G. Bertrand, *Compt. Rend.*, 121 (1895) 166.
25 H. Yoshida, *J. Chem. Soc.*, 43 (1883) 472.
26 D. Keilin and T. Mann, *Nature*, 143 (1939) 23.
27 G. Bertrand, *Compt. Rend.*, 124 (1897) 1032.
28 G. Bertrand, *Compt. Rend.*, 124 (1897) 1355.
29 G. Bertrand, *Compt. Rend.*, 123 (1896) 463.
30 G. Bertrand, *Ann. Chim. 7th Ser.*, 12 (1897) 115.
31 L. J. Thenard, *Ann. Chim.*, 9 (1818) 314.
32 L. J. Thenard, *Ann. Chim.*, 11 (1819) 85.
33 O. Loew, *Rept. U.S. Dept. Agric.*, No. 68 (1901).
34 G. Linossier, *C.R. Soc. Biol.*, 50 (1898) 373.
35 R. Chodat and A. Bach, *Ber. deut. Chem. Ges.*, 36 (1903) 606.
36 F. Schardinger, *Z. Unters. Nahr. Genussm.*, 5 (1902) 22.
37 R. Kuhn, D. B. Hand and M. Florkin, *Z. physiol. Chem.*, 201 (1931) 255.
38 K. A. C. Elliott and D. Keilin, *Proc. Roy. Soc. (London), Ser. B.*, 114 (1934) 210.
39 D. Keilin and T. Mann, *Proc. Roy. Soc. (London), Ser. B*, 122 (1937) 119.
40 W. Pfeffer, *Landw. Jahrb.*, 7 (1878) 807.
41 H. Devaux, *Compt. Rend.*, 128 (1899) 1346.
42 M. Berthelot, *Compt. Rend.*, 128 (1899) 1366.
43 P. Mazé, *Compt. Rend.*, 128 (1899) 1608.
44 P. Mazé, *Ann. Inst. Pasteur*, 16 (1902) 195.
45 P. Mazé and A. Perrier, *Ann. Inst. Pasteur*, 18 (1904) 721.
46 E. Godlewski, *Bull. Acad. Cracovie*, 13 (1904) 15.

47 S. Kostychev, *Zentralbl. Bakteriol.*, 13 (1904) 490.
48 W. Palladin, *Z. physiol. Chem.*, 47 (1906) 407.
49 W. Palladin and S. Kostychev, *Z. physiol. Chem.*, 48 (1906) 214.
S. Kostychev, *Biochem. Z.*, 15 (1908) 164.
51 S. Kostychev, *Biochem. Z.*, 23 (1909) 137.
52 S. Kostychev, *Z. Physiol. Chem.*, 67 (1910) 116.
53 W. Palladin, *Z. physiol. Chem.*, 55 (1908) 209.
54 W. Palladin, *Biochem. Z.*, 18 (1909) 151.
55 H. Wieland, *On the Mechanism of Oxidation*, New Haven, 1932.
56 P. Portier, *Les Oxydases dans la Série Animale*, Paris, 1897 (Thesis).
57 H. S. Mason, *Advan. Enzymol.*, 19 (1957) 79.
58 J. B. Conant, *Harvey Lect.*, 28 (1933) 159.
59 J. B. Conant and L. F. Fieser, *J. Biol. Chem.*, 62 (1925) 595.
60 R. Hill and H. F. Holden, *Biochem. J.*, 20 (1926) 1326.
61 R. Hill and H. F. Holden, *Biochem. J.*, 21 (1927) 625.

Chapter 27

Iron Catalysis in Biological Oxidations and the Identification of Warburg's Atmungsferment as an Iron-Porphyrin Compound

1. Role of metals in the activity of oxidases

Enzymes catalyzing color changes in the presence of oxygen or hydrogen peroxide have been studied since the observation that plant tissues turn guaiacum blue[1–4].

We have mentioned in Chapter 26 the experiment in which P. Ehrlich injected several dyes into a rabbit and observed that they were appearing in the tissues in the oxidized or in the reduced state. P. Ehrlich had observed that when a mixture of α-naphthol and of dimethyl-*p*-phenylenediamine was injected into a rabbit, an oxidative synthesis of indophenol was observed to take place in tissues. Ehrlich never thought that this was due to intracellular catalysts.

To quote Ehrlich[5]

"Ich stelle mir das so vor ... wie die Meteore durch den Dunstkreis der Erde dringen, so werden im Kleinsten die Moleküle in das Plasma der Zelle geschleudert ... sie bleiben dort ... irgendwie nutzbringend ... werden zum Beispiel als Sauerstoffträger verwendet und dann wieder ... ausgeschieden."

It must be stated that these experiments of P. Ehrlich confirmed his opinion of the existence of affinities between certain chemicals and the substance of cells and played a seminal role in the development of his work in chemotherapy (see Marquardt[6]).

Other authors[7,8] tested different tissues for the so-called *indophenol reaction* produced by the addition of the so-called "Nadi"-reagent (an abbreviation of the names of two components α-naphthol and dimethyl-*p*-phenylenediamine, which did not prevent some authors from attributing

References p. 212

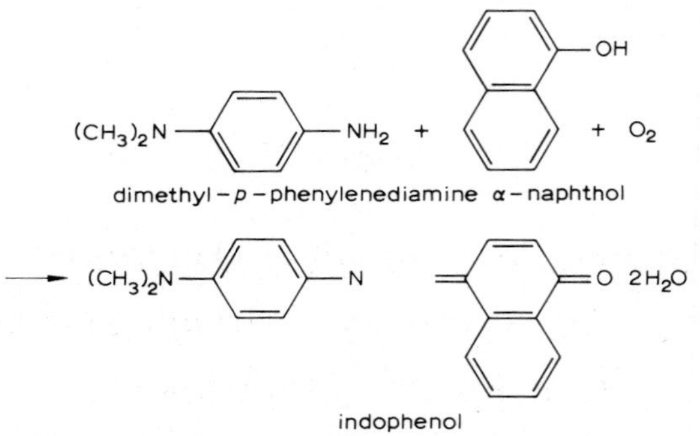

$$(CH_3)_2N-C_6H_4-NH_2 + \alpha\text{-}C_{10}H_7OH + O_2$$

dimethyl-*p*-phenylenediamine α-naphthol

$$\longrightarrow (CH_3)_2N-C_6H_4-N=C_{10}H_6=O \quad 2H_2O$$

indophenol

Fig. 9. The oxidative synthesis of indophenol catalyzed by indophenol oxidase (indophenol reaction).

its discovery to a mysterious Mr. Nadi). Röhmann and Spitzer[7] attributed the oxidative synthesis of indophenol to an intracellular enzyme. Spitzer[8] fractionated several tissue extracts and obtained preparations showing marked activity. He found a certain degree of proportionality between iron content and activity of what he considered as nucleoprotein preparations, and concluded that the enzyme-catalyzing indophenol synthesis was an iron-nucleoprotein compound. Keilin[9] refers to Spitzer's contribution as follows:

"This was the first clearly formulated conception of a role for iron in the catalysis of intracellular oxidation."

Kastle[10] introduced the name indophenol oxidase for the enzyme concerned (corresponding to what was later called Atmungsferment and today cytochrome oxidase) and Vernon carried out quantitative studies on the reaction[11–13].

2. Reversible cyanide inhibition of respiration

That cyanide poisoning was in fact an asphyxia of the tissues in the presence of excess of oxygen came from the observations of Hoppe-Seyler[14] who correctly observed that by inhibiting the respiration processes of tissues, cyanide prevents them from using oxygen so that oxyhemoglobin retains both its colour and its oxygen when it passes from the arterial to the venous state. Battelli and Stern[15] demonstrated that cyanide poisoning of tissue respiration was a reversible process.

3. Warburg's *Atmungsferment* and the theory of oxygen activation

The philosophical attitude of Warburg which appears in his first studies on cell respiration was an approach to cell physiology commanded by his interest in physical chemistry and particularly in surface phenomena as suggested by his blood-charcoal model. Warburg[16] wrote, in 1921:

> "Die Zellatmung ist ein capillar-chemischer Vorgang, der an den eisenhaltigen Oberflächen der festen Zellbestandteile abläuft. Durch Adsorption an diesen Oberflächen werden die trägen organischen Verbindungen aus dem gleichen Grunde gegenüber Sauerstoff reaktionsfähig, wie die Aminosäuren an der Oberfläche der Blutkohle. Die Zellatmung ist damit nicht physikalisch erklärt, jedoch zurückgefuhrt auf Phänomene der unbelebten Welt.
>
> "Narkotica hemmen die Zellatmung, indem sie—selbst an den Oberflächen nicht oxydabel—die Oberflächen bedecken und dadurch die Brennstoffe verdrängen".

In 1925, Warburg proposed new views on cellular respiration, based on a series of observations[17–19]. Warburg states that all cells contain iron (from 0.01 to 0.1 mg/g), and that the respiratory activity of cells and tissues varies between 7000 and 100 000 μl O_2/mg Fe/h. In the oxidation of cysteine to cystine in the presence of iron, at 37°C, about 400 000 μl O_2/mg Fe/h are involved and we must therefore conclude that the iron content of the cells is sufficient to account for cell respiratory activities. Cyanide, at the low concentrations at which it inhibits the catalysis by iron salts of oxidation reactions *in vitro*, also inhibits cell respiration. Warburg also observed that the oxygen uptake of granular (mitochondrial) extracts of unfertilized sea urchin eggs (later shown to contain mitochondria) is increased by the addition of iron. Warburg has realized a number of artificial models in which a study was made of the catalysis of the oxidation of amino acids such as leucine or cysteine by iron-containing charcoals.

From these observations Warburg concluded that the "combustion" of organic molecules involved in cell respiration is brought about by an *Atmungsferment* containing atoms of iron as oxygen-transporting components; in its bivalent state this reacts with molecular oxygen and passes into a higher valency state. The oxidized iron reacts with organic substances and reverts to the bivalent state. The combustible substances are bound by unspecific surface forces to the solid constituents of the cells but these unspecific surface forces are insufficient to bring about the reaction with oxygen unless the specific chemical force of the *Atmungsferment* comes into play. As, according to the concepts of the time,

References p. 212

Plate 88. Frédéric Battelli.

respiration was considered essentially a surface reaction, it is inhibited by narcotics which displace the combustible molecules from the surface, while cyanide inhibits respiration by its action on the *Atmungsferment* which was first conceived by Warburg as the sum of the catalytic iron compounds of the cell. This was later characterized by him as a well-defined enzyme which he henceforth called "oxygen-transferring enzyme of respiration" and which was finally recognized as the auto-oxidizable component of the respiratory chain, Keilin's cytochrome oxidase (corresponding to the old indophenoloxidase), the most essential piece in the respiratory chain.

4. Battelli and Stern's oxydones

Studies of the oxidation of *p*-phenylenediamine and of succinate by several tissues and extracts of tissues were performed in Geneva by Battelli and Stern[20–30]. Determining the rate of oxidation of both substrates, measured by oxygen uptake, and determining colorimetrically the rate of oxidation of *p*-phenylenediamine, they obtained, from disintegrated tissue extracts, which, after acid precipitation and resuspension, were highly active in the oxidation of both substrates. This they attributed to insoluble oxidases, which they called *oxydones* to differentiate them from soluble oxidases. The important contribution of Battelli and Stern lies in their conclusion that there is a definite relationship between cell respiration in the tissues and catalysis of the oxydones obtained after the disintegration of the tissues. Cyanide, which inhibits the tissue respiration reversibly, also reversibly inhibits the indophenol reaction[31,32] as well as the aerobic oxidation of succinate by tissues[33].

From their observations, Battelli and Stern concluded that the activity of oxydones was linked with the main respiration of tissues ("Haupt-atmung"), by which they meant what we recognize now as the respiratory chain, as distinct from what they called accessory respiration, and considered as accomplished by enzymes which can be separated from the tissues.

5. The inhibitory effect of carbon monoxide

It was in 1926 that Warburg started a study of the effect of carbon monoxide on the respiration of yeast. Contrary to hemoglobin, the active group for which CO and O_2 compete in yeast has a much greater affinity

References p. 212

for oxygen than for carbon monoxide. The rate of respiration of yeast in a mixture of nitrogen and oxygen is inhibited if the nitrogen is replaced by carbon monoxide and the inhibition is reversible. A very high partial pressure of CO is required to cause the inhibition which is diminished by an increase of pO_2. Different pressures of CO may produce the same degree of inhibition if the CO/O_2 ratio remains constant.

While engaged in this study, Warburg, as he relates[34], was made aware by A. V. Hill of an old observation of Haldane and Smith[35] who had shown the effect of light, with shifting of the equilibrium towards the left, on the reversible reaction

$$HbO_2 + CO \rightleftharpoons HbCO + O_2$$

due to the fact that HbCO is sensitive to light.

To quote Sir Hans Krebs[36]

> "I was present when A. V. Hill visited Warburg's laboratory. Warburg showed him the inhibition of yeast respiration by carbon monoxide and Hill then told him of the old work of Haldane and Smith on the light sensitivity of CO-haemoglobin. Within 24 hours Warburg tested the light sensitivity of the inhibition of yeast respiration and found the well known effect."

This led Warburg[37] to a detailed study of the effect of light on yeast, the respiration of which was inhibited by CO. Warburg's laboratory was particularly well equipped for these studies, as he had been working on the determination of quantum efficiency of photosynthesis. By measuring the inhibitory effect of carbon monoxide on aerobic fermentation and the reversal of this inhibition by light of certain wavelengths, as a result of the dissociation of CO, Warburg in a most brilliant study obtained indications showing that his *Atmungsferment*, combined with CO, shows absorption bands at 436, 546 and 578 mμ. In 1927, using as controls gas mixtures in which CO was replaced by nitrogen, Warburg[37], from the percentage of inhibition of yeast respiration in different CO/O_2 gas mixtures, calculated the relative affinity of the *Atmungsferment* for oxygen and for carbon monoxide. At this time he concluded that the *Atmungsferment* was related to haemoglobin and was a single definite substance.

In 1928, Warburg and Negelein[38,39] obtained a photochemical absorption spectrum from the effect of light at 8 wavelengths for yeast cells. In 1929, they described the methods and developed the theories applied to the determination at 15 different wavelengths, of the photochemical absorption spectrum of a yeast *(Torula utilis)*. The absorption bands of the carbon monoxide compound of the *Atmungsferment* Warburg and Negelein[40] found to be situated nearer the red end of the spectrum than in the case of CO-protohaem. Later in the same year, Warburg and Negelein[41] concluded that there is a close resemblance between the photochemical absorption spectra of *Atmungsferment* and *Spirographis* CO-haemoglobin, prepared by combining the haem of *Spirographis* chlorocruorin with globin. It was made clear that *Atmungsferment* is an iron-porphyrin compound, a major advance, as this compound figures in our schemes of biological oxidation under the name of cytochrome oxidase, the only auto-oxidizable component of cytochrome. Its nature and function was therefore recognized by Warburg as an iron-porphyrin compound and an oxidase, major discoveries which became part of the paradigm of our present theory of biological oxidation.

REFERENCES

1 W. H. Wollaston, *J. Nat. Philos.*, 8 (1804) 293.
2 W. T. Brande, *Phil. Trans.*, (1806) 89.
3 L. A. Planche, *Bull. Pharm.*, 2 (1810) 578.
4 L. A. Planche, *J. Pharm. Chim.*, 6 (1820) 16.
5 P. Ehrlich, *Das Sauerstoff-Bedürfnis des Organismus*, Berlin, 1885.
6 M. Marquardt, *Paul Ehrlich*, Berlin, 1951.
7 F. Röhmann and W. Spitzer, *Ber. deut. chem. Ges.*, 28 (1895) 567.
8 W. Spitzer, *Arch. Ges. Physiol.*, 67 (1897) 615.
9 D. Keilin, *The History of Cell Respiration and Cytochrome*, Cambridge, 1966.
10 J. H. Kastle, *Bull. U.S. Hyg. Lab.*, *No. 59*, Washington, 1910.
11 H. M. Vernon, *J. Physiol.*, 42 (1911) 402.
12 H. M. Vernon, *J. Physiol.*, 43 (1911) 96.
13 H. M. Vernon, *J. Physiol.*, 45 (1912) 197.
14 F. Hoppe-Seyler, *Arch. Ges. Physiol.*, 12 (1876) 1.
15 F. Battelli and L. Stern, *J. Physiol. Pathol. Gen.*, 9 (1907) 228.
16 O. Warburg, *Biochem. Z.*, 119 (1921) 134.
17 O. Warburg, *Festschr. Ges. Förd. Wiss.*, (1921) 224.
18 O. Warburg, *Biochem. Z.*, 152 (1924) 479.
19 O. Warburg, *Ber. deut. chem. Ges.*, 58 (1) (1925) 1001.
20 F. Battelli and L. Stern, *J. Physiol. Path. Gen.*, 30 (1911) 172.
21 F. Battelli and L. Stern, *Biochem. Z.*, 30 (1911) 172.
22 F. Battelli and L. Stern, *Ergeb. Physiol.*, 12 (1912) 96.
23 F. Battelli and L. Stern, *Biochem. Z.*, 46 (1912) 317.
24 F. Battelli and L. Stern, *Biochem. Z.*, 46 (1912) 343.
25 F. Battelli and L. Stern, *Biochem. Z.*, 52 (1913) 226.
26 F. Battelli and L. Stern, *Biochem. Z.*, 52 (1913) 253.
27 F. Battelli and L. Stern, *Biochem. Z.*, 56 (1913) 35.
28 F. Battelli and L. Stern, *Biochem. Z.*, 56 (1913) 50.
29 F. Battelli and L. Stern, *Biochem. Z.*, 56 (1913) 59.
30 F. Battelli and L. Stern, *Biochem. Z.*, 67 (1914) 443.
31 W. Spitzer, *Arch. Ges. Physiol.*, 60 (1895) 303.
32 S. Gräff, *Beitr. Pathol. Anat.*, 70 (1922) 1.
33 F. Battelli and L. Stern, *J. Physiol. Path. Gen.*, 9 (1907) 228.
34 O. Warburg, *Biochem. Z.*, 177 (1926) 471.
35 J. Haldane and J. L. Smith, *J. Physiol.*, 20 (1896) 497.
36 H. A. Krebs, *Letter to Author*, 24th July, 1972.
37 O. Warburg, *Biochem. Z.*, 189 (1927) 354.
38 O. Warburg and E. Negelein, *Biochem. Z.*, 193 (1928) 339.
39 O. Warburg and E. Negelein, *Biochem. Z.*, 202 (1928) 202.
40 O. Warburg and E. Negelein, *Biochem. Z.*, 214 (1929) 64.
41 O. Warburg and E. Negelein, *Biochem. Z.*, 214 (1929) 101.

Chapter 28

The Wieland-Thunberg Theory

"Biological oxidations are, according to Wieland[1–4] catalysed by specific enzymes which he called dehydrases, later known as dehydrogenases, which activate certain hydrogen atoms of substrate molecules with the result that they become labile and can be transferred to a suitable hydogen acceptor such as methylene blue, quinone or oxygen, which does not require to be activated". (Keilin[5])

The hydrogen donor is dehydrogenated (*i.e.* oxidized) and the hydrogen acceptor becomes reduced (methylene blue to its leuco-base, oxygen to H_2O_2, *etc.*). Wieland, who was a chemist, and used palladium surface as a model, was not as impressed by the complete inhibition of cell respiration by cyanide as was Warburg. He considered it to be the result of an accumulation of H_2O_2 resulting from the poisoning of catalase. Wieland's theory was developed from purely chemical experiments, and was extended by him to the oxidation of alcohol by acetic bacteria, as well as to the oxidation of aldehydes by the Schardinger enzyme of milk. This is the reason for which it was neglected by those interested in animal biochemistry until Thunberg extended and developed it[6,7].

In order to perform rapid tests, Thunberg imagined a convenient "vacuum tube" in which the oxidation of metabolites is brought about anaerobically in the presence of methylene blue. The hydrogen donor is oxidised at the expense of methylene blue, which becomes reduced to the leuco-compound. Thunberg[8] had, in 1909, noticed that addition of succinic acid to finely chopped muscle increases its oxygen consumption. Battelli and Stern[9] later showed that succinic acid was oxidized by muscle. This was a break-through, as it was generally admitted, on chemical grounds, that succinic acid is very resistant to oxygen. With respect to the following developments we may cite Thunberg[10] himself:

"Some years later, when Wieland published his hydrogen activation theory, it was possible for me to explain the mechanism of biological oxidation of succinic acid. Oxidation is caused by an agent in the muscle which has the power of giving to hydrogen groups in succinic

References p. 218

Plate 89. Heinrich Otto Wieland.

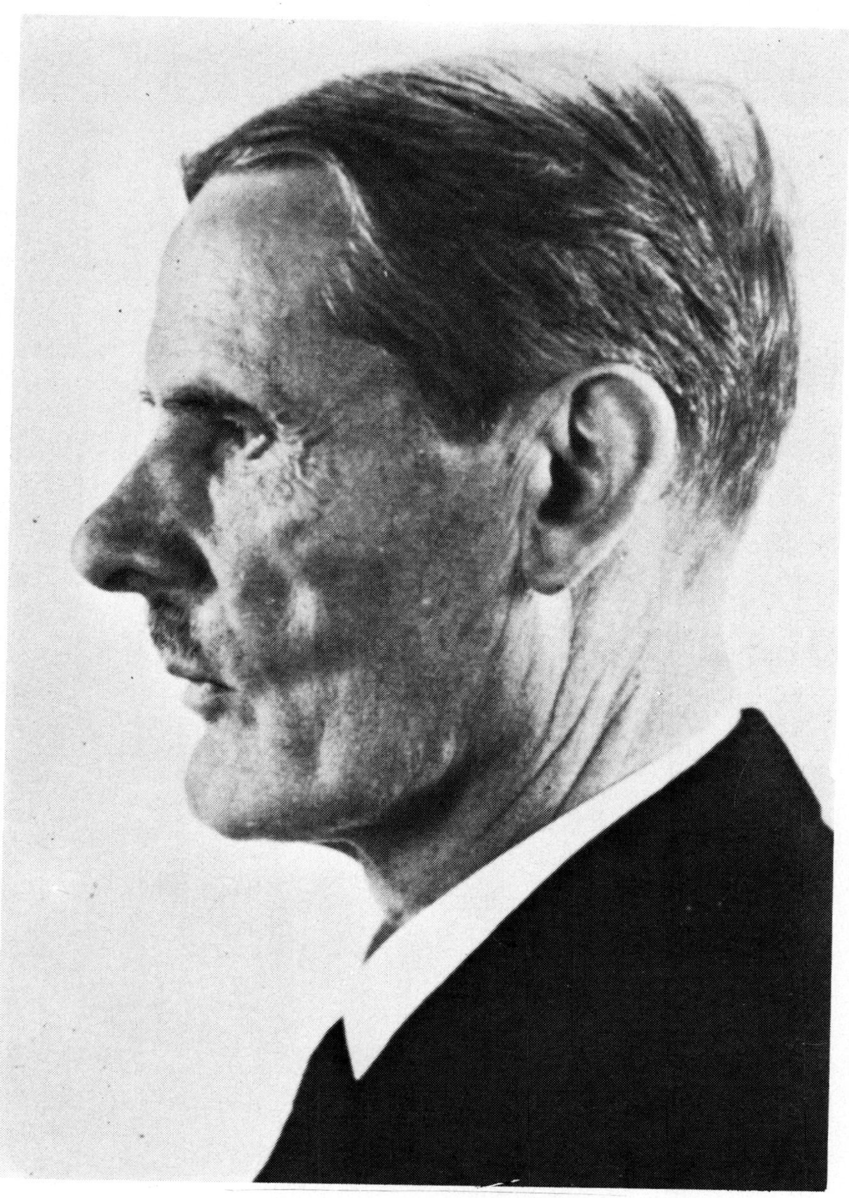

Plate 90. Torsten Ludwig Thunberg.

Plate 91. Thomas P. Singer

acid a reactivity that it had not had before. This reactivity was readily demonstrated by the use of methylene blue. Succinic acid and methylene blue (in water solution) do not react with one another. Thoroughly washed muscle and methylene blue do not react with one another either. However, if succinic acid is added to an aqueous solution (an homogenate) of methylene blue in which muscle is suspended, there will occur a reaction that is readily observed, because it is accompanied by a decolorization of methylene blue. Under the influence of this agent in muscle, succinic acid thus appears as a hydrogen donor because it gives off a hydrogen group to methylene blue (a hydrogen acceptor) which is thereby reduced to leuko-methylene blue. It is this very agent in the muscle which I christened succinic dehydrogenase".

That the enzyme was able to convert succinic acid to fumaric acid was first shown by Einbeck[11,12]. It must be noted that at that time the enzyme was not isolated. Particles of diverse degrees of complexity were used and no reliable assay method was available. The difficulty of solubilizing the enzyme was solved by the brilliant work of Singer and his collaborators who isolated the enzyme from several sources.

This essential development came about in a fortuitous way, as Singer and Kearney[13] recall it:

"During the study of L-cysteinsulfinate metabolism in mammalian mitochondria and in *Proteus vulgaris* in 1952–1953, Singer and Kearney found that soluble enzyme preparations from these sources, when supplemented with phenazine methosulfate, catalyzed the coupled oxidation of this amino acid and a substance of unknown structure to yield pyruvate, aspartate and sulfite or sulfate[14,15]. The cosubstrate was isolated from yeast extracts by ion-exchange procedures and identified as succinic acid[14]. The reaction sequence leading to the over-all formation of pyruvate and aspartate is given in Eqns. 2–4.

$$\text{Succinate} \rightarrow \text{Fumarate} \rightarrow \text{Malate} \rightarrow \text{Oxaloacetate} \quad (2)$$

$$\text{Oxaloacetate} + \text{Cysteine sulfinate} \rightarrow \text{Aspartate} + \beta\text{-Sulfinylpyruvate} \quad (3)$$

$$\beta\text{-Sulfinylpyruvate} \rightarrow \text{Pyruvate} + SO_3^{--} \quad (4)$$

Owing to the irreversibility of reaction (4), the reaction sequence proceeded to completion, and, by preventing oxaloacetate accumulation, cysteinsulfinate permitted the stoichiometric oxidation of succinate. The same soluble extract catalyses the one-step oxidation of succinate to fumarate in the presence of phenazine methosulfate[14,15]. Thus the discovery of satisfactory methods for the extensive solubilization was a fortuitous event."

Succinate dehydrogenase, as it is now called, is a flavoprotein enzyme. The developments of its knowledge are retraced by Singer in Chapter 3 of Vol. 14 of this Treatise, to which the reader is referred.

But let us turn back to Thunberg's theory[7–16] of biological oxidations. The essential point of it is that according to him, the free oxygen does not directly combine with carbon as understood in the combustion theory, but reacts with hydrogen, not to form H_2O_2, but to form water. According

References p. 218

to the theory, the oxygen of carbon dioxide is not derived from the molecular oxygen of respiration, but represents "endo-oxygen" from the substrate. Experiments with ^{18}O confirm that the oxygen in respiration is converted to water[17].

REFERENCES

1 H. Wieland, *Ber. deut. chem. Ges.*, 45 (1912) 489.
2 H. Wieland, *Ber. deut. chem. Ges.*, 45 (1912) 2606.
3 H. Wieland, *Ber. deut. chem. Ges.*, 46 (1913) 3327.
4 H. Wieland, *Ergeb. Physiol.*, 20 (1922) 477.
5 D. Keilin, *The History of Cell Respiration and Cytochrome*, Cambridge, 1966.
6 T. Thunberg, *Quart. Rev. Biol.*, 5 (1930) 318.
7 T. Thunberg, *Ergeb. Physiol.*, 39 (1937) 76.
8 T. Thunberg, *Skand. Arch. Physiol.*, 22 (1909) 430.
9 F. Battelli and L. Stern, *Biochem. Z.*, 30 (1911) 172.
10 T. Thunberg, *Kungl. Fysiografiska Sällskapets i Lund Forhandlingar*, 20 (1950) 1.
11 H. Einbeck, *Z. physiol. Chem.*, 90 (1914) 301.
12 H. Einbeck, *Biochem. Z.*, 95 (1919) 296.
13 T. P. Singer and E. B. Kearney, in P. D. Boyer, H. Lardy and K. Myrbäck (Eds.), *The Enzymes*, Vol. 7, New York and London, 1963.
14 E. B. Kearney and T. P. Singer, *Biochim. Biophys. Acta*, 14 (1954) 572.
15 T. P. Singer and E. B. Kearney, in W. D. McElroy and B. Glass (Eds.), *Amino Acid Metabolism*, Baltimore, 1954.
16 T. Thunberg, *Bull. Soc. Chim. Biol.*, 21 (1939) 887.
17 N. Lifson, G. Gordon, M. B. Visscher and A. O. Nier, *J. Biol. Chem.*, 180 (1949) 803.

Chapter 29

Keilin's Rediscovery of Histohematin (Cytochrome)

1. MacMunn's histohematin

The first of the components of the system of biological oxidations appearing in our present theory was discovered by MacMunn and described by him in papers published between 1884 and 1887 (refs. 1–3). The pigment presents a characteristic four-banded absorption spectrum and MacMunn, who considered it as a single substance, gave it different names according to its origin: myohematin, histohematin, actiniohematin. As the pigment is also present in plants and in unicellular organisms, Keilin[4], in 1925, in order to recognize its wide distribution in aerobic cells, proposed calling it cytochrome (cellular pigment). In spite of the fact that it was shown, as we shall see later, to be composed of several hemochromogens, the name cytochrome in the singular remains in use to designate MacMunn's histohematin and signifies that its components are interdependent in their oxidation–reduction activity, an aspect to which we shall return later on.

MacMunn, the author of a well-known treatise on spectroscopy[5], had an extensive medical practice and with the help of the Royal Society, he had built a spectroscopical laboratory in the garden of his house at Wolverhampton, near Birmingham, where he later became a Life Governor of the University. He was inclined towards comparative studies and spent his summer holidays at the Marine Biological Laboratory at Plymouth.

During spectroscopic examinations of tissues of a number of animal species, MacMunn observed a characteristic four-banded spectrum. He found that when he added H_2O_2 the spectrum disappeared, and reappeared spontaneously on reduction. About these pigments, he wrote:

"they appear to be capable of oxidation and reduction and are therefore respiratory". (ref. 2)

A whole chapter (Chapter 6) of Keilin's book[6] on the history of cell respiration is devoted to MacMunn's work and we refer the reader to that chapter to which the following lines owe much.

References. p. 234

Plate 92. Charles Alexander MacMunn.

MacMunn had always been careful to remove blood hemoglobin by perfusion, when present. He had carefully distinguished between muscle hemoglobin (later called myoglobin), when present, and myohematin or histohematin. But the technicalities of the problem were difficult and he evidently made a few errors. In pigeon breast muscle, for instance, he was unable to separate the myoglobin spectrum, and the spectrum he described was in reality a bulk spectrum of myoglobin and myohematin, which he considered to be myohematin, and he therefore wrongly concluded that pigeon breast muscle pigment was deprived of myoglobin and made entirely of myohematin[1]. In a great number of species of invertebrates well known to contain neither hemoglobin, nor myoglobin, nor chlorocruorin, he showed the presence of histohematin, which, on the other hand, is very widely distributed in the tissues, from sponge to man.

A few years after his first publication, the unfortunate statement made by MacMunn[1] that myohematin is the only pigment present in pigeon breast muscle, was criticized in Hoppe-Seyler's Institute by one of his students, Levy, in a doctoral dissertation[7]. Levy did not make any pigment examination on the tissues themselves, as he was not equipped to do so. He worked on extracts and found myoglobin in extracts of pigeon breast muscles. Myohematin, as defined by MacMunn, he considered a decomposition product of hemoglobin.

MacMunn's reply[8] to Levy's criticism was that myohematin can be seen directly by microspectroscopic examination of pigeon breast muscle freed from blood, that myohematin cannot be a derivative of hemoglobin because it is present in the muscles of invertebrates such as insects, which are devoid of hemoglobin, and finally that the "modified myohematin", obtained by Struve's method differs from an ordinary hemochromogen in the appearance and position of its absorption bands as well as in the fact that it cannot be oxidized even by oxygen.

This went above the head of the neophyte Levy and was answered by Hoppe-Seyler himself[9]. He showed that when a slice of pigeon breast was treated with CO, the absorption spectrum was replaced by that of CO-myoglobin. Having proved beyond doubt that pigeon breast muscle contains myoglobin, Hoppe-Seyler rather unwittingly rejected MacMunn's claim that myohematin is present in the tissues of lower organisms which are devoid of hemoglobin, the argument being that observations made on lower animals are not applicable to higher organisms. For him, the four-banded spectrum attributed by MacMunn to myohematin was in fact the

References p. 234

result of a mixture of bands contributed by oxyhemoglobin and the surface-reduced hemoglobin in the interior, and perhaps a little hemochromogen. MacMunn[10] replied to Hoppe-Seyler by reaffirming the argument in favour of the existence of myohematin and histohematin. To this last attempt of MacMunn to defend his views, Hoppe-Seyler, as editor of the journal, appended a foot-note in which he firmly expressed his wish not to discuss the question any more nor to accept the presence of any special colouring matter in pigeon muscle[11].

When Mörner[12] resumed the important study of the relations between muscle hemoglobin and blood hemoglobin, he was able to map the absorption bands of the two proteins more exactly, and showed that in myochrome, as he called the muscle protein, they lie nearer to the red end of the spectrum. Nevertheless he did not confuse myochrome (myoglobin) with MacMunn's myohematin.

The confusion between myohematin and myoglobin starts later with a review by von Fürth[13] who stated that Mörner had been led to confirm MacMunn's (erroneous) view on the muscle pigment, a statement which is far from the truth.

An important paper on muscle hemoglobin, to which he gave the name of myoglobin, was published by Günther[14] in 1921. This author confirmed the specific characteristics of the muscle protein established by Mörner. He further identified reduced myohematin with metmyoglobin (oxidized myoglobin).

This confusion between myoglobin and myohematin (cytochrome of muscle) contaminated other authors who obtained their information from Günther, for instance Hans Fischer[15]*. Halliburton[16] considered myohematin as derived from hemoglobin. Some authors considered the existence of histohematin as uncertain while others accepted it as being different from hemoglobin or its derivatives, but considered that its respiratory function had not been demonstrated[17,18].

2. Keilin's rediscovery of histohematin

MacMunn died frustrated and discouraged in 1911. It was only in 1925 that MacMunn's histohematin, rechristened cytochrome, was led by Keilin[4] into the Hall of Fame.

* "Fischer's opinion that MacMunn's findings were correct was based upon an erroneous identification of myohematin with myoglobin and on the false belief that what MacMunn had demonstrated was the existence of myoglobin as distinct from blood hemoglobin."[6]

We are fortunate in having recently been able to read Keilin's own memories of the events that led him to the study of the properties and of the respiratory function of cytochrome[6]. This text shows that Keilin did not merely revive MacMunn's forgotten discovery of which he was unaware, but was led to the same subject by the purpose of solving problems of an entirely different nature.

David Keilin, who had left Poland to seek a medical education abroad, studied first in Liège and later in Paris. There, on account of the asthma of which he was suffering, he turned to biological studies and he received the degree of D.Sc. in Paris. Soon after the beginning of the First World War, in 1915, Keilin received an invitation to join the University of Cambridge where he remained for the rest of his life[19]. For Keilin, who, from being an entomologist and a parasitologist, became a biochemist, it was natural to study an insect in order to obtain information on biological oxidations, while this would not have occurred to Warburg, who used charcoal surfaces, or to Wieland, who used palladium surfaces. We advise the reader to peruse pages 140 to 165 of Keilin's book which is paraphrased in what follows. From 1919 onwards, Keilin tells how actively engaged he had been on a study of the anatomy of respiratory systems in dipterous larvae and pupae. During the years 1922-1924 he was preoccupied with two closely linked problems. The first was related to the long quiescent period of the *Gasterophilus intestinalis* larva and the fate of hemoglobin during and after the period of metamorphosis outside the equine host. He had to abandon the first problem owing to difficulties of culture.

The second problem was less difficult to tackle. Early in the summer of 1923, Keilin obtained a first batch of *Gasterophilus* pupae. He detected in these pupae a gradual disappearance of larval hemoglobin, especially during the latter stages of metamorphosis. Thoracic muscles of adult flies emerged from the pupae and those which had died in captivity showed a spectrum with four distinct absorption bands.

Early in the summer of 1924, Keilin obtained a large number of *Gasterophilus* pupae, from which several perfect adult flies emerged. Examining their thoracic muscles, Keilin observed the same four-band spectrum. Keilin then took several other insects and in particular an adult wax-moth *(Galleria mellonella)* which had been bred in the laboratory from a culture of its caterpillars fed on honey comb. Keilin had cultures of *Bacillus subtilis* in his laboratory. He removed the bacilli from the surface

References p. 234

Plate 93. David Keilin.

of the culture in the form of a thick cheesey mass and examined it under the microspectroscope. The same four-banded absorption spectrum appeared again. Keilin took a small lump of fresh baker's yeast and compressed it between two glass slides: there was the four-banded spectrum once more.

One day, while examining a suspension of yeast freshly prepared from a few bits of compressed yeast shaken vigourously with some water in a test tube, Keilin failed to observe the four-banded spectrum. But before he had time to remove the suspension from the field of vision, the four bands reappeared. They disappeared again on shaking the tube, to reappear within a few seconds of standing. He wrote:

> "I must admit that this first visual perception of an intracellular respiratory process, was one of the most impressive spectacles I have witnessed in the course of my work. Now I had no doubt that cytochrome is not only widely distributed in nature and completely independent of hemoglobin, but that it is an intracellular respiratory pigment which is much more important than hemoglobin. On the other hand, the nature of the pigment showing four distinct absorption bands when it was in the reduced state, but none when it was oxidized, still remained very obscure". (ref. 6)

However, a very careful study of yeast cells and of the thoracic muscles of insects under diverse conditions showed Keilin that the complex spectrum of cytochrome was in fact that of a mixture of three chromogens, the a–b and c-bands being their α-bands, while d- represents their β-bands fused into one complex band. Cytochrome was thereby resolved into three compounds: cytochrome *a*, cytochrome *b* and cytochrome *c*. At that stage, Keilin abandoned for the time being his comparative studies on insect respiration in order to devote himself entirely to the problems of cell respiration.

A survey of a large number of species immediately convinced him of the general occurrence of cytochrome in aerobic organisms. At that time Keilin had not heard of MacMunn's work. It was made known to him, he recalls in his book, by Günther's paper[14] on muscle hemoglobin, in which it is mentioned. As he writes, Keilin[6]

> "having already gained a first-hand knowledge of this pigment, had no difficulty of disentangling MacMunn's correct observations from some of its own erroneous conclusions and, above all, from the confusion of myohematin with myoglobin which had subsequently arisen."

For his studies on cytochrome, Keilin used two simple pieces of apparatus, the microspectroscope ocular and the Hartridge reversion spectroscope,

References p. 234

both instruments being attached to a microscope in place of the eyepiece. Owing to its low dispersion, the Zeiss microspectroscope ocular used by Keilin was ideal for the study of plant or animal tissues, suspensions of cells (yeasts and bacteria) and opalescent fluids.

But besides these technical advantages properly selected, Keilin's memoir of 1925 (ref. 4) is an admirable example of the use of the comparative method in solving such a baffling problem as that of the four-banded spectrum of the unknown compound concerned. Keilin solved it in a spectroscopic study of baker's yeast and of insect wing muscles in a great variety of conditions.

He himself summarized the evidence as follows:

"(*1*) Although the positions of the absorption bands (a, b, c and d) of cytochrome in different organisms are fairly constant, their relative intensities show marked variations. This is most readily explained by postulating variations in the relative concentrations of the three components.

(*2*) Band d, unlike the other bands, has a complex structure showing several maxima of intensity which I ascribed to the β-bands of the three compounds.⋆

(*3*) When urethane is added to a suspension of yeast cells shaken with air, bands a and c and the main part of band d disappear and only return after a long delay, whereas band b and the remaining portion of band d not only remain distinct but appear to be reinforced. Thus, in the presence of a narcotic and air, cytochromes *a* and *c* undergo oxidation and cannot subsequently be easily reduced, while cytochrome *b* remains reduced and cannot undergo oxidation.

(*4*) The spectroscopic examination of thoracic muscles of bees, which had been spread on a slice and allowed to dry, showed, according to the area examined, one of the following combinations of absorption bands: a, b, c, d; a, c, d; c, d; c, d or no bands at all. The missing bands were those of the components which had undergone oxidation.

(*5*) Cells heated to above 70°C showed only absorption bands c and d, while band a and b as well as a portion of d disappeared permanently. This showed that while component c is thermostable, components a and b are thermolabile.

(*6*) Clear, aqueous extracts of air-dried or acetone-dried yeast showed only absorption band c and that portion of band b belonging to cytochrome *c*."

Not only did Keilin in his paper of 1925 demonstrate the existence and the composition of cytochrome, but he was led to the view that cytochrome plays a fundamental part in cellular respiration. The results on which this conclusion was based are listed by him in the last lines he wrote of the manuscript his daughter J. Keilin[6] published in 1966:

⋆ "This complex structure of band d led me to assume that one of the maxima of intensity was the β-band of cytochrome *a*. Although this assumption was later shown to be incorrect, at the time it was made it was consistent with the available knowledge of hemochromogen spectra and was indeed helpful in demonstrating that the four-banded absorption spectrum of reduced cytochrome belongs to three hemochromogen-like compounds".

"(*1*) Cytochrome, which is composed of three hemochromogen-like compounds designated as cytochrome *a*, *b* and *c* is very widely, if not universally, distributed in the cells of aerobic organisms.

(*2*) There is in nature a marked parallelism between the concentration of cytochrome *c* and the respiratory activity or oxygen requirement of cells; highly respiring cells of baker's yeast, aerobic bacteria, rhythmically contracting and very active vertebrate heart muscle; the pectoral muscles of flying birds and the wing muscles of flying insects are particularly rich in cytochrome.

(*3*) Cytochrome not only has the property of reversible oxidation and reduction, but it can actually be seen in living cells (in yeast and in insect wing muscles) to undergo continual oxidation and reduction.

(*4*) The activity of cytochrome, in other words its oxidation and reduction in living cells is affected reversibly by respiratory inhibitors in the same way and to the same degree as the respiratory activity of the cell is affected. However, while cyanide inhibits the oxidation of all the three components (*a*, *b* and *c*) of cytochrome, narcotics such as ethylurethane inhibit the reduction of cytochromes *a* and *c* and the oxidation of cytochrome *b*.

(*5*) In the living organism the state of cytochrome, as seen spectroscopically, denotes only a difference between the activities of the process of respiration responsible for its oxidation and reduction at that particular time. Thus, whatever may be the spectroscopic appearance of cytochrome in living, normally respiring cells, it is not a static condition but represents a dynamic equilibrium (or kinetic steady state) between the processes of oxidation and reduction.

(*6*) This explains why cytochrome, which can be seen in the oxidized state for a long time in an aerated suspension of yeast or in the wing muscle of living Galleria, suddenly becomes reduced on the addition of cyanide, although cyanide does not act as a reducing agent.

(*7*) The continual reduction of cytochrome within cells must naturally be accompanied by the simultaneous oxidation of other substrates within the cells; in other words, it is a measure of intracellular respiration. Thus in cytochrome an intracellular reversible oxidation–reduction system was discovered which links the molecules of substrate or metabolite with molecular oxygen. It was also the first respiratory system the activity of which could be observed in living intact cells, so that it was possible to localize the effects of different respiratory inhibitors."

3. Cytochrome studies of Keilin, from 1926–1933*

During the period from 1926 to 1933, Keilin's work on cytochrome was mainly devoted to a precision of the nature of its three compounds, and to the mechanism of its oxidation and reduction. With improved methods, the spectrum of oxidized cytochrome was found to be of a parahematin type, consisting of a narrow band at 566.5 nm and a stronger and wider β-band at 528.7 nm[20].

As cytochromes are not auto-oxidizable, their rapid oxidation in the cells requires the presence of a catalyst. The indophenoloxidase system seemed to be a possible candidate. But at that time yeast cells, the best source of

*Drawn from Chapter 9 of Keilin[6].

cytochrome, were considered as being devoid of indophenoloxidase[21,22]. In fact, baker's yeast as well as the other cells showing a typical cytochrome system, presents an indophenol reaction as long as the activity of their dehydrogenases is lowered to a greater extent than their oxidase system[23]. Comparing indophenoloxidase with other oxidizing systems (polyphenol oxidase, xanthine oxidase, uricase), Keilin concluded that indophenol oxidase was the only one which could be linked directly with the cytochrome system and with the respiratory activity of the cell.

The fundamental properties of cytochrome *c* had been established by Keilin[24] on a sample isolated from yeast and purified to a hematin content of about 1.3% of the dry weight consisting mainly of proteins. In the reduced state, cytochrome *c* showed the spectrum of a hemochromogen, except that it was not auto-oxidizable nor did it combine with CO. In solutions below pH 4.0 and above pH 12 it acquired these properties, losing them again if brought back to its native state by neutralization of the solutions. It was this preparation which was used by Keilin to reconstitute a yeast cytochrome *c*–succinic acid dehydrogenase system of heart muscle preparation. It was also used for the determination of the absorption spectrum by Dixon, Hill and Keilin[25].

Zeile and Reuter[26] purified cytochrome *c* to a hematin content of 3.6% and suggested a minimal molecular weight. Cytochrome was therefore shown to belong to an intracellular system the constituents of which are associated in definite relationships. Unfortunately the reconstruction of the complete system from its constituents isolated from tissues was made difficult by the impossibility of separating them. Only cytochrome *c* was isolated from yeast. A number of properties of this constituent[24] were

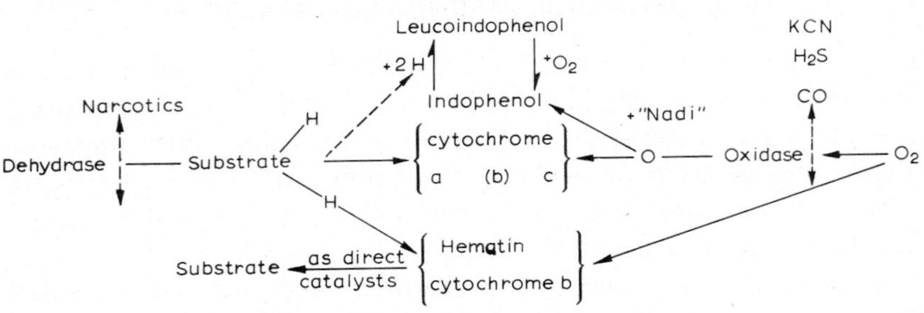

Fig. 10. (Fig. 14 of Keilin[23]) The functional relationship between the intracellular haematin compounds and the two kinds of respiratory enzymes (dehydrogenases, oxidases). In this system, the oxidized cytochromes *a*, *b* and *c* are reduced by the dehydrogenases, and the reduced cytochrome is reoxidized by the oxidase and oxygen.

determined and a cytochrome *c*-oxidase system was obtained which could oxidize cysteine. This was a respiratory chain in which oxygen transferred was largely replaced by hydrogen transfer. It must be recalled that, in 1927, Warburg[27] still considered cytochrome as *transferring oxygen* from the *Atmungsferment* to the substrate.

Commenting on Fig. 10 Nicholls writes:

"Here it may be noted that cytochrome b had been in part removed from the catalytic system dependent on activations by dehydrogenases and indophenol oxidase. This was because cytochrome b appeared to be oxidized in the presence of cyanide and reduced in the presence of narcotics. The cytochrome oxidase is referred to as indophenol oxidase; indophenol was found to be oxidized by the same system, and Keilin regarded the cytochromes as "carriers" rather than as "activators". He did not consider that the cytochromes themselves provided at least part of the indophenol oxidase and thus concluded that an unknown enzyme was responsible for both effects. He compared its behavior with that of polyphenol oxidase." (ref. 28)

4. Indophenoloxidase and *Atmungsferment* (oxygen-transferring enzyme)

Warburg's views on this subject are found in his book published in 1946 (ref. 29).

At that time it was clear to many workers in the field that the work of Battelli and Stern (1907–1914) and of Vernon (1912) on indophenol oxidase related above, and Warburg's studies on his *Atmungsferment*, were concerned with the same oxidation system of cells, which is, like respiration, strongly and reversibly inhibited by cyanide. In 1927, as we have said, Keilin suggested that the enzyme responsible for the oxidation of cytochrome by molecular oxygen is indophenoloxidase. Keilin also showed that the indophenol reaction of cells, tissues and cell-free preparations is also, like respiration, inhibited by carbon monoxide, and that this inhibition is reversible and light-sensitive. Further work of Keilin[30,31] brought forward more evidence in favour of the kinship of indophenoloxidase and *Atmungsferment*. Warburg published in 1929 his views on the subject in a paper on *Atmungsferment* and oxidases[32], a point of view he stated again in 1929 in his Herter Lecture on the enzyme problem and biological oxidation[33].

In this lecture he considered the apparent contradiction between the identification of a number of different oxidases, and the unity of his *Atmungsferment*. His explanation is that the different oxidases are not

References p. 234

preformed in cells but are "decomposition products of a substance uniform in life"[33].

As we have said, Keilin's work on the oxidation of cytochrome by indophenoloxidase had led him to the concept according to which cytochrome is linked on the one hand with reducing systems composed of enzymes and of their substrates undergoing oxidation and on the other hand, with an oxidase, the well-known indophenoloxidase which catalyzes the oxidation of the reduced cytochrome.

In a short note of 1927 (ref. 34), Warburg for the first time suggested that the *Atmungsferment* was a pigment probably related to hemoglobin. He even agreed to consider both cytochrome and *Atmungsferment* as hemoglobin-like respiratory pigments. However, in the following year, in another note[35], he applied to cytochrome the term of "degenerate ferment" depriving it of function in cellular respiration, a view which expressed his conviction that the implication of oxygen in respiration was not controlled by cytochrome but by the *Atmungsferment*.

5. Cytochrome oxidase: the signified and the significant, in a historical perspective

The name "cytochrome oxidase" was proposed by Malcolm Dixon[36] in 1929 to designate a "signified" corresponding to the biological activity of oxygen activation in biological oxidations.

As stated above, this biological activity was discovered by Warburg who called the "significant" a "respiratory ferment" (*Atmungsferment*). As described in Chapter 27, Warburg, influenced by the prevailing theories of biocolloidology, conducted a series of studies on whole cells and models and described, in the framework of his membrane theory, the significant corresponding to the signified of this respiratory ferment as a ferric iron colloidal complex. An inquiry into the background of Warburg's theories leads to his studies on the respiration of sea urchin eggs. At the time it was widely believed (by Loeb[37], amongst others) that respiration occurs in the cell nucleus. In a paper dated 1908, Warburg[38] reported on measurements of oxygen consumption by suspensions of fertilized and unfertilized sea urchin eggs. He observed a 6- to 7-fold increase in the rate of respiration, coupled with a rapid synthesis of nuclear material. Comparing the stages of 8 and of 32 cells, however, he observed no difference in respiration intensity. He also observed that fertilized eggs, even if prevented from dividing, showed the same high respiration rate in comparison with unfertilized eggs. He concluded that the nucleus was not the seat of cellular respiration. In 1910, Warburg[39] reported on a number of observations which led him to locate cell respiration in the cell membrane. Warburg's approach in these studies stems from the bias characteristic of studies on the complete biological system. Experimenting with erythrocytes, Warburg[40] showed that the effect of narcotics on respiration was proportional to their solubility in lipids, a concept he[41] extended to bacteria, spermatozoa and liver cells. All these studies appeared to be in agreement with Overton's theory that narcotics act on the lipid membrane only. In 1912, Warburg[42] extended his inquiry to cell-free systems and observed that narcotics give rise to a coagulation of yeast extracts. As a result he abandoned Overton's theory, exchanging it — as a 1914 paper indicates[43] — for that of biocolloidal adsorption. His observations of yeast extracts suggested to him the possibility of cell proteins being involved in the mechanism of narcosis.

To analyze cell-surface activity Warburg[42] adopted the activated-charcoal model. Planning a study of respiration in cell extracts, Warburg visited the Physiological Laboratory at King's College, London to learn the method of cell-grinding with the mechanical mill developed by Macfadyen and Rowland. As stated in Chapter 27, Warburg[44,45] recognized that what he had described in the framework of his membrane theory as a ferric iron colloid complex, was in truth an iron-containing enzyme reduced by metabolites to the ferrous iron state and inhibited by cyanide.

It was in 1925 that Keilin (see Chapter 29) described the spectral properties of the heme compounds known as cytochromes *a*, *b* and *c*. In 1927, Warburg proved that his respiratory enzyme was a heme compound (see Chapter 27).

As stated by Slater *et al.*[46] in a 1964 review paper (published in 1965):

> "Thus by 1926 Warburg had strong evidence that a CO- and cyanide-sensitive respiratory enzyme containing heme plays a central role in intracellular respiration, while Keilin had directly demonstrated the respiratory function of three heme compounds, cytochromes *a*, *b* and *c*. The difficulty was that none of these cytochromes showed any spectroscopically visible alteration of the absorption spectrum when cyanide or CO was added, as would have been expected, had these respiratory inhibitors combined with one of the cytochromes."

Nevertheless, Keilin showed in 1925 that cyanide inhibits the oxidation of the cytochromes. In 1927, however, Keilin[47] demonstrated that indophenol oxidase (see Chapter 27) is inhibited, with slight reversibility, by CO. In 1930, he showed[48] that the oxidation of ferrocytochrome *c* by oxygen,

References p. 234

catalyzed by a heart muscle preparation, is also inhibited by cyanide and, again with slight reversibility, by CO. Keilin concluded that Warburg's oxygen-transferring enzyme and indophenol oxidase display the same properties with respect to inhibition by cyanide or by CO and that indophenol oxidase is responsible for oxygen activation. Though based only on consideration of the "signified" and its inhibitions, these conclusions won wide recognition.

The "significant" of the oxidase-activating oxygen had only been approached with the help of Warburg's photochemical adsorption method (Chapter 27). In 1933, Keilin and Hartree[49] proposed to identify the significant of cytochrome oxidase as a component of the cytochrome a_3 which differs from cytochrome *a*. This concept did not meet with general approval; indeed, as related by Okunuki in Chapter 5 of Volume 14 of the present Treatise, the very existence of cytochrome a_3 was subject to doubt. Assumptions concerning a spectral differentiation between cytochromes *a* and a_3 have been considered unfounded. Then again, there is no evidence that cytochrome a_3 itself is autoxidizable as well as able to oxidize "ferrocytochrome *a*" without the cooperation of cytochrome *c*. To quote Okunuki (in Volume 14 of this Treatise):

"From this evidence it seems probable that cytochrome *a* is not autoxidizable and requires native cytochrome *c*, but not the acylated form, for cytochrome oxidase activity. From the above facts, and the results of Yakushiji and Okunuki showing that electrons are transferred from cytochrome *c* and *vice versa*, the following scheme is proposed to interpret the mechanism of action of cytochrome oxidase:

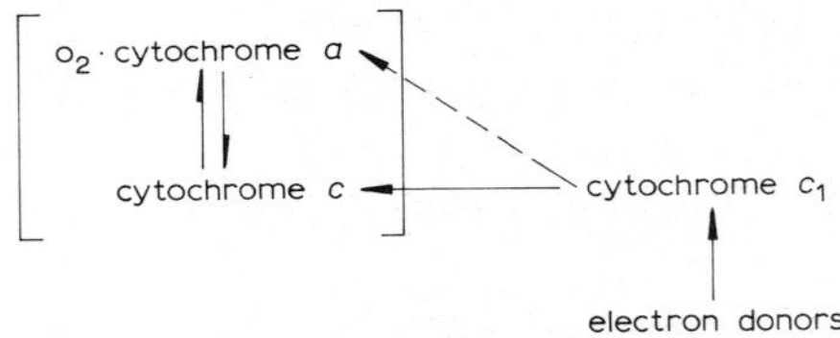

Here the bracket indicates a ternary complex, corresponding to the cytochrome oxidase, while O_2 represents molecular oxygen, though the reaction scheme is not stoichiometric. As to the mode of action of the reconstituted cytochrome oxidase composed of highly purified cytochromes *a* and c_1, it will be described in detail later that the haem of cytochrome *a* functions as an electron acceptor for reduced cytochrome *c* and as an activator of oxygen in cooperation with cytochrome *c*, whereas cytochrome *c* acts not only as an electron donor but also as an essential constituent of cytochrome oxidase forming an active complex with cytochrome *a*." (literature in Okunuki's chapter).

The oxygen-transferring enzyme renamed "cytochrome oxidase" by Dixon, whose photochemical absorption spectrum was mapped by Warburg, has thus been recognized as "significant" in the sense of a complex between cytochromes *a* and *c*. Wainio[50] details the long-lasting controversies that raged before this (still disputed) conclusion was reached. As a later chapter will show the term "cytochrome oxidase" (as "signified") is now identified by a number of authors with the significant consisting of particle IV isolated from mitochondria. This may serve to illustrate the need to keep in mind both aspects of the "sign" (significant and signified) in unravelling the traits of recurrent history (on biosemes and biosemiotics, see Chapter 1 of Volume 29 of this Treatise). It is, of course, at the level of the relation between significant and signified that the interplay of chemistry and biology can best be grasped.

REFERENCES

1 C. A. MacMunn, *J. Physiol.*, 5 (1884) XXIV.
2 C. A. MacMunn, *Phil. Trans.*, 177 (1886) 267.
3 C. A. MacMunn, *J. Physiol.*, 8 (1887) 51.
4 D. Keilin, *Proc. Roy. Soc. (London), Ser. B*, 98 (1925) 312.
5 C. A. MacMunn, *The Spectroscope in Medicine*, London, 1880.
6 D. Keilin, *The History of Cell Respiration and Cytochrome*, Cambridge, 1966.
7 L. Levy, *Z. physiol. Chem.*, 13 (1889) 309.
8 C. A. MacMunn, *Z. physiol. Chem.*, 13 (1889) 497.
9 F. Hoppe-Seyler, *Z. physiol. Chem.*, 14 (1890) 106.
10 C. A. MacMunn, *Z. physiol. Chem.*, 14 (1890) 328.
11 F. Hoppe-Seyler, Footnote added to MacMunn's paper[10].
12 K. A. H. Mörner, *Nord. Med. Ark.*, 30 (1896) 1.
13 O. von Fürth, *Ergeb. Physiol.*, 1 (1902) 110.
14 H. Günther, *Virchow's Arch.*, 230 (1921) 146.
15 H. Fischer, *Strahlentherapie*, 18 (1924) 185.
16 W. D. Halliburton, in E. A. Schäfer (Ed.), *Textbook of Physiology*, Edinburgh and London, 1898.
17 Ch. Dhéré, *J. Physiol. Pathol. Gen.*, 18 (1919) 221.
18 J. Verne, *Les Pigments dans l'Organisme Animal*, Paris, 1926.
19 T. Mann, *Biogr. Mem. of Fellows of the Roy. Soc.*, 10 (1964) 183.
20 D. Keilin, *Proc. Roy. Soc. (London), Ser. B*, 100 (1926) 129.
21 A. Harden and S. S. Zilva, *Biochem. Z.*, 8 (1914) 217.
22 A. Bach, *Fermentforsch.*, 1 (1916) 197.
23 D. Keilin, *Proc. Roy. Soc. (London), Ser. B*, 104 (1929) 206.
24 D. Keilin, *Proc. Roy. Soc. (London), Ser. B*, 106 (1930) 418.
25 M. Dixon, R. Hill and D. Keilin, *Proc. Roy. Soc. (London), Ser. B*, 109 (1931) 29.
26 K. Zeile and F. Reuter, *Z. physiol. Chem.*, 221 (1933) 101.
27 O. Warburg, *Naturwissenschaften*, 15 (1927) 546.
28 P. Nicholls, in P. D. Boyer, H. Lardy and K. Myrbäck (Eds.), *The Enzymes*. 2nd ed., Vol. 8, New York and London, 1963.
29 O. Warburg, *Schwermetalle als Wirkungsgruppen von Fermenten*, 2nd ed., Berlin, 1948.
30 D. Keilin, *Proc. Roy. Soc. (London), Ser. B*, 104 (1929) 206.
31 D. Keilin, *Proc. Roy. Soc. (London), Ser. B*, 106 (1930) 418.
32 O. Warburg, *Biochem. Z.*, 214 (1919) 1.
33 O. Warburg, *Johns Hopkins Hosp. Bull.*, 46 (1930) 341.
34 O. Warburg, *Naturwissenschaften*, 15 (1927) 546.
35 O. Warburg, *Naturwissenschaften*, 16 (1928) 345.
36 M. Dixon, *Biol. Rev.*, 4 (1929) 352.
37 J. Loeb, *The Dynamics of Living Matter*, New York, 1906.
38 O. Warburg, *Z. physiol. Chem.*, 57 (1908) 1.
39 O. Warburg, *Z. physiol. Chem.*, 66 (1910) 305; 328.
40 O. Warburg, *Z. physiol. Chem.*, 69 (1910) 452.
41 O. Warburg, *Z. physiol. Chem.*, 71 (1911) 479.
42 O. Warburg, *Arch. ges. Physiol.*, 145 (1912) 277.
43 O. Warburg, *Arch. ges. Physiol.*, 155 (1914) 547.
44 O. Warburg, *Biochem. Z.*, 152 (1924) 479.
45 O. Warburg, *Ber. d. deutsch. chem. Ges.*, 58 (1925) 1001.

46 E. C. Slater, B. F. van Gelder and K. Minnaert, in T. E. King, H. S. Mason and M. Morrison (Eds.) *Oxidases and Related Redox Systems*, Vol. 2, New York, 1965.
47 D. Keilin, *Nature*, 119 (1927) 670.
48 D. Keilin, *Proc. Roy. Soc. (London) Ser. B*, 106 (1930) 418.
49 D. Keilin and E. F. Hartree, *Nature*, 141 (1938) 870.
50 W. W. Wainio, *The Mammalian Mitochondrial Respiratory Chain*, New York, 1970.

Chapter 30

The Theory of Szent-Györgyi

There are many varieties of biochemists. Some follow the pathways opened by acceptance of paradigms, and add details to the picture. Others go deeper into the chemistry of single metabolic steps. The leaders, nevertheless, are the explorers, discovering new territories of biochemical reality. Szent-Györgyi is one in the last category of researchers, the pioneer variety. He says:

"I make the wildest theories, connecting up the test tube reaction with broadest philosophical ideas, but spend most of my time in the laboratory, playing with living matter, keeping my eyes open, observing and pursuing the smallest detail. The current fashion is to avoid making theories (they may be wrong!) and limit one's observations to reading pointers. I think that an intimate finger-tip friendship with living matter is still important for the biologist. By working in this way, usually something crops up, some small discrepancy, which, if followed up, may lead to basic discoveries. The theories seem to satisfy the mind, prepare it for an "accident", and keep one going. I must admit that most of the new observations I made were based on wrong theories. My theories collapsed, but something was left afterwards." (ref. 1)

The first topic studied by Szent-Györgyi, then at Hamburger's laboratory in Groningen, was the problem of biological oxidations. This took place at the time of violent controversies between Wieland's theory of hydrogen activation and Warburg's theory of oxygen activation. The opposition became so sharp that in 1924 Warburg stated that it was impossible to go on in this way and declared that he would accept no more discussion[2]. In an address to the Physiological Congress in Stockholm in 1926, Hopkins nevertheless expressed the view that the two theories

"are mutually incompatible only when either is expressed in too dogmatic form" (ref. 3)

Szent-Györgyi had at that time shown by using *p*-phenylenediamine or succinate as substrates, that both processes are involved[4]. As he states

References p. 247

Plate 94. Albert Szent-Györgyi.

"I simply knocked out O_2 activation (and with it respiration) by cyanide and then added methylene blue to the minced tissue. The dye restored respiration, replacing O_2 activation. It was reduced by activated H and then reoxidized spontaneously". (ref. 1)

Simultaneously, at Cambridge, Fleisch[5] also showed that the aerobic oxidation of both substrates was inhibited by cyanide, while the oxidation of succinate in the presence of methylene blue was insensitive to cyanide.

While other researchers would deny the importance of oxygen activation, since respiration could take place without it if another electron acceptor was added, Szent-Györgyi[6] agreed with Warburg on the point that, as cyanide completely abolishes cell respiration without influencing hydrogen activation, the oxygen activation should be considered as a fundamental aspect of cell respiration.

Szent-Györgyi, when he recalls his memories of Groningen, likes to recount that, as his wife contributed to the household expenditures by working as a gymnastics teacher, he used to help her in the preparation of the meals. He became interested in the fact that potatoes or apples would turn black when exposed to air while white cabbage did not. This aroused his interest in theories of plant respiration. Palladin (see Ch. 26) had emphasized the importance of hydrogen carriers such as hydroquinone which is oxidized to quinone in the presence of the "oxidase" and again reduced by the cell to hydroquinone. Starting from the theories of Palladin[7,8] and from the researches of Onslow and Robinson[9], Szent-Györgyi proposed a theory of plant oxidation in which a preparation isolated from the potato and called *tyrin* by him is capable of "oxidation" to a red compound *(oxytyrin)* by *orthoquinone* or by the system oxidase catechol. According to him, preparations similar to *tyrin* could be obtained from various mammalian and other tissues, and he concluded that *tyrin* was a member of Palladin's group of respiratory pigments. Platt and Wormall[10] showed that the oxidative properties attributed to *tyrin* by Szent-Györgyi can be explained by the presence of free or combined amino acids. The theory collapsed, but as we shall see in a subsequent chapter, it eventually led Szent-Györgyi to the identification of ascorbic acid with vitamin C, one of the most brilliant of his many brilliant achievements.

Szent-Györgyi recalls his Groningen studies on biological oxidation as follows:

"I also became interested in vegetable respiration, being convinced that there is no basic difference between man and the grass he mows. Plants, at that time, were divided into

References p. 247

two groups: the "catechol oxidase" and "peroxidase" plants. I started with the catechol oxidase plants which contain catechol and a strong catechol oxidase. I simplified the accepted, rather complex ideas about this oxidation system. Then I shifted to "peroxidase" plants which are called so because they contain peroxidase in high concentration. If peroxidase is added to a mixture of peroxidase and benzidine, immediately an intense blue colour appears, due to the oxidation of benzidine. I found that if the reaction was performed with plant juice, instead of purified peroxidase, there was a very short delay, of a second or so, in the benzidine reaction. This fascinated me. There had to be present a reducing agent which reduced the oxidized benzidine, the delay corresponding to the time necessary to oxidize away this unknown reducing agent, later to be known as ascorbic acid."

In 1932, Szent-Györgyi returned to his native country, Hungary, as professor of medical chemistry at the University of Szeged, where he stimulated the formation of a group of active and able disciples and was supported by the Josiah Macy Jr. Foundation. During his studies in Groningen, he had been impressed by the special position of succinic acid and other four-carbon acids as metabolites and the existence in animal tissues, of powerful enzymes catalyzing their conversion. "Nature is not extravagant": this thought led Szent-Györgyi to suspect that these metabolites and enzymes must play an important role in biological oxidations.

At the time, dicarboxylic acids had been the subject of a number of biochemical studies. Battelli and Stern[11,12], as well as Thunberg[13,14], had studied their oxidation in preparations from muscles, and it was known that succinic acid was oxidized to fumaric acid, and that this took up the elements of water to form L-malic acid.

The presence of succinic and fumaric acids in fresh muscle had been demonstrated[15,16]. In 1930, Needham[17] published a paper, now classical, on muscle succinic acid. She had demonstrated in a previous paper[18] that formation of succinic, malic and fumaric acids takes place in minced muscle. She had also previously shown an increased formation of these acids when glutamic and aspartic acids were added[19]. She shows in her paper of 1930

"'that glutamic and aspartic acid when added to minced muscle under anaerobic conditions give rise to an increased amount of succinic, fumaric and malic acids",

and that

"an increased oxygen uptake was found when glutamic or aspartic acid were added to minced muscle." (ref. 17)

She also showed that

"after incubation with muscle in air or oxygen, one half to two thirds of the aspartic acid had disappeared."

... "Washing the muscle with distilled water reduces the oxygen uptake of the control to about one-tenth of the original value, but it reduces still more the excess oxygen uptake in the samples containing amino-acids." (ref. 17)

It was known at that time that an increased oxygen uptake results from the addition of fumaric acid to muscle after various degrees of washing[20–23]. It had been shown that in the presence of muscle, L-malic acid was further oxidized to oxaloacetic acid[24]. An important step was Quastel's demonstration of the inhibitory action of malonate on succinate oxidation in bacteria[25] and in muscle or brain[26], and of the strong inhibitory action of malonate on succinodehydrogenase[27].

Such was the situation when Szent-Györgyi attacked the problem. He writes[28]:

"I began to suspect that something must be wrong about the WK (Warburg–Keilin)–Wieland theory. It might be true, but must be incomplete, and the C_4 dicarboxylic acids must play some very important catalytic role in respiration. So I investigated two things: (*1*) what happens to the respiration if we cut out the oxidation of succinic acid, and (*2*) what happens if we increase the minute quantity of fumarate normally present in the tissue."

In 1934, Göszy and Szent-Györgyi[29] confirmed that the rate of respiration of pigeon breast muscle suspensions* is greatly increased by the addition of small amounts of fumaric acid, while small amounts of malonic acid decrease it. An addition of fumarate to the suspension treated with malonate restores the respiration rate. The authors proposed to consider succinate as a catalytic agent of hydrogen transfer in respiration.

With a series of collaborators, Szent-Györgyi[30] extended these observations. Considering the dehydrogenases known at that time in animal tissues which react with a speed comparable to the rate of respiration of these tissues (succinic, lactic, phosphoglyceraldehyde, glycerophosphate, hexose monophosphate, glutamic acid and citric dehydrogenases), he concludes that succinic dehydrogenase is unique among them in its capacity to reduce the cytochrome–cytochrome oxidase system. When fumarate is added, it prevents the rate of respiration from decreasing with time.

When fumarate or malonate is added to tissues suspended in Ringer, no effect is observed. The stimulation or inhibition effects are developed only if phosphate is present.

---

* The method of preparing these suspensions preserved in a certain measure the integrity of the cells, as the Latapie machine used was deprived of the "Scheibe" which is situated behind the sieve plate and which completes the mincing of the tissue.

References p. 247

When fumarate is added to tissues suspended in a solution containing phosphate, in the presence of malonate, the fumarate is reduced to succinate, if the experiment is accomplished in anaerobic conditions. Oxaloacetate is readily reduced to malate[31].

These experimental data were the basis of the theory of the fumaric catalysis of respiration[30], the basic concepts of which (Fig. 11) were confirmed by Boyland and Boyland[32] and by Greville[33].

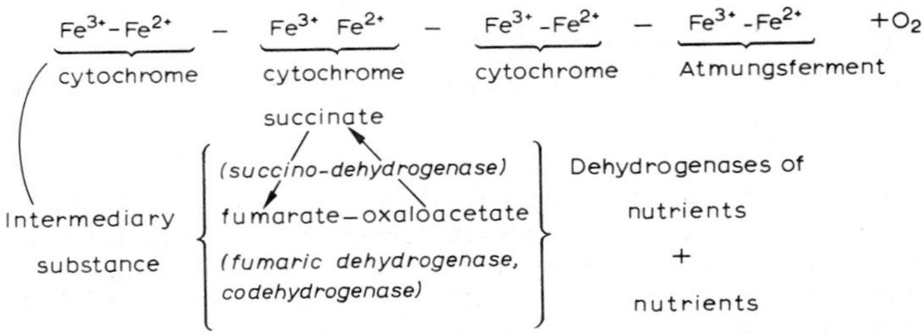

$Fe^{3+}-Fe^{2+}$ — Fe^{3+} Fe^{2+} — $Fe^{3+}-Fe^{2+}$ — $Fe^{3+}-Fe^{2+}$ $+O_2$
cytochrome cytochrome cytochrome Atmungsferment
succinate
(succino-dehydrogenase) Dehydrogenases of
Intermediary fumarate—oxaloacetate nutrients
substance *(fumaric dehydrogenase, codehydrogenase)* +
nutrients

Fig. 11. (Szent-Györgyi, in Annau *et al.*[30]). Respiration cycle as proposed by Szent-Györgyi in 1935.

Szent-Györgyi and his collaborators[34] published in 1936 more data, confirming the "fumaric acid theory". In the same paper they propose a theory of respiration considered as an oxidation of trioses by oxaloacetic acid, a view they place parallel with Meyerhof's views according to which glycolysis is an oxidation of trioses by pyruvate. According to these views, respiration and fermentation are identical processes, with the proviso that in fermentation the hydrogen acceptor is pyruvate, while in respiration the hydrogen acceptor is a carboxy-pyruvate, oxaloacetate.

"Die Natur scheint bei der Atmung das System der Gärung beibehalten zu haben, sie hat nur das als Acceptor dienende Brenztraubensäure-Molekül mit einer Carboxylgruppe beschwert, hierdurch der Resynthese zu Kohlenhydrat entzogen und somit zu einer katalytischen Funktion befähigt." (ref. 34)

At that time it was still believed that the lactic acid produced in muscle in anaerobic conditions was resynthetized to carbohydrate.

The authors also derive from their theory an explanation of the Pasteur effect: pyruvate and oxaloacetate are in competition for activated hydrogen. In anaerobiosis, no oxaloacetic acid is produced, and as pyruvate is the only substance produced, the carbohydrate can only be metabolized by fermentation; while in aerobic conditions, there is production of oxaloacetate, competing with pyruvate for activated hydrogen.

In 1937, Straub[35] observed that the rate of reduction of cytochrome is more rapid after adding succinate, fumarate or malate than after adding lactic acid or α-glycerophosphoric acid.

In pigeon breast muscle, the substance ultimately oxidized is considered as being triose-phosphate by Banga[36].

Laki, Straub and Szent-Györgyi[37], pointing to the fact that in pigeon breast one observes a rapid and reversible reduction of oxaloacetate to malate and a rapid oxidation of succinate to fumarate, suggested associating the two systems in a hydrogen-transporting system. (Fig. 12)

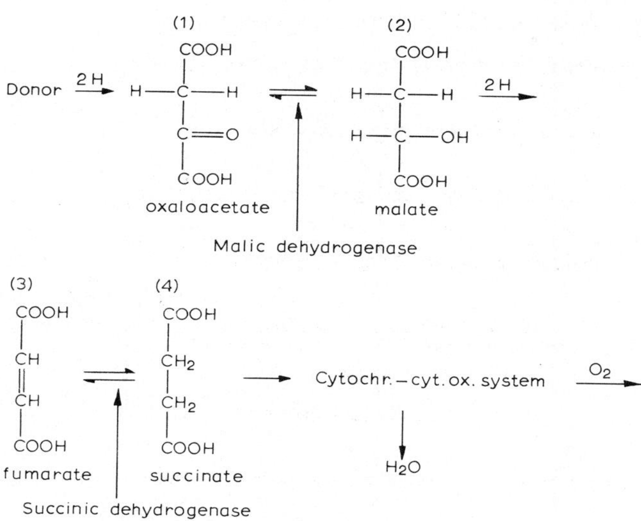

Fig. 12. (Laki, Straub and Szent-Györgyi[37]).

At that time, the oxidation of succinate by the cytochrome–cytochrome oxidase system was supposed to be realized directly without a mediator.

The link between the oxaloacetete malate system and the succinate–fumarate system was not explicated in the scheme. A completed theory was presented by Szent-Györgyi[38] in 1937. He took into account a number of observations. Green[39] had formulated the possibility of establishing a link between the oxaloacetete–malate system and the succinate–fumarate system. He had found that in the presence of NAD, methylene blue and the appropriate dehydrogenase preparation, malic acid could be oxidized by fumaric acid

malate + methylene blue → oxaloacetate + reduced methylene blue

fumarate + reduced methylene blue → succinate + methylene blue

the addition to these two reactions giving

malate + fumarate → oxaloacetate + succinate

Fischer and Eysenbach[40] found that fumaric acid could be reduced by certain dyes in the presence of a flavoprotein from yeast. Banga[41,42] and Laki[43] concluded that a similar system was present in pigeon breast muscle. The unknown connecting link was proposed to consist of the "old yellow enzyme" of Warburg and Christian. Szent-Györgyi's theory[38], proposed in 1937, is represented in Fig. 13.

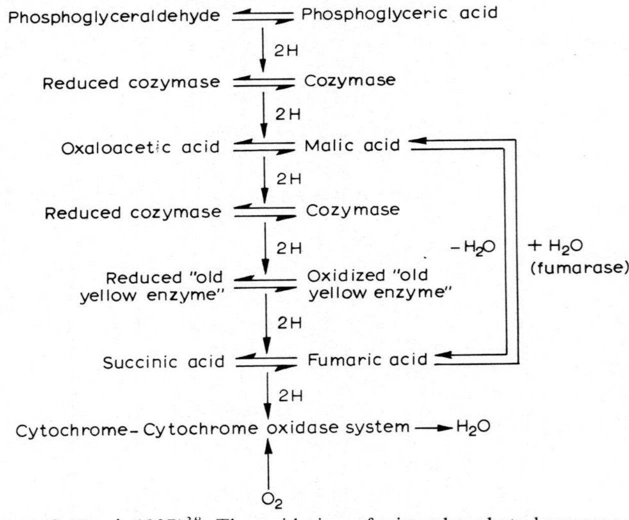

Fig. 13. (Szent–Györgyi, 1937)[38]. The oxidation of triosephosphate by means of the Szent-Györgyi C_4-dicarboxylic acid cycle.

Szent-Györgyi's theory (oxidation of triosephosphate by means of the C_4-dicarboxylic acid cycle) received the support of a number of authors (literature in Sumner and Somers[44], p. 360). On the other hand, many objections were raised against it. Stare and Baumann[45] considered it unlikely that the "old yellow enzyme" could act as a mediator between the oxalo-acetate–malate and the succinate–fumarate systems, though they did not deny that another flavoprotein could play that part.

Keilin and Hartree[46], Stern[47] and Stotz[48] raised objections concerning the direct reaction of succinodehydrogenase with the cytochrome system. Hogness[49] questioned the possibility of a mediation between the "old yellow enzyme" and cytochrome *c* by the fumarate–succinate system.

The great merit of Szent-Györgyi's theory was that it considered the C_4 dicarboxylic acids not as muscle fuels, as was the tendency after the contributions of Battelli and of Thunberg, but as parts of a system of oxidation. By considering triose as the starting-point of his scheme, he clarified the common basis of glycolysis and respiration which had been recognized by Kostychev (see Chapter 26).

He wrote in 1937:

"Now, with this new knowledge, let us write side by side the simplified schemes of both processes, fermentation and oxidation:

$$\text{Triose} \xrightarrow{2\text{ H}} \text{pyruvic acid} \rightarrow \text{lactic acid}$$

$$\text{Triose} \xrightarrow{2\text{ H}} \text{oxaloacetic acid} \rightarrow \text{malic acid} \rightarrow$$

$$\text{fumaric acid} \rightarrow \text{succinic acid} \xrightarrow{2\text{ H}} \text{WK–O}_2$$

"This scheme reveals at once a great similarity of plan in the two processes. In both, two hydrogen atoms are taken off the triose, which is thus oxidized into pyruvic acid. The only difference is that in fermentation these two hydrogen atoms are taken over by pyruvic acid while in oxidation it is oxaloacetic acid which takes them over. In oxidation there is also an annexe to this, the Warburg–Keilin system, which has the function of unloading the hydrogen from the dicarboxylic acids.

"The fundamental unity of both processes is not revealed until we incorporate the real chemical formulae into the scheme:

$$\text{Triose} \xrightarrow{-2\text{H}} \underset{\left[\text{pyruvic acid}\right]}{\begin{matrix}CH_3\\|\\CO\\|\\COOH\end{matrix}} \xrightarrow{+2\text{H}} \underset{\left[\text{lactic acid}\right]}{\begin{matrix}CH_3\\|\\HCOH\\|\\COOH\end{matrix}}$$

$$\text{Triose} \xrightarrow{-2\text{H}} \underset{\left[\text{oxaloacetic acid}\right]}{\begin{matrix}COOH\\|\\CH_3\\|\\CO\\|\\COOH\end{matrix}} \xrightarrow{+2\text{H}} \underset{\left[\text{malic acid}\right]}{\begin{matrix}COOH\\|\\CH_2\\|\\HCOH\\|\\COOH\end{matrix}} \xrightarrow{-H_2O} \text{fumaric acid} \xrightarrow{+2\text{H}} \text{succinic acid} \xrightarrow{-2\text{H}} \text{WK—O}_2$$

References p. 247

"If we compare the formulae of oxaloacetic and pyruvic acid on the one side, malic acid and lactic acid on the other side, we realize the remarkable fact that oxaloacetic acid is but a carboxy-pyruvic acid, and malic acid is but a carboxy-lactic acid, and we also realize what happened millions of years ago, when Nature discovered oxidation; it simply fitted some of the pyruvic acid molecules with carboxyl groups and completed the old system of fermentation with a new one, the WK chain, having no other function than to free the reduced oxaloacetic acid from its two hydrogen atoms, transferring them on to oxygen." (ref. 28)

REFERENCES

1 A. Szent-Györgyi, *Ann. Rev. Biochem.*, 32 (1963) 1.
2 O. Warburg, *Biochem. Z.*, 142 (1924) 518.
3 F. G. Hopkins, *Skand. Arch. Physiol.*, 49 (1926) 33.
4 A. Szent-Györgyi, *Biochem. Z.*, 150 (1924) 195.
5 A. Fleisch, *Biochem. J.*, 18 (1924) 294.
6 A. Szent-Györgyi, *Biochem. Z.*, 157 (1925) 50.
7 W. Palladin, *Ber. deut. bot. Ges.*, 23 (1905) 240.
8 W. Palladin, *Z. Physiol. Chem.*, 47 (1906) 407.
9 M. W. Onslow and M. E. Robinson, *Biochem. J.*, 14 (1920) 1138.
10 B. S. Platt and A. Wormall, *Biochem. J.*, 21 (1927) 26.
11 F. Battelli and L. Stern, *Biochem. Z.*, 30 (1911) 172.
12 F. Battelli and L. Stern, *Biochem. Z.*, 63 (1914) 369.
13 T. Thunberg, *Skand. Arch. Physiol.*, 35 (1918) 163.
14 T. Thunberg, *Skand. Arch. Physiol.*, 43 (1923) 275.
15 H. Einbeck, *Z. physiol. Chem.*, 87 (1913) 145.
16 H. Einbeck, *Z. physiol. Chem.*, 90 (1914) 301.
17 D. Moyle Needham, *Biochem. J.*, 24 (1930) 208.
18 D. Moyle Needham, *Biochem. Z.*, 21 (1927) 739.
19 D. Moyle, *Biochem. J.*, 18 (1924) 351.
20 T. Thunberg, *Skand. Arch. Physiol.*, 22 (1909) 431.
21 T. Thunberg, *Skand. Arch. Physiol.*, 25 (1911) 37.
22 O. Meyerhof, *Arch. Ges. Physiol.*, 175 (1919) 20.
23 H. Grönvall, *Skand. Arch. Physiol.*, 45 (1924) 303.
24 A. Hahn, W. Haarman and E. Fischbach, *Z. Biol.*, 88 (1929) 587.
25 J. H. Quastel and M. D. Whetham, *Biochem. J.*, 18 (1924) 519.
26 J. H. Quastel and A. H. M. Wheatley, *Biochem. J.*, 25 (1931) 117.
27 J. M. Quastel and W. R. Woolridge, *Biochem. J.*, 22 (1928) 689.
28 A. Szent-Györgyi, in J. Needham and D. E. Green (Eds.), *Perspectives in Biochemistry*, London, 1937.
29 B. Göszy and A. Szent-Györgyi, *Z. physiol. Chem.*, 224 (1934) 1.
30 E. Annau, I. Banga, B. Göszy, St. Huszak, K. Laki, B. Straub and A. Szent-Györgyi, *Z. Physiol. Chem.*, 236 (1935) 1.
31 K. Laki, F. B. Straub and A. Szent-Györgyi, *Z. physiol. Chem.*, 247 (1937) 1.
32 E. Boyland and M. E. Boyland, *Biochem. J.*, 30 (1936) 224.
33 G. D. Greville, *Biochem. J.*, 30 (1936) 877.
34 E. Annau, I. Banga, A. Blaszo, B. Bruckner, K. Laki, F. B. Straub and S. Szent-Györgyi, *Z. physiol. Chem.*, 244 (1936) 105.
35 F. B. Straub, *Z. physiol. Chem.*, 249 (1937) 189.
36 I. Banga, *Z. physiol. Chem.*, 249 (1937) 183.
37 K. Laki, F. B. Straub and A. Szent-Györgyi, *Z. physiol. Chem.*, 247 (1937) 1.
38 A. Szent-Györgyi, *Studies on Biological Oxidations*, Leipzig, 1937.
39 D. E. Green, *Biochem. J.*, 30 (1936) 2095.
40 F. G. Fischer and H. Eysenbach, *Ann. Chem.*, 530 (1937) 99.
41 I. Banga, *Z. physiol. Chem.*, 249 (1937) 200.
42 I. Banga, *Z. physiol. Chem.*, 249 (1937) 205.
43 K. Laki, *Z. physiol. Chem.*, 249 (1937) 61.
44 J. B. Sumner and G. F. Somers, *Chemistry and Methods of Enzymes*, 2nd ed., New York, 1947.

45 F. J. Stare and C. A. Baumann, *Cold Spring Harbor Symp.*, 7 (1931) 227.
46 D. Keilin and E. F. Hartree, *Proc. Roy. Soc., (London) Ser. B*, 129 (1940) 277.
47 K. G. Stern, *Cold Spring Harbor Symp.*, 7 (1939) 312.
48 E. Stotz, in *Symposium on Respiratory Enzymes*, University of Wisconsin Press, 1942.
49 T. R. Hogness, in *Symposium on Respiratory Enzymes*, University of Wisconsin Press, 1942.

Chapter 31

The Conversion of Citrate into Succinate

Citric acid was discovered in 1784 by Scheele who isolated it from lemon juice. Citrate was soon recognized as a component of many plants and considered to be a typical vegetable compound. After Soxhlet and Henkel[1] had shown that milk contains citrate and Allen[2] that it contains 1–4 g of this acid per liter, it was believed that its physiological importance in animals was related to lactation.

In 1936, it was known to be a constituent of the urine of man[3–9], horse[10], cow[10], guinea pig[10,6], rabbit[11,6], dog[6,8,12], rat[10] and hog[13,6]. It had also been detected in cerebrospinal fluid[14], amniotic fluid[15,8], aqueous humor[15] and saliva[12], as well as in animal tissues. Data had also been collected, within the limits of the analytical methods of the time, on the concentration of citric acid in these different sites (see for literature, Sherman *et al.*[17]).

Such a general distribution in animal organisms suggested a metabolic significance.

A theory was proposed according to which citric acid and citrate acted as fixed acid-conserving devices during alkalosis[10]. Other authors could not see a relation between citrate excretion and acid-base requirements, and they suggested that citrate must play a role in normal metabolism[7,8]. Considering the large amount of citrate excreted by dogs on a citrate-low diet during prolonged alkalosis, and the absence of stores of preformed citrate in blood, liver, muscle and kidney, Sherman *et al.*[17] came to the conclusion that the dog can synthesize citric acid.

The production of citric acid from sugar by certain microorganisms was discovered by Wehmer in 1893 (ref. 18). He prepared the acid with the use of moulds, particularly *Penicillium*, and this became the basis of an industrial procedure.

The synthesis of citric acid was accomplished by Grimaux and Adam in 1881 (ref. 19).

References p. 253

Plate 95. Carl Martius.

In the course of his experiments on the behaviour of different organic acids when added to muscle homogenates, Thunberg[20] found that citric acid increased the respiration of the homogenate, from which it could be concluded that citric acid was a metabolite, dehydrogenated in the presence of citricodehydrogenase.

From some evidence (not confirmed later) about the oxidation of citric acid by bacteria[21] and by yeast[22], it was admitted that the primary oxidation product of citric acid was β-ketoglutaric acid

$$\text{COOH–CH}_2\text{–C(OH)COOH–CH}_2\text{–COOH–2H} \rightarrow \text{COOH–CH}_2\text{–CO–CH}_2\text{–COOH} + \text{CO}_2$$

A difficulty was raised by the fact that β-ketoglutaric acid does not act as donor in respiration[23].

The rupture with this theory came from important work accomplished by Martius in Knoop's laboratory in Tübingen[24, 25]. Martius showed that citric acid is not catabolized to β-ketoglutaric acid but to α-ketoglutaric acid and then to succinic acid. He showed that between citric acid and α-ketoglutaric acid, a reversible system of tricarboxylic acids was at work.

Martius describes the process as involving two phases. In the first (anaerobic) phase, water is removed from citric acid and the *cis*-aconitic acid which results is again hydrated into isocitric acid.

$$\underset{\text{citric acid}}{\text{HOOC—CH}_2\text{—C(OH)(COOH)—CH}_2\text{—COOH}} \underset{+H_2O}{\overset{-H_2O}{\rightleftharpoons}} \underset{\textit{cis}\text{-aconitic acid}}{\text{HOOC—CH}_2\text{—C(COOH)=CH(COOH)}} \underset{-H_2O}{\overset{+H_2O}{\rightleftharpoons}} \underset{\text{isocitric acid}}{\text{HOOC—CH}_2\text{—CH(COOH)—CH(OH)—COOH}}$$

In the second oxidative phase, the aconitic or the isocitric acid is dehydrated with the formation of oxalosuccinic acid which, probably spontaneously, loses CO_2 producing α-ketoglutaric acid. This, as was known, is dehydrated and decarboxylated, a process leading to succinic acid

$$\underset{\text{isocitric acid}}{\text{HOOC—CH}_2\text{—CH(COOH)—CH(OH)—COOH}} \xrightarrow{-H_2O} \underset{\text{oxalosuccinic acid}}{\text{HOOC—CH}_2\text{—CH(COOH)—CO—COOH}} \longrightarrow \underset{\alpha\text{-ketoglutarate}}{\text{HOOC—CH}_2\text{—CH}_2\text{—CO—COOH}}$$

Martius shows that the first phase of his scheme is reversible. Breusch[26] confirmed the results obtained by Martius and showed that cellular respiration (in muscle as well as in liver) is increased to about the same

References p. 253

degree by citric acid, *cis*-aconitic acid and isocitric acid which can replace each other and that α-ketoglutaric acid also brings about a respiration increase.

In his classical paper, Martius considers the enzymatic aspects of his pathway of citrate catabolism and concludes that the pathway leading from citric acid to α-ketoglutaric acid cannot be catalyzed by a single citricodehydrogenase. At least two enzymes are required, a dehydrogenase (our isocitrate dehydrogenase) and a hydratase (our aconitase).

REFERENCES

1 F. Soxhlet and T. Henkel, *Münch. med. Wschr.*, 35 (1888) 328.
2 L. A. Allen, *J. Dairy Res.*, 3 (1931) 1.
3 S. Amberg and W. B. McClure, *Am. J. Physiol.*, 44 (1917) 453.
4 H. Fasold, *Z. Biol.*, 90 (1930) 192.
5 H. Sullman and E. Schaerer, *Schweiz. med. Wschr.*, 13 (1932) 619.
6 O. Furth, H. Minnibeck and E. Edel, *Biochem. Z.*, 269 (1934) 379.
7 A. C. Kuyper and H. A. Mattil, *J. Biol. Chem.*, 103 (1933) 51.
8 W. M. Boothley and M. Adams, *Am. J. Physiol.*, 107 (1934) 471.
9 C. Schuck, *J. Nutr.*, 7 (1934) 679.
10 O. Ostberg, *Skand. Arch. Physiol.*, 62 (1931) 81.
11 H. Langecker, *Arch. exptl. Pathol. Pharmakol.*, 171 (1933) 744.
12 G. W. Pucher, C. C. Sherman and H. B. Vickery, *J. Biol. Chem.*, 113 (1936) 235.
13 E. B. Woods, *Am. J. Physiol.*, 79 (1927) 321.
14 B. Benni, *Skand. Arch. Physiol.*, 63 (1931) 84.
15 I. I. Nitzescu and I. D. Georgescu, *Compt. Rend.*, 190 (1930) 1325.
16 C. L. Gemmill, *Skand. Arch. Physiol.*, 67 (1934) 201.
17 C. C. Sherman, L. B. Mendel and A. Smith, *J. Biol. Chem.*, 113 (1936) 247.
18 C. Wehmer, *Beiträge zur Kenntnis einheimischer Pilze*, No. 1, Hannover, 1893.
19 E. Grimaux and P. Adam, *Blood*, (2) 36 (1881) 21.
20 T. Thunberg, *Skand. Arch. Physiol.*, 25 (1911) 37.
21 J. Butterworth and T. K. Walker, *Biochem. J.*, 23 (1929) 926.
22 H. Wieland and R. Sonderhof, *Ann. Chem.*, 503 (1933) 61.
23 T. Wagner-Jauregg, *Z. physiol. Chem.*, 242 (1936) 1.
24 C. Martius and F. Knoop, *Z. physiol. Chem.*, 246 (1937) 1.
25 C. Martius, *Z. physiol. Chem.*, 247 (1937) 104.
26 F. L. Breusch, *Z. physiol. Chem.*, 250 (1937) 262.

Chapter 32

The Original "Citric Acid Cycle"

At the time of the formulation of the Szent-Györgyi's theory, a number of researchers had been interested in the sequence of intermediates in the biological oxidation of carbohydrates. In 1920, the leader in the field of the theory of hydrogen activation, Thunberg, proposed a theory which remained current for twenty years, and which obtained the blessing of Knoop[2] and of Wieland[3]. It postulated an oxidation and condensation of two molecules of acetate to succinate, and the oxidation of succinate, *via* fumarate, malate, oxaloacetate and pyruvate to a single molecule of acetic acid, as shown in Fig. 14. It has sometimes been recorded (for

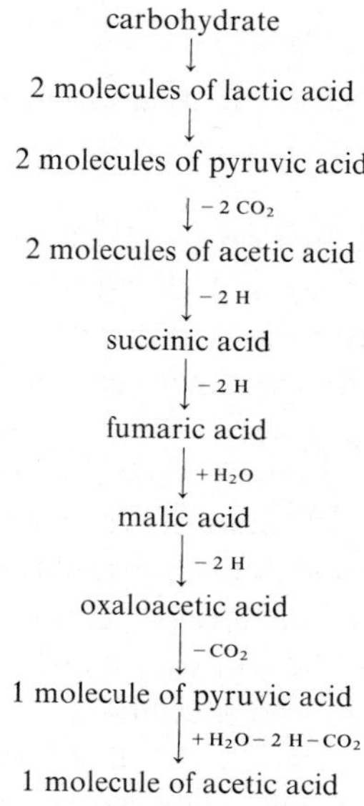

Fig. 14. Thunberg's[1] pathway of carbohydrate metabolism.

References p. 262

instance by Thannhauser[4]) that the dehydrogenation of acetic acid to succinate had been confirmed by several authors[5,6]. What was observed by them is the dehydrogenation of artificially introduced succinic acid into fumaric acid by minced muscle after an addition of methylene blue.

The weakness of Thunberg's theory was the entire lack of experimental evidence for the formation of succinate from acetate.

The metabolic fate of pyruvate was first studied on the whole animal. Toenniessen[7,8], after subcutaneously injecting a rabbit with neutralized pyruvic acid solution, isolated from the urine an organic acid which, though the evidence was scarce, he tentatively identified as succinic acid. Better data were obtained in perfusion experiments on the rabbit's surviving legs, and the fact was established by Toenniessen and Brinkmann[9] with perfusion experiments alone.

These authors found that when pyruvate was added to the perfusion fluid, the perfused muscle gave some succinate, together with formate, while the addition of acetate yielded no succinate. They assumed that a condensation occurs at the pyruvate stage, the "polymerization product" being assumed to be 1,4-diketo-adipic acid:

$$\underset{\text{pyruvic acid}}{CH_3\text{–}CO\text{–}COOH} + \underset{\text{pyruvic acid}}{CH_3\text{–}CO\text{–}COOH} - 2H \rightarrow \underset{\text{1,4-diketo-adipic acid}}{COOH\text{–}CO\text{–}CH_2\text{–}CH_2\text{–}CO\text{–}COOH}$$

$$\underset{\text{1,4-diketo-adipic-acid}}{COOH\text{–}CO\text{–}CH_2\text{–}CH_2\text{–}CO\text{–}COOH} + 2H_2O \rightarrow \underset{\text{succinic acid}}{COOH\text{–}CH_2\text{–}CH_2\text{–}COOH} + \underset{\text{formic acid}}{H\text{–}COOH}$$

The whole scheme of carbohydrate catabolism was considered, as had been proposed by Kostychev, as a follow-up to the anaerobic scheme of fermentation, and the latter was conceived according to Neuberg's scheme (see Chapter 19), in which pyruvate is derived from methylglyoxal (Fig. 15). Wille[10] submitted the proposed pathway to an experimental test. He synthetized 1,4-diketo-adipic acid, added the substance to respiring tissues (muscle, kidney, liver), and found its decomposition rate to be very low compared to the pyruvate catabolism. The substance was therefore discarded as a possible intermediary step, Krebs[11] suggested that the succinate formed in the experiments after the infusion of pyruvate was derived *via* transamination from aspartate and glutamate, and that the formic acid was the result of bacterial activity. At this point it may be emphasized that in the theories of carbohydrate oxidation formulated above, as well as in the theory of Szent-Györgyi, hexoses are not oxidized as such, but,

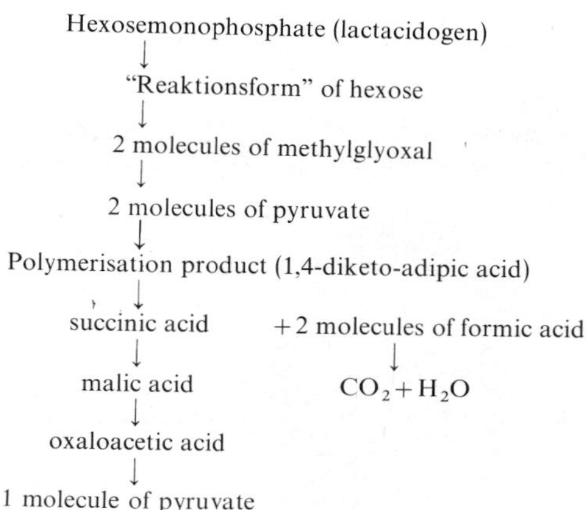

Fig. 15. The pathway of carbohydrate oxidation according to Toenniessen and Brinkmann[9].

according to Kostychev's views, only after the anaerobic process of glycolysis. This view, according to which anaerobic glycolysis precedes the oxidation is confirmed by the observation that the inhibitors of glycolysis (iodoacetate, fluoride) also inhibit the biological oxidations, although they do not inhibit pyruvate oxidation[12–15].

The pioneer concept included in Szent-Györgyi's theory is that the malate, fumarate and oxaloacetate in muscle were not fuels undergoing "combustion" but catalytic hydrogen carriers, "between foodstuff and cytochrome". The way is longer than was realized at the time, but the dicarboxylic acids have kept a place in the completed pathway. The theory could not explain a series of observations reported by Krebs and Johnson[16] in 1937. Citrate exerts on the respiration of pigeon breast muscle a "catalytic effect" of the same order of magnitude as the effect of succinate observed by Szent-Györgyi. Citrate, isocitrate, *cis*-aconitate and α-ketoglutarate are rapidly oxidized by the same preparation, as had already been shown in 1911 by Battelli and Stern[36] working with frog muscle. Citrate is formed in muscle tissue from oxaloacetate[16]. Fumarate and oxaloacetate can be converted into succinate, provided that oxygen is available. There are therefore two pathways to succinate: direct reduction, and an *oxidative* reaction. This last contention was criticized by Barron and Stare[17] as being unsubstantiated. It had nevertheless been quite well documented by Szent-Györgyi[18] as well as in the paper of Krebs and Johnson[16]. Later on it was confirmed by the isotopic method[19,20].

References p. 262

Plate 96. Sir Hans Adolf Krebs.

The discovery, by Krebs and Johnson, of the formation of citrate from oxaloacetate and pyruvate is the decisive, elegant and simple contribution which gave the start to the original citric acid cycle. It was combined by these authors with both a reinterpretation of the cyclic part of Szent-Györgyi's scheme and the pathway of the conversion of citrate to succinate revealed by Martius, into a "citric acid cycle" shown in Fig. 16. This metabolic scheme, which "in a master stroke", created a "new order"

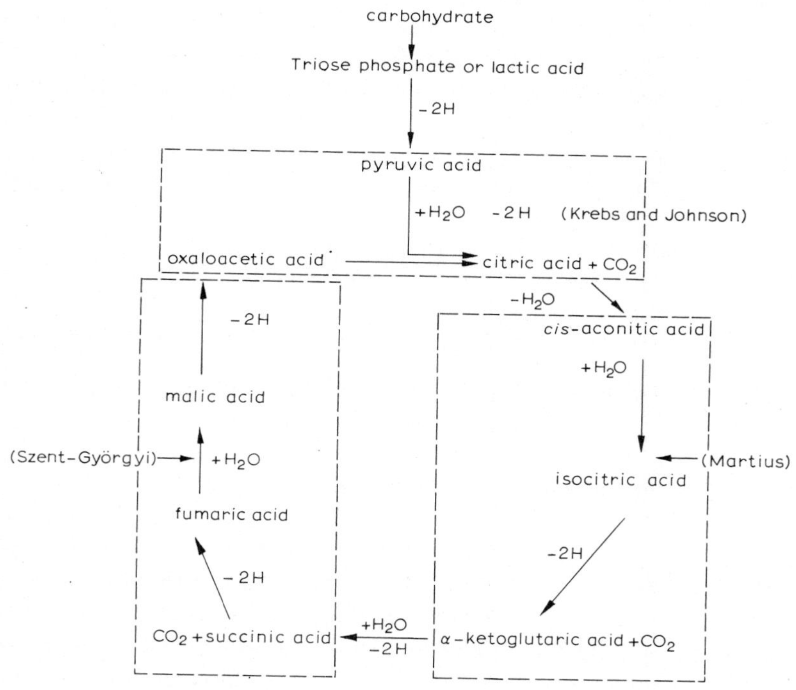

Fig. 16. The original "citric acid cycle" as proposed by Krebs and Johnson[16].

exerted a decisive influence on the development of modern biochemistry, and was the subject of a great deal of discussion. Heart muscle studies[22,23] as well as experiments on citric acid formation in different animal species[24,25] supported it. It was observed by Banga, Ochoa and Peters[27] that an addition of citric acid and α-ketoglutaric acid to brain preparations stimulated pyruvate oxidation, but the authors reported that these acids were less active than C_4-dicarboxylic acids.

References p. 262

While confirming the catalytic effect of citrate, α-ketoglutarate, fumarate, succinate, and malate, Stare and his collaborators[28,29], on the basis that citric acid could not remove malonate inhibition, concluded that the citric acid cycle was not functional, a criticism which was opposed, on kinetic grounds, by Evans[30]. While admitting the reality of the citric acid cycle, several authors held the view that it was only a side reaction without real importance in the picture of biological oxidations. Breusch[31,32] expressed the view that the concentration of oxaloacetate had been unphysiologically high in the experiments, resulting in the formation of citrate. As, in his experiments, the yield of citrate never exceeded 2% of the oxaloacetate, he suggested that the origin of the citrate was to be found in an impurity. Thomas[33] also got very low yields of citrate. Both authors concluded that the catalytic function of the C_4-dicarboxylic acids was fully explained by the theory of Szent-Györgyi according to the reaction

$$\text{fumarate} \underset{+2\,H}{\overset{-2\,H}{\rightleftharpoons}} \text{oxaloacetate}$$

whilst, according to the original citric acid cycle, the oxaloacetate reacts as follows:

$$\text{oxaloacetate} + \text{pyruvate} \rightarrow \text{citrate} \rightarrow \alpha\text{-ketoglutarate} \rightarrow \text{succinate} \rightarrow \text{fumarate}$$

In answering these objections, Krebs[34] stresses that in Szent-Györgyi's theory, the oxidative formation of succinate from oxaloacetate in the presence of malonate remained unexplained, and he defined the conditions under which, although the direct reduction of oxaloacetate was virtually blocked, large amounts of succinic acid were formed from oxaloacetate. To the objection relating to the alleged "unphysiological" concentrations of oxaloacetate, he answers that citrate is also formed when predecessor oxaloacetate arose physiologically from the addition of fumarate. To the contention of Breusch that pigeon breast muscle "has only slight ability to break down citric acid", Krebs refers to the repeated demonstration of this ability by a number of researchers[35,36], Thomas[33] included. Krebs also underlines the misunderstanding of the citric acid cycle by its opponents. Thomas states that if the citric acid cycle is correct, malonate must inhibit the formation of malate from oxaloacetate, forgetting that the reaction

$$\text{succinate} \rightleftharpoons \text{fumarate} \rightleftharpoons \text{malate} \rightleftharpoons \text{oxaloacetate}$$

is reversible.

In Krebs and Johnson's paper of 1937 (ref. 16), the order of formation of the dicarboxylic acids had been rather arbitrarily chosen as follows:

$$\text{oxaloacetate} + \text{pyruvate} \rightarrow \text{citrate} \rightarrow \text{cis-aconitate}$$

However, tracer experiments[37,38] suggested that this order was incorrect and that the results would be better interpreted by a theory according to which the condensation of oxaloacetate with pyruvate or with a pyruvate derivative yields primarily *cis*-aconitate which is directly converted into isocitrate, whereas the formation of citrate is due to a side reaction. This (erroneous) concept, due to the overvalorization of the isotopic method at that time, gained recognition and remained current between 1941 and 1948.

References p. 262

REFERENCES

1 T. Thunberg, *Skand. Arch. Physiol.*, 40 (1920) 1.
2 F. Knoop, *Klin. Wschr.*, 60 (1923) 1.
3 H. Wieland, in C. Oppenheimer (Ed.), *Handbuch der Biochemie*, 2nd ed., Vol. 2, Jena, 1924.
4 S. J. Thannhauser, *Stoffwechsel und Stoffwechsel-Krankheiten*, Munich, 1929.
5 B. Fischer, *Ber. deut. Chem. Ges.*, 60 (1927) 2257.
6 A. Hahn and W. Haarmann, *Z. Biol.*, 87 (1928) 107.
7 E. Toenniessen, *Verh. deut. Ges. inn. Med.*, (1927) 213.
8 E. Toenniessen, *Verh. deut. Ges. inn. Med.*, (1928) 252.
9 E. Toenniessen and E. Brinkmann, *Z. physiol. Chem.*, 187 (1930) 137.
10 F. Wille, *Ann. Chem.*, 538 (1939) 237.
11 H. A. Krebs, *Advan. Enzymol.*, 3 (1943) 191.
12 H. A. Krebs, *Biochem. Z.*, 234 (1931) 278.
13 O. Meyerhof and E. Boyland, *Biochem. Z.*, 237 (1931) 406.
14 E. G. Holmes and C. A. Ashford, *Biochem. J.*, 24 (1930) 1119.
15 S. P. Colowick, M. S. Welch and C. F. Cori, *J. Biol. Chem.*, 133 (1940) 641.
16 H. A. Krebs and W. A. Johnson, *Enzymologia*, 4 (1937) 148.
17 E. S. G. Barron and F. J. Starc, *Ann. Rev. Biochem.*, 10 (1941) 20.
18 A. Szent-Györgyi, *Z. physiol. Chem.*, 235 (1936) 1.
19 H. G. Wood, C. H. Werkman, A. Hemingway and A. O. Nier, *J. Biol. Chem.*, 139 (1941) 377.
20 H. G. Wood, C. H. Werkman, A. Hemingway and A. O. Nier, *J. Biol. Chem.*, 142 (1942) 31.
21 H. M. Kalckar, *Biological Phosphorylations. Development of Concepts*, Englewood Cliffs, N.J., 1969.
22 N. Hallman and P. E. Simola, *Science*, 90 (1939) 594.
23 D. H. Smyth, *Biochem. J.*, 34 (1940) 1046.
24 H. A. Krebs, E. Salvin and W. A. Johnson, *Biochem. J.*, 32 (1938) 113.
25 J. M. Orten and A. H. Smith, *Biochem. J.*, 117 (1937) 555.
26 A. H. Smith and E. M. Curtis, *J. Biol. Chem.*, 131 (1939) 45.
27 I. Banga, S. Ochoa and R. A. Peters, *Biochem. J.*, 33 (1939) 1980.
28 C. A. Baumann and F. J. Stare, *J. Biol. Chem.*, 133 (1940) 183.
29 F. J. Stare, M. A. Lipton and J. M. Goldfinger, *J. Biol. Chem.*, 141 (1941) 981.
30 E. A. Evans, Jr., *Bull. Johns Hopkins Hosp.*, 69 (1941) 225.
31 F. L. Breusch, *Z. Physiol. Chem.*, 250 (1937) 262.
32 F. L. Breusch, *Biochem. J.*, 33 (1939) 1757.
33 Q. Thomas, *Enzymology*, 7 (1937) 231.
34 H. A. Krebs, *Biochem. J.*, 34 (1940) 460.
35 T. Thunberg, *Skand. Arch. Physiol.*, 24 (1910) 23.
36 F. Battelli and L. Stern, *Biochem. Z.*, 31 (1911) 478.
37 H. G. Wood, C. H. Werkman, A. Hemingway and A. O. Nier, *J. Biol. Chem.*, 139 (1941) 483.
38 E. A. Evans, Jr. and L. Slotin, *J. Biol. Chem.*, 141 (1941) 439.

Chapter 33

The Oxidative Decarboxylation of Pyruvate and the Biosynthesis of Citrate

1. Rehabilitation of citrate

The roadblock which confined citric acid to a lateral pathway, and which persisted for a decade, was due to the results obtained, as we said, by the isotopic method. In their book on biological oxidation, published in 1939, Oppenheimer and Stern[1] still present the reader with the Wieland–Thunberg scheme as a paradigm.

The citric acid cycle, as the authors say

"constitutes a modification and extension of Szent-Györgyi's C_4-acid theory."

To reject this new development, Oppenheimer and Stern base their argument on a study by Breusch[2] in Szent-Györgyi's laboratory. Breusch concludes that in physiological conditions, citric acid is not formed from oxaloacetic acid and furthermore he concludes that there is no proof that the C_6-tricarboxylic acid system fulfils a function in hydrogen transport comparable to that performed by the C_4-dicarboxylic acid system.

The rehabilitation of citrate in the metabolic pathway of biological oxidations was the result of two new developments which accomplished the decisive rupture with the Wieland–Thunberg system.

The first of these developments came in 1948 from a paper of Ogston[3] who pointed out a fallacy in the interpretation of the isotopic experiments of Wood, Werkman, Hemingway and Nier[4] and of Evans and Slotin[5]. In his Harvey lecture delivered in March 1949[6], as well as in the chapter he wrote in the first edition of Greenberg's *Metabolic Pathways*[7], Krebs developed a number of considerations necessary to appreciate Ogston's argument. The difficulty in explaining the asymmetrical distribution of the isotopic carbon was removed by the concept that even if citrate itself

References p. 281

Plate 97. Sir Rudolph Albert Peters.

is a symmetrical molecule, it does not necessarily react symmetrically. The direct proof that citrate reacts asymmetrically in the presence of enzymes was afforded by several isotopic experiments during the following year[8–11]. But the removal of the roadblock did not point to the true way of pyruvate oxidation.

2. The connection between pyruvate oxidation and thiamine

This development has its source in the work of Sir Rudolph Peters and his group on the biochemical lesion involved in an avitaminosis, vitamin B_1 deficiency. In his work on the isolation of thiamine from yeast, Peters used pigeons as test animals. It has long been known in nutrition studies that the withdrawal of vitamin B_1 (thiamine) induced the condition of "opisthotonus" as manifested by head retraction and convulsions of a special type. Using birds in a definite stage of thiamine deficiency, in which the pigeons are cured by thiamine only, Peters could follow the vitamin in the different stages of isolation and finally obtained it, from yeast*, in the crystalline state[12,13]. It was the purpose of determining the biochemical lesion presented by the pigeons in the acute signs of thiamine deficiency which led Peters to focus his attention on the central nervous system and on possible enzyme changes[12]. This was a rupture with current views which put the emphasis on the peripheral nervous system, as shown by the wide use of the word "polyneuritic" in the description of the avitaminosis state. Kinnersley and Peters[14], using a mince of pigeon brain, always found more lactate in the avitaminous tissues. That the alteration was not at the level of glycolysis from glucose to lactate, but at the level of the removal of lactate was shown by Fisher[15] and by Hayasaka[16].

Gavrilescu, Meiklejohn, Passmore and Peters[17] found that a specific lowering of oxygen uptake could be observed in the mince of the brain of pigeons suffering from thiamine deficiency. If the pigeons were suffering from acute opisthotonus, a concentrate of thiamine was able to restore the normal respiration, as shown by Gavrilescu and Peters[18]. That *in vitro*, it was pyruvate which accumulated and which disappeared in the presence of the vitamin, was demonstrated by Peters and Sinclair[19] and by Peters and Thompson[20]. Thompson and Johnson[21] found that when, in the avitaminous pigeons pyruvate accumulates in tissues, some leaks into the blood where its concentration increases. This happens also in patients

* See *Note added in proof*, p. 280.

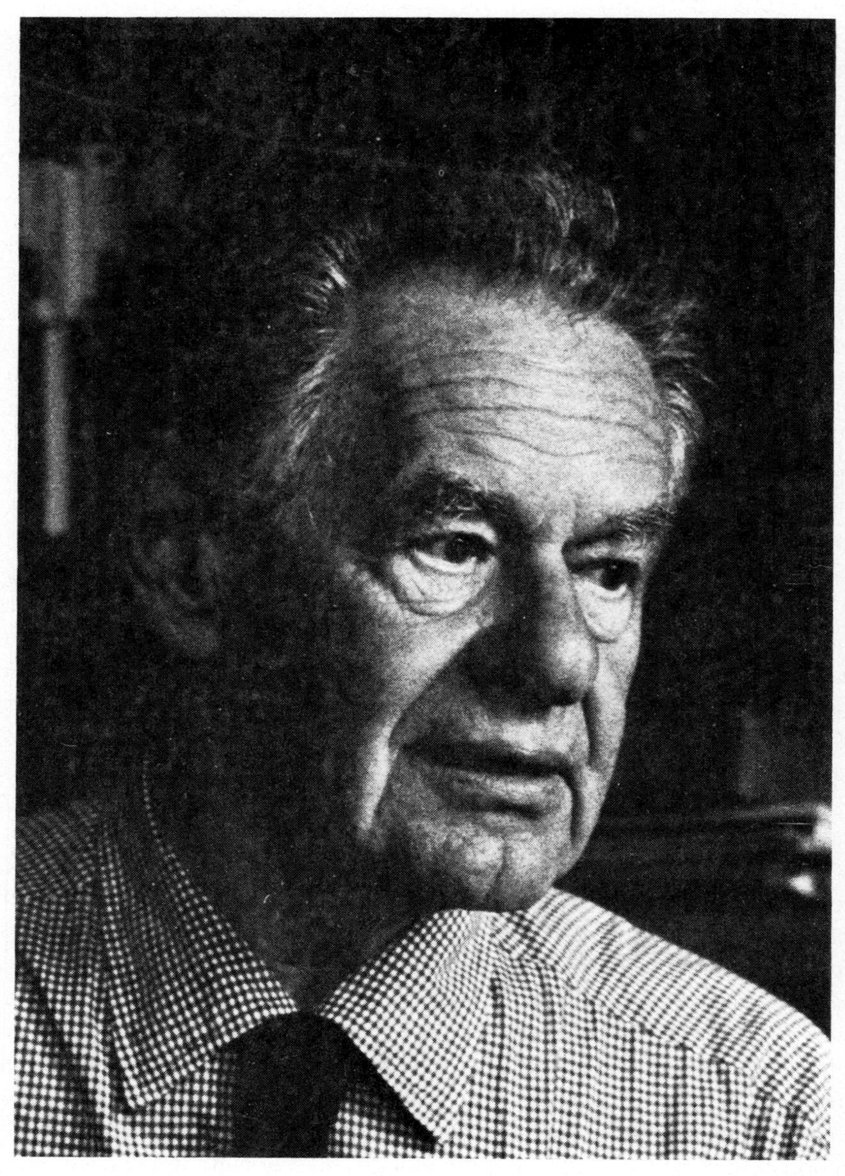

Plate 98. Fritz Albert Lipmann.

suffering from beri-beri (Platt and Lu[22,23]). It must be stressed that the experiments of Thompson and Johnson provided the demonstration that in the whole organism, pyruvate is a normal metabolite.

In the meantime, in the field of alcoholic fermentation studies, cocarboxylase was shown by Lohmann and Schuster[24], to be thiamine pyrophosphate

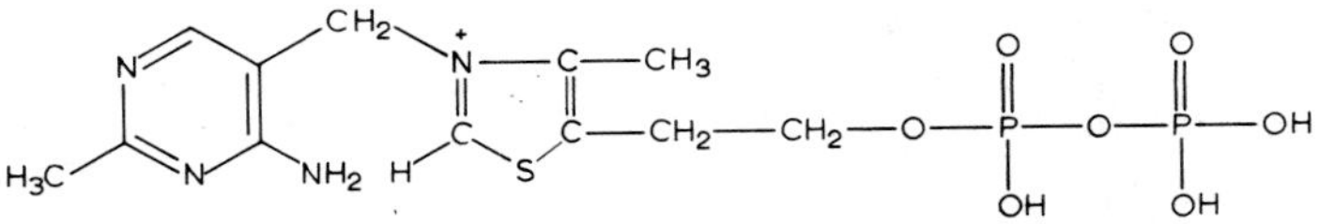

That thiamine pyrophosphate was also the active form in the brain respiration was demonstrated by Banga, Ochoa and Peters[25] when they showed that a dispersion of brain tissue of an avitaminous pigeon would only oxidize pyruvate when thiamine pyrophosphate was present, and not in the presence of thiamine only. So the biochemical lesion was traced to the lack of thiamine pyrophosphate.

3. Acetylphosphate

Lipmann[26] had gone with Meyerhof to Heidelberg but for personal sentimental reasons, he wished to go back to Berlin. Warburg had become interested in the technique of tissue culture, and a laboratory unit had been given in Dahlem to Albert Fischer, a Dane who had been initiated in the technique by Carrel at the Rockefeller Institute. Fischer wanted a collaborator knowledgeable in tissue metabolism, in order to study the metabolism of growing cultures and to use metabolism as a measure of tissue proliferation. There, Lipmann, who had lived on fellowships until that time, got his first job as an assistant in Fischer's laboratory. After a year, Fischer started moving to Copenhagen where a laboratory was built for him, and while waiting for the completion of the Institute, Lipmann took advantage of a Rockefeller Foundation fellowship to pay a visit to the U.S.A. and work in Levene's department at the Rockefeller Institute, where he arrived in 1931. There he started on his first self-chosen project. During his work on tissue cultures, Lipmann had become familiar with the presence of phosphoproteins in the food for growing tissue cultures (egg yolk and milk). On the other hand, Lipmann conceived the idea that phosphate could be bound in protein in an energy-rich N$\sim$P linkage similar to that in creatine phosphate. From egg yolk, Levene and

References p. 281

Alsberg[27] had isolated a vitellinic acid (now named phosvitin) containing 10–12% phosphate. It must be noted that Lipmann, at the time, was not aware of the work of Pasternak who had provided evidence for the attachment of phosphate to serine in phosphoproteins. With Levene, Lipmann published a paper describing the isolation of serine phosphate from vitellinic acid[28]. Lipmann soon realized that the characteristics of protein phosphate did not suggest an N ~ P link.

In the autumn of 1932, Lipmann went back to Copenhagen where he stayed in Fischer's laboratory, at the Carlsberg Institute, until 1939 (ref.26). His interest in the metabolic patterns of fibroblasts, which display a strong anaerobic glycolysis paired with a high respiration, led him to study the Pasteur effect, *i.e.* the repression of glycolysis by respiration. This work focused his interest on the crossroad in pyruvate, reduced to lactate in muscle glycolysis and oxidized to acetate and CO_2 in respiration. As shown by the work of Peters retraced above, pyruvate piled up in the thiamine deficiency.

To quote Lipmann[26]:

"At that time I was quite unhappy that, in contrast to the fermentation, the respiration system was refractive to solubilization; its mitochondrial nature was not then known. One of the lessons I had learned in Meyerhof's laboratory was that if mammalian systems show difficulties, one should turn to microorganisms. This made me look for pyruvic acid oxidation in bacteria. A pyruvate oxidation had been reported with acetone powder suspensions of a *Lactobacillus delbrueckii acidificans longissimus*, which I obtained from Germany. This preparation seemed promising and I began, hopefully expecting it to yield a soluble enzyme. The pyruvic acid oxidation system, which I soon learned to extract from dry preparations of *Lactobacillus delbrueckii*, was first used to verify further the connection between pyruvate oxidation and thiamine."

The coenzyme of pyruvate decarboxylation in alcoholic fermentation had been identified by Lohmann and Schuster[24] as thiamine pyrophosphate. In this work, an extraction with slightly alkaline phosphate buffers had been applied.

Lipmann[29] observed that the pyruvate oxidase of *L. delbrueckii* could be inactivated by the same procedure, and the oxidation reactivated by thiamine pyrophosphate, which pointed to the fact that thiamine pyrophosphate, the coenzyme of pyruvate decarboxylation, was also a coenzyme of pyruvate oxidation. When acid ammonium sulphate precipitation was applied to the pyruvate oxidase of *L. delbrueckii* (a technique used by Warburg and Christian in their fractionation of D-amino acid oxidase) the enzyme was inactivated and reactivation took place by the addition of thiamine pyrophosphate together with flavin–adenine dinucleotide (which had been identified by Warburg and Christian[30] as a coenzyme).

That a chemical correlation existed between pyruvate oxidation and acetate activation was suggested by an experiment accomplished by Lipmann in the last period of his Copenhagen episode.

"At that time a chemical connection of pyruvate oxidation with acetate activation was already suggested by the accidental finding of a role for inorganic phosphate in pyruvate oxidation with the Lactobacillus enzyme. This happened when, in order to employ the Warburg technique for following methylene blue reduction manometrically, I had to replace otherwise routinely used phosphate with a bicarbonate buffer. To my surprise the system became virtually inactive in bicarbonate, although methylene blue was easily reduced in phosphate. It was natural to try to put the phosphate back, which proved it to be the missing factor. This was interpreted as indicating that acetyl phosphate was likely to be an intermediary." (ref. 26)

In the paper[31], in which he described these experiments, Lipmann concluded that the dehydrogenation of pyruvic acid proceeds according to the following formulation

pyruvate + phosphate-2 H → *acetylphosphate* + CO_2
acetylphosphate + H_2O → acetic acid + phosphate
(or adenylic acid) (or adenosinepolyphosphate)

"In investigations of pyruvic acid metabolism it has long been puzzling that, although the oxidative breakdown of pyruvic acid has been found to go through a stage corresponding to acetic acid, added acetic acid is completely inactive in systems oxidizing pyruvic acid. The intermediate formation of acetylphosphate provides a reasonable explanation for the "active acetate", and for various acetylation processes connected with the active breakdown of pyruvic acid and carbohydrate, for example, acetylation of amino acids and of choline." (ref. 31)

After Lipmann had established that in *Lactobacillus delbrueckii* the oxidation product of pyruvate was acetylphosphate[32], the same demonstration was made for *E. coli* by Utter and Werkman[33] and for *Clostridium butylicum* by Koepsell and Johnson[34]

$CH_2COCOOH + HOPO_3^{=}$ → $CH_3COO{\sim}PO_3^{=} + H_2O_2 + CO_2$
→ $CH_2COO{\sim}PO_3^{=} + H_2 + CO_2$ *Lactobacillus delbrueckii*
→ $CH_2COO{\sim}PO_3^{=} + HCOOH$ *(Clostridium butylicum)* *Escherichia coli*

Even if these studies may be concerned with specific cases and be more interesting from the viewpoint of comparative biochemistry than from that of general biochemistry, and even if it turned out that acetylphosphate is not the intermediary in animals, the appearance of acetylphosphate, the mixed anhydride of acetic acid and phosphoric acid as a product of pyruvate oxidation was of great heuristic value in the unravelling of the general pathway of pyruvate oxidation. The acetylphosphate was, as Lipmann emphasized, a donor of phosphoryl as well as a donor of acetyl

"Like the head of Janus, this molecule appeared to be able to face in two directions:

$$\begin{array}{c} \text{Acetyl} \rightarrow \text{acetyl} \sim \\ \wr \\ \sim\text{Phosphoryl} \leftarrow \text{phosphate} \end{array}$$
" (ref. 34 a)

To quote Lipmann[26]:

"The appearance of the mixed anhydride of acetic acid and phosphoric acid as a product of pyruvate oxidation became a prime mover in my thinking and in the planning of further work. It furnished the first example of the functioning of an energy-rich phosphate bond in group activation. So far such bonds had been found to function in energy transformations to carry $\sim$P to ADP and fill the ATP pool. On the receiving side only mechanical energy production had appeared as a utilizing device for phosphoryl in ATP, but I had visualized earlier that $\sim$P was to be used in biosynthesis. Here, then, was the first representative of a group of phosphoryl acceptors whose phosphoryl potential was used in group activation".

The note in *Nature* in which the role of acetylphosphate as intermediary is suggested, is dated July 21, 1939 from the Carlsberg Foundation in Copenhagen, and the Department of Biochemistry, Cornell University Medical College, New York City. At the time, Lipmann had left Denmark for the United States.

4. A trail-blazing review

When the Nazi pressure made itself felt in Denmark, the Lipmanns left early in July, 1939, for the United States where Burk, who had been a colleague of Lipmann in Meyerhof's laboratory took him into a cancer-oriented group temporarily located in Du Vigneaud's laboratory at Cornell during the completion of the building of a Cancer Institute in Washington. During the two years he stayed in this team, Lipmann contributed to the refutation of the presence of D-amino acids in cancers and to the demonstration of D-amino acids in a number of antibiotics, but his thoughts had not been taken away from acetylphosphate.

When F. F. Nord prepared the first volume of the *Advances in Enzymology*, he asked Lipmann, whom he had known in Germany, to prepare a review on the Pasteur effect. But Lipmann, whose experiments with acetylphosphate had led him to develop thoughts on the role of phosphate bonds as means to parcel and transfer energy in energy transformations and in biosynthesis, suggested writing on that subject rather than on the Pasteur effect.

"My paper caused a stir when it appeared, although at first it was given what one might call a mixed reception. The talk of energy-rich bonds aroused forceful, sometimes virulent antagonism. Unwittingly, I had stepped into a hornet's nest by using bond energy to express the potential energy derivable from a bond, brushing aside its accepted use for energy expended to form a bond. I was interested in the amount of free energy that could be derived from a bond and was groping for a definition of its capacity to carry potential chemical energy that could be used for synthesis. To amplify this I proposed alternatively the use of group potential, best expressed, probably, by the absolute value of the ΔF of hydrolysis." (ref. 26)

If the squiggle ($\sim$) introduced by Lipmann to denote energy-carrying bond is occasionally adopted by organic chemists, the physical chemists have remained antagonists. To quote Lipmann[26]:

"The physical chemist remains aloof. He may be forced to accept the usage, but he usually refrains from referring to the dilettante who originated it. Worldwise Du Vigneaud, from whom I learned a great deal, once said to me: You always have to see too that your union card is in good order."

Lipmann's review in the first volume of *Advances in Enzymology*[35] is one of the major landmarks of the development of bioenergetics. It contains the classical statement of the ATP–ADP cycle as well as the basis for future developments in the biochemistry of biosynthesis.

5. Coenzyme for acetylation (CoA)

After two years spent with the Burk's group, Lipmann who was not inclined to take up cancer research as a full-time occupation, won a CIBA fellowship to work in the Department of Surgery of the Massachusetts General Hospital and returned to his work on acetylphosphate[26].

In his classical paper on high energy bonds[35], Lipmann reports on the synthesis of acetylphosphate and its production as a compound of the energy-rich series during pyruvate oxidation (oxidative decarboxylation) in cell-free preparations of bacteria, in the presence of inorganic phosphate.

$$\underset{\text{pyruvate}}{CH_3COCOO^-} + (HO)_2PO_2^- + O_2 \rightarrow \underset{\text{acetylphosphate}}{CH_3COOPO_3^{2-}} + CO_2 + H_2O$$

References p. 281

Lipmann also showed that acetylphosphate, in the bacterial preparations, could transfer its phosphoryl group to ADP. He suspected that he had found the general definition of "active acetate". Unfortunately the animal experiments did not confirm these hopes and tests with added acetylphosphate were not encouraging. Lipmann continued to search for a general definition of "active acetate" and he turned to the enzymatic acetylation of sulphanilamide in liver. This choice was justified by earlier experiments of Klein and Harris[36] showing with rabbit liver slices that this acetylation was coupled with respiration.

"This coupling indicated a chain of reactions transferring oxidation-derived energy to the energy acceptor, the acetylation system." (ref. 37)

Rabbit liver homogenates did not prove very able to carry the acetylation reaction, but examining different species, Lipmann found that extracts of pigeon liver catalyzed a very strong reaction[38]. Lipmann recognized that a coenzyme was involved in the sulphonamide acetylation in pigeon liver and prepared partially purified samples, which he sent to several laboratories for vitamin tests.

This decisive initiative rested on a general concept, the notion of the metabolic role of vitamins after they are built into coenzymes. The induction was that the coenzyme for acetylation was probably a derivative of one of the B vitamins, the functions of which was still undefined[26].

While other vitamins were only present in traces, it was found in the laboratory of Roger Williams, by Miss B. Guirard, that the samples, after hydrolysis, had a high concentration of β-alanine, which suggested the possible presence of pantothenic acid. This was confirmed, and the name of "coenzyme for acetylation" (CoA) was coined[39]. In his Harvey Lecture, delivered in December, 1948, Lipmann[34a] records the events as follows:

"On dialysis as well as on ageing the enzyme solution lost the ability to acetylate, which was regained on addition of boiled extracts. None of the known coenzymes could replace this factor and the isolation of this apparently new coenzyme was therefore attempted. On purification it appeared that the new coenzyme, coenzyme A was a pantothenic acid derivative. The same coenzyme was found to activate the acetylation of choline[40]. Concurrently, Nachmansohn and Berman[41] and Feldberg and Mann[42] observed the need for an activator in choline acetylation. We find the activators of Nachmansohn and of Feldberg and Mann to be identical with coenzyme A."

Coenzyme A is also involved in the formation of acetoacetate from acetate[43], of citrate from oxaloacetate and acetate[44,45] and therefore became considered as the universal enzyme of acetylation, and the concept of "active acetate" became identified with the chemical reality of acetyl-CoA[46,47] (the history of this development is retraced by Goldman and Vagelos in Vol. 15, Chapter 3, of this Treatise).

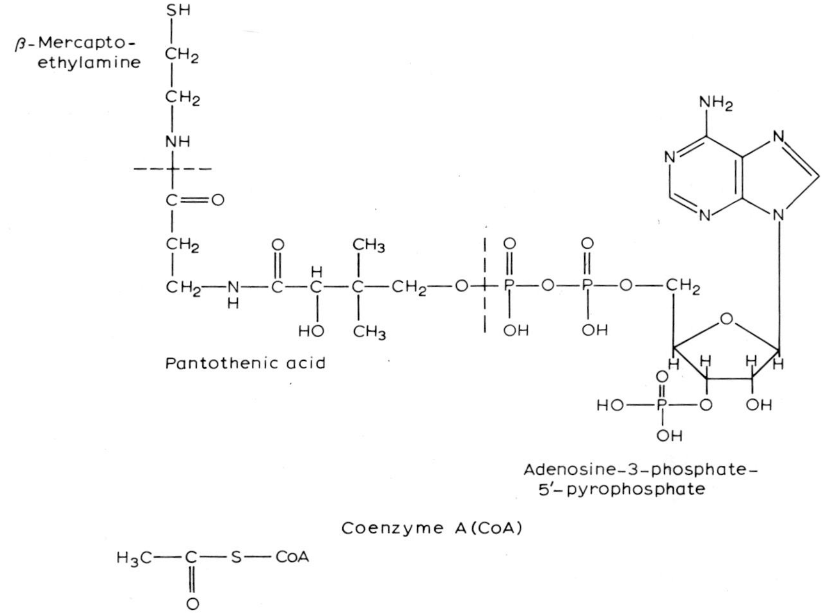

β-Mercapto-ethylamine

Pantothenic acid

Adenosine-3-phosphate-5'-pyrophosphate

Coenzyme A (CoA)

$H_3C-C(=O)-S-CoA$

Fig. 17. Coenzyme A and Acetyl-CoA.

The concept of active acetate[48–52] had evolved from the fact that if citrate synthesis can be considered from the chemical viewpoint as a condensation of acetic acid and oxaloacetic acid, no biochemical argument could be recognized for the participation of acetic acid as such in the condensation reaction, and there was evidence that "active acetate" can be obtained from sugars through pyruvate (as well as from fatty acids, and from acetate itself).

Coenzyme A, which was recognized[39] as containing pantothenic acid in combination with adenylic acid, was found to be widely distributed in organisms[53] and to account for almost all the pantothenate present[54]. Snell *et al.*[55] recognized mercaptoethanolamine as a component of coenzyme A which led to the brilliant work of Lynen[56], who demonstrated that the functional group of CoA was the third group of mercaptoethanolamine (Fig. 17). The participation of pantothenate in carbohydrate metabolism had already been suspected by Teague and Williams[57] in 1942, a view which was made more interesting by the demonstration of the participation, in *Proteus inorgani*, of pantothenic acid in pyruvate oxidation[58,59].

6. Citric acid biosynthesis

Lipmann, believing from its universal distribution that coenzyme A has an importance in metabolic pathways besides acetylation, thought that while citric acid biosynthesis is not a true acetylation, perhaps acetate activation might occur in both cases by the same mechanism.

He was led by the work of Hills[59] on *Proteus inorgani*

"to seek as common factor the acetyl activation, to carry pyruvate into the citric acid cycle". (ref. 34a)

Novelli and Lipmann[60] showed in *Proteus inorgani* the correlation between the content of coenzyme A and the oxidation of pyruvate, an observation which was extended by Olson and Kaplan[61] who worked with liver slices and showed a remarkable parallelism between coenzyme A level and rate of pyruvate oxidation. Lipmann says:

"We therefore turned our attention to an organism with an outspoken acetate metabolism, choosing a strain of yeast that metabolizes acetate rather rapidly[62]. Earlier, the use of isotopic acetate had shown that in yeast the acetate is metabolized through the citric acid cycle[63]. Yeast thus appeared to be a very suitable organism for a further testing of the proposition that coenzyme A is involved in citric condensation." (ref. 34a)

Experiments by Novelli and Lipmann[62] on pantothenic deficient yeast gave support to the concept of the involvement of coenzyme A in citrate synthesis.

On the other hand Kaplan and Lipmann[64] observed that an "active acetate" can be generated from acetate and ATP in liver extracts.

These ideas of Lipmann were confirmed when Stern and Ochoa found that

"ammonium sulfate fractions from extracts of acetone-dried pigeon liver, prepared and aged as described by Kaplan and Lipmann[64] readily forms citrate from acetate (or acetoacetate) and oxaloacetate in the presence of ATP, CoA and Mg^{++}". (ref. 65)

Similar results were obtained with preparations from yeast and from

Escherichia coli by Novelli and Lipmann[45] and with kidney acetone powder extracts by Elliott and Kalnitsky[66]. Acetoacetate[65] and other β-keto acids[67] were also found to serve as acetyl donors in pigeon liver preparation.

The synthesis of citrate proceeds through the steps

$$\text{acetate} + \text{ATP} \rightarrow \text{active acetate}$$

$$\text{active acetate} + \text{oxaloacetate} \rightarrow \text{citrate}^{67}$$

The first acetate-activating enzyme could be extracted from only a few sources, including pigeon liver, *E. coli* and lyophilized yeast[67,68]. The second could be extracted from many sources. It is the condensing enzyme which was crystallized from pig heart by Ochoa, Stern and Schneider[69], which gave the final proof that citric acid is the product of the condensation catalyzed by this enzyme, the first enzyme to be crystallized among those taking part in what became known as the tricarboxylic acid cycle. On the other hand, the acetate-activating enzyme was shown by Stern, Shapiro, Stadtman and Ochoa[70] to correspond with transacetylase discovered by Stadtman[71] in *Clostridium kluyveri.* Stern, Shapiro, Stadtman and Ochoa[70] drew the scheme of citric acid synthesis as follows:

$$\text{(1) acetylphosphate} + \text{CoA} \rightleftharpoons \text{acetyl-CoA} + \text{phosphate (transacetylase)}$$

$$\text{(2) acetyl-CoA} + \text{oxaloacetete} \rightleftharpoons \text{citrate} + \text{CoA(condensing enzyme)}$$

$$\text{(3) Sum: acetyl phosphate} + \text{oxaloacetate} \overset{\text{CoA}}{\rightleftharpoons} \text{citrate} + \text{phosphate}$$

In the same paper, the authors present proofs of the reversibility of these reactions.

The final demonstration that the condensing enzyme catalyzes reaction (2) was the use of Lynen's acetyl-CoA with crystalline condensing enzyme. This was described in a joint paper by Stern, Ochoa and Lynen[71a].

But how does the oxidation of pyruvate generate acetyl groups for citrate synthesis?

Korkes, Stern, Gunsalus and Ochoa reported in 1950, in a preliminary note[72] that soluble enzyme preparations from *Escherichia coli* and *Streptococcus faecalis* containing lactic dehydrogenase catalyse the conversion of two molecules of pyruvate to acetylphosphate, carbon dioxide and lactate in the presence of orthophosphate and NAD. They formulate the reactions as follows:

$$\text{Pyruvate} + \text{phosphate} + \text{NAD} \rightarrow \text{acetylphosphate} + CO_2 + \text{NADH} \quad (1)$$

$$\text{pyruvate} + \text{NADH} \rightarrow \text{lactate} + \text{NAD (lactic dehydrogenase)} \quad (2)$$

$$\text{Sum: 2 pyruvate} + \text{phosphate} \rightarrow \text{acetylphosphate} + CO_2 + \text{lactate} \quad (3)$$

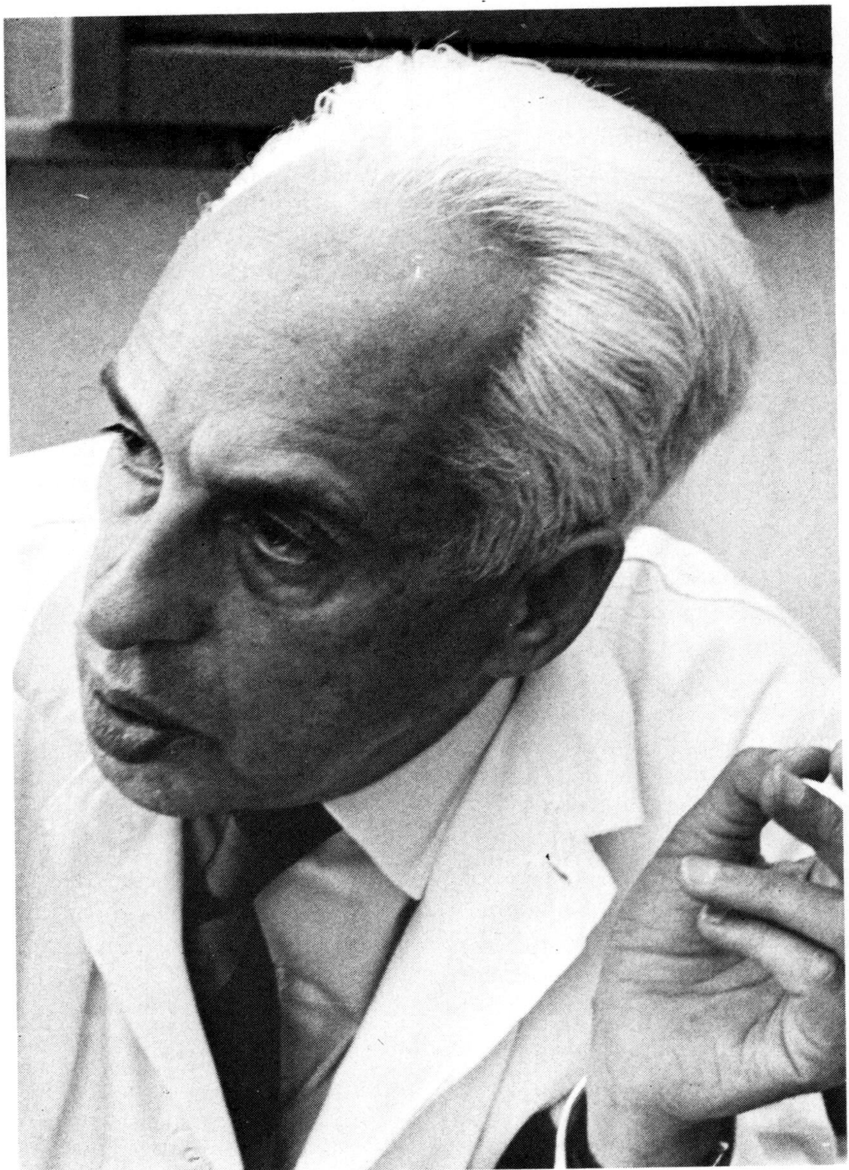

Plate 99. Severo Ochoa.

In the absence of phosphate, no acetylphosphate is formed. If oxaloacetate and condensing enzyme are added, citrate is formed instead of acetylphosphate.

In these conditions the reaction becomes

$$2 \text{ pyruvate} + \text{oxaloacetate} \rightarrow \text{citrate} + CO_2 + \text{lactate} \quad (4)$$

Purifying, at least partly, the pyruvate oxidation system of *E. coli*, Korkes, Del Campillo, Gunsalus and Ochoa[73] isolated two enzyme fractions (A and B)

"which, in the presence of NAD and CoA catalyze the conversion of pyruvate either to acetylphosphate, provided transacetylase and orthophosphate are added, or to citrate provided condensing enzyme and oxaloacetate are present." (ref. 73)

No reaction is observed in the absence of an acetyl acceptor system, either transacetylase–orthophosphate or condensing enzyme-oxaloacetate. Diphosphothiamine is also required. The authors concluded that the enzyme fractions A and B from *E. coli* catalyze the overall reaction.

$$\text{pyruvate} + NAD^+ + \text{CoA} \rightarrow \text{acetyl CoA} + CO_2 + NADH + H^+ \quad (5)$$

Acetyl-CoA now replaces the "active acetate" of earlier schemes (Ochoa[74]).

In his pioneering work on pyruvate oxidation in animal tissues, Peters, as stated above, showed that thiamine is one of the coenzymes involved. Thiamine pyrophosphate was found to be a cofactor of many ketoacid decarboxylases[75–78].

In 1947, O'Kane and Gunsalus[79] identified another cofactor, lipoic acid. This brings us to the development related by L. J. Reed in Vol. 14 (Chapter 2) of this Treatise, where the reader will find the development of the knowledge of pyruvate and α-ketoglutarate decarboxylation respectively leading to acetyl-CoA and to succinyl-CoA after the discovery of lipoic acid as a cofactor, including the structural organization of *a*-keto-acid dehydrogenation complexes of mitochondria.

7. The present theory of the tricarboxylic acid cycle

The unravelling of the mechanism of citrate synthesis from pyruvate and oxaloacetate and of the conversion of α-ketoglutarate to succinyl-CoA was included in the present theory of the tricarboxylic acid cycle as shown in Fig. 18. This cycle, a development of the original citric acid cycle formulated by Krebs in 1937 as a scheme "describing the intermediary stages

References p. 281

of the oxidation of carbohydrate in animals tissues"[6] was at the time essentially a theory, the chief aim of which was

"to describe, step by step, the chemical changes of the carbohydrate molecule leading to the formation of carbon dioxide and water."[80]

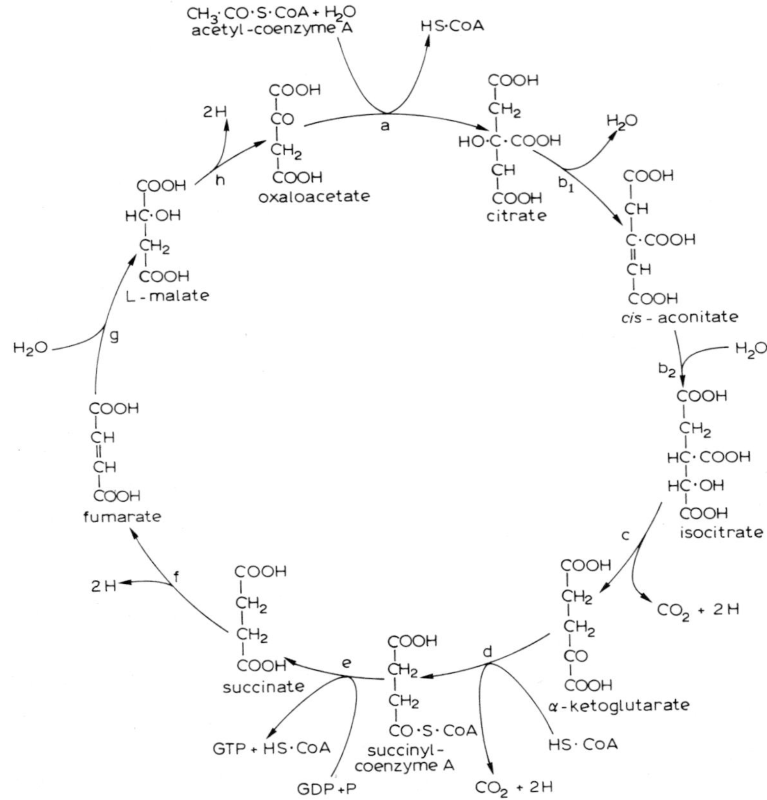

Fig. 18. The tricarboxylic acid cycle.

Set in a new context, the theory of Szent-Györgyi originally conceived as a system for hydrogen transfer, is inserted in what we now call the respiratory chain. The development inaugurated by Martius concerning the metabolism of citrate, and by Lipmann on one hand and Stern and Ochoa on the other, on the synthesis of citrate from pyruvate have also been among the basic elements included in the cycle. The great merit of Krebs, besides his experiments on the stimulation of respiration by citrate and his demonstration of the oxidative formation of succinate from oxaloacetate or fumarate, has been to conceive a cyclic scheme of oxidations (dehydrogenations) defining the stages through which the carbon atoms of the carbohydrates pass in a cycle of acetyl transfer.

The originality of the theory of Krebs is attested by the objections of those biochemists who reproached him for not demonstrating that his "citric acid cycle" was a chain of hydrogen transfer, what it precisely is not.

While all previous theories of biological oxidations went from the nutrient to oxygen through a sequential multienzyme system accepting electrons from the nutrient, (*i.e.* they described the path of hydrogen), Krebs provided the so far missing link through which all the different nutrients were brought to generate the same starting reaction of the electron transfer chains derived from the dehydrogenation in the cycle (*i.e.* he clarified the path of carbon). The tricarboxylic acid cycle also clarified for the first time the functional significance of the occurrence of succinate in muscle and of the ready oxidation of malate and citrate, phenomena which had been known for nearly 30 years[81,82].

8. Generality of the Krebs cycle

As stated by Green and McLennan[83],

"the tricarboxylic acid cycle may be looked upon as a combustion chamber in which 'hot' electrons are generated by oxidation. (ref. 83)

Each major advance in biochemical theory must be tested for its generality among living cells. This always involves a lot of painstaking and unglorified, though necessary activity. For the literature on the general occurrence of the tricarboxylic acid cycle in cells the reader is referred to Lowenstein's chapter in the 3rd edition of Greenberg's *Metabolic Pathways*[7].

A large amount of work has been performed on the isolation, compartmentalization in cells, characterization, mechanism, stereospecificity and regulation of the reactions of the cycle. This recent work is presented by Lowenstein in Vol. 18 S, Chapter 1, of this Treatise, to which the reader is referred.

According to the method of modern biochemistry, the endeavour has been towards isolating and purifying the enzymes of the cycle. It is necessary to confirm the reconstructed scheme by experimentation on cells or organisms.

References p. 281

In animals, the injection of malonate has been shown to cause an accumulation of succinate in urine[84] and in several tissues[85]. An accumulation of citrate in several tissues follows the injection of fluoroacetate[86,87].

But it is the isotopic method which, through the use of labelled compounds, has afforded the final proof of the operative nature of the tricarboxylic acid cycle in living animals[88–91], higher plants[92–95], algae[96–100] and microorganisms[101–105].

Note added in proof

The isolation of crystalline thiamine from rice polishing had been accomplished in 1925 by Jansen and Donath (B. C. P. Jansen and W. F. Donath, *Koninkl. Ned. Akad. Wetenschap. Proc.*, 29 (1926) 1390) who used a protective test and not a curative one. The importance of this great achievement accomplished on Java, where Eykmann worked, will be underlined in a further section of this History.

REFERENCES

1 C. Oppenheimer and K. G. Stern, *Biological Oxidation*, The Hague, 1939.
2 F. L. Breusch, *Z. physiol. Chem.*, 250 (1937) 262.
3 A. G. Ogston, *Nature*, 162 (1948) 963.
4 H. G. Wood, C. H. Werkman, A. Hemingway and A. O. Nier, *J. Biol. Chem.*, 139 (1941) 483.
5 E. A. Evans, Jr. and L. Slotin, *J. Biol. Chem.*, 141 (1941) 439.
6 H. A. Krebs, in *Harvey Lectures* Ser. 44, 1948–1949, Springfield, Ill., 1950.
7 H. A. Krebs, in D. M. Greenberg (Ed.), *Chemical Pathways of Metabolism*, New York, 1954 (2nd ed., 1960, title *Metabolic Pathways;* 3rd ed., 1969). In the 2nd ed. the chapter is signed by H. A. Krebs and J. M. Lowenstein; the revised chapter in the 3rd ed. is signed by J. M. Lowenstein.
8 V. R. Potter and C. Heidelberger, *Nature*, 164 (1949) 180.
9 P. E. Wilcox, C. Heidelberger and V. R. Potter, *J. Am. Chem. Soc.*, 72 (1950) 5019.
10 V. Lorber, M. F. Utter, H. Rudney and M. Cook, *J. Biol. Chem.*, 185 (1950) 689.
11 C. Martius and G. Z. Schorre, *Z. Naturforsch.*, 5b (1950) 170.
12 Sir R. A. Peters, *Biochemical Lesions and Lethal Synthesis*, Oxford, 1963.
13 H. W. Kinnersley, R. A. Peters and V. Reader, *Biochem. J.*, 22 (1929) 276.
14 H. W. Kinnersley and R. A. Peters, *Biochem. J.*, 23 (1929) 1126.
15 R. B. Fisher, *Biochem. J.*, 25 (1931) 1410.
16 E. Hayasaka, *Tohoku J. exptl. Med.*, 14 (1929) 85.
17 N. Gavrilescu, A. P. Meiklejohn, R. Passmore and R. A. Peters, *Proc. Roy. Soc. (London), Ser. B*, 110 (1932) 431.
18 N. Gavrilescu and R. A. Peters, *Biochem. J.*, 25 (1931) 1937, 2150.
19 R. A. Peters and H. M. Sinclair, *Biochem. J.*, 27 (1933) 1671.
20 R. A. Peters and R. S. H. Thompson, *Biochem. J.*, 28 (1934) 916.
21 R. S. H. Thompson and R. E. Johnson, *Biochem. J.*, 29 (1935) 694.
22 B. S. Platt and G. D. Lu, *Quart. J. Med.*, 5 (1946) 355.
23 B. S. Platt and G. D. Lu, *Biochem. J.*, 33 (1939) 1525.
24 K. Lohmann and P. Schuster, *Biochem. Z.*, 294 (1937) 188.
25 I. Banga, S. Ochoa and R. A. Peters, *Biochem. J.*, 33 (1939) 1109, 1980.
26 F. Lipmann, *Wanderings of a Biochemist*, New York, 1971.
27 P. A. Levene and C. L. Alsberg, *J. Biol. Chem.*, 2 (1906–1907) 127.
28 F. A. Lipmann and P. A. Levene, *J. Biol. Chem.*, 98 (1932) 109.
29 F. A. Lipmann, *Nature*, 143 (1939) 436.
30 O. Warburg and W. Christian, *Biochem. Z.*, 298 (1938) 150.
31 F. Lipmann, *Nature*, 144 (1939) 381.
32 F. Lipmann and L. C. Tuttle, *J. Biol. Chem.*, 161 (1945) 415.
33 M. F. Utter and C. H. Werkman, *J. Biol. Chem.*, 146 (1942) 289.
34 H. J. Koepsell and M. J. Johnson, *J. Biol. Chem.*, 145 (1942) 379.
34a F. Lipmann, *Harvey Lectures*, 44 (1950) 99.
35 F. Lipmann, *Advan. Enzymol.*, 1 (1941) 99.
36 J. R. Klein and J. S. Harris, *J. Biol. Chem.*, 124 (1938) 613.
37 F. Lipmann, *Advan. Enzymol.*, 6 (1946) 231.
38 F. Lipmann, *J. Biol. Chem.*, 160 (1945) 173.
39 F. Lipmann, N. O. Kaplan, G. D. Novelli, L. C. Turtle and B. M. Guirard, *J. Biol. Chem.*, 167 (1947) 869.
40 F. Lipmann and N. O. Kaplan, *J. Biol. Chem.*, 162 (1946) 743.
41 D. Nachmansohn and M. Berman, *J. Biol. Chem.*, 165 (1946) 551.

42 W. Feldberg and T. Mann, *J. Physiol.*, 104 (1946) 411.
43 M. Soodak and F. Lipmann, *J. Biol. Chem.*, 175 (1948) 449.
44 J. R. Stern, B. Shapiro and S. Ochoa, *Nature*, 166 (1950) 403.
45 G. D. Novelli and F. Lipmann, *J. Biol. Chem.*, 182 (1950) 213.
46 S. Korkes, J. R. Stern, I. C. Gunsalus and S. Ochoa, *Nature*, 166 (1950) 439.
47 E. R. Stadtman, G. D. Novelli and F. Lipmann, *J. Biol. Chem.*, 191 (1951) 365.
48 C. Martius, *Z. physiol. Chem.*, 279 (1943) 96.
49 A. L. Lehninger, *J. Biol. Chem.*, 164 (1946) 291.
50 F. L. Breusch, *Science*, 97 (1943) 490.
51 H. Wieland and C. Rosenthal, *Ann. Chem.*, 554 (1943) 241.
52 G. Kalnitsky, *J. Biol. Chem.*, 160 (1945) 1015.
53 N. O. Kaplan and F. Lipmann, *J. Biol. Chem.*, 174 (1948) 37.
54 G. D. Novelli, N. O. Kaplan and F. Lipmann, *J. Biol. Chem.*, 177 (1949) 97.
55 E. E. Snell, G. M. Brown, V. J. Peters, J. A. Craig, E. L. Wittle, J. A. Moore, V. M. McGlohon and O. D. Bird, *J. Am. Chem. Soc.*, 72 (1950) 5349.
56 F. Lynen and E. Reichert, *Z. Angew. Chem.*, 63 (1951) 47.
57 P. Teague and R. J. Williams, *J. Gen. Physiol.*, 25 (1942) 777.
58 A. Dorfman, S. Berkman and S. A. Koser, *J. Biol. Chem.*, 144 (1942) 393.
59 G. M. Hills, *Biochem. J.*, 37 (1943) 418.
60 G. D. Novelli and F. Lipmann, *Arch. Biochem.*, 14 (1947) 23.
61 R. E. Olson and N. O. Kaplan, *J. Biol. Chem.*, 175 (1948) 515.
62 G. D. Novelli and F. Lipmann, *J. Biol. Chem.*, 171 (1947) 833.
63 S. Weinhouse and R. H. Millington, *J. Am. Chem. Soc.*, 69 (1947) 3089.
64 N. O. Kaplan and F. Lipmann, *Federation Proc.*, 7 (1948) 163.
65 J. R. Stern and S. Ochoa, *J. Biol. Chem.*, 179 (1949) 491.
66 W. B. Elliott and G. Kalnitsky, *J. Biol. Chem.*, 186 (1950) 477.
67 J. R. Stern and S. Ochoa, *Federation Prod.*, 9 (1950) 234.
68 J. R. Stern and S. Ochoa, *J. Biol. Chem.*, 191 (1951) 161.
69 S. Ochoa, J. R. Stern and M. C. Schneider, *J. Biol. Chem.*, 193 (1951) 691.
70 J. R. Stern, B. Shapiro, E. R. Stadtman and S. Ochoa, *J. Biol. Chem.*, 193 (1951) 703.
71 E. R. Stadtman, *Federation Proc.*, 9 (1950) 233.
71a J. R. Stern, S. Ochoa and F. Lynen, *J. Biol. Chem.*, 198 (1952) 313.
72 S. Korkes, J. R. Stern, I. C. Gunsalus and S. Ochoa, *Nature*, 166 (1950) 439.
73 S. Korkes, A. Del Campillo, I. C. Gunsalus and S. Ochoa, *J. Biol. Chem.*, 193 (1951) 721.
74 S. Ochoa, *Physiol. Rev.*, 31 (1951) 56.
75 D. E. Green, D. Herbert and V. J. Subrahmanyan, *J. Biol. Chem.*, 135 (1940) 795.
76 D. E. Green, W. W. Westerfield, B. Vennesland and W. E. Knox, *J. Biol. Chem.*, 145 (1942) 69.
77 S. Kobayashi, *J. Biochem. (Tokyo)*, 33 (1941) 301.
78 S. Ochoa, in E. A. Evans (Ed.), *The Biological Action of Vitamins*, Chicago, 1942.
79 D. J. O'Kane and I. Gunsalus, *J. Bacteriol.*, 54 (1947) 20.
80 H. A. Krebs, *Advan. Enzymol.*, 3 (1943) 191.
81 T. Thunberg, *Skand. Arch. Physiol.*, 24 (1910) 23.
82 F. Battelli and L. Stern, *Biochem. Z.*, 31 (1910) 478.
83 D. E. Green and D. H. MacLennan, in D. M. Greenberg (Ed.), *Metabolic Pathways*, 3rd ed. Vol. 1, New York and London, 1967.
84 H. A. Krebs, E. Salvin and W. A. Johnson, *Biochem. J.*, 32 (1938) 113.
85 H. Bush and V. R. Potter, *J. Biol. Chem.*, 198 (1952) 71.
86 P. Buffa and R. A. Peters, *J. Physiol.*, 110 (1949) 488.
87 A. Lindenbaum, M. R. White and J. Schubert, *J. Biol. Chem.*, 190 (1951) 585.

88 N. Lifson, V. Lorber, W. Sakami and H. G. Wood, *J. Biol. Chem.*, 176 (1948) 1263.
89 J. S. Lee and N. Lifson, *J. Biol. Chem.*, 193 (1951) 253.
90 V. Lorber, N. Lifson, H. G. Wood, W. Sakami and W. W. Shreeve, *J. Biol. Chem.*, 183 (1950) 517.
91 E. O. Weinman, E. H. Strisower and I. L. Chaikoff, *Physiol. Rev.*, 37 (1957) 252.
92 P. Saltman, G. Kunitake, H. Spolter and C. Stills, *Plant Physiol.*, 31 (1956) 464.
93 R. S. Bandurski, *J. Biol. Chem.*, 217 (1955) 137.
94 R. S. Bandurski and C. M. Greiner, *J. Biol. Chem.*, 204 (1953) 781.
95 P. Saltman, V. H. Lynch, G. Kunitake, C. Stills and H. Spolter, *Plant Physiol.*, 32 (1957) 197.
96 M. Calvin and A. A. Benson, *Science*, 107 (1948) 476.
97 M. Calvin and A. A. Benson, *Science*, 109 (1949) 140.
98 M. Calvin and P. Massini, *Experientia*, 8 (1952) 445.
99 V. Moses, O. Holm-Hansen, J. A. Bassham and M. Calvin, *J. Mol. Biol.*, 1 (1959) 21.
100 A. A. Benson, J. A. Bassham, M. Calvin, T. C. Goodale, V. A. Haas and W. Stepka, *J. Am. Chem. Soc.*, 72 (1950) 1710.
101 H. Y. Saz and L. O. Krampitz, *J. Bacteriol.*, 67 (1954) 409.
102 H. E. Swim and L. O. Krampitz, *J. Bacteriol.*, 67 (1954) 419.
103 H. E. Swim and L. O. Krampitz, *J. Bacteriol.*, 67 (1954) 426.
104 J. A. Demoss and H. E. Swim, *J. Bacteriol.*, 74 (1957) 445.
105 A. O. M. Stoppani, L. Couches, S. L. S. de Favelukes and F. L. Sacerdote, *Biochem. J.*, 70 (1958) 438.

Chapter 34

From Fatty Acids to Acetyl-CoA

Introduction

A consideration of the process whereby biochemistry developed into an autonomous discipline, shows that, if it is true that physiological chemistry finally merged with biochemistry, it was in microbiological and immunological research that biochemistry—in the form of studies on glycolysis—was born. Similarly, the catabolism of fatty acids also became the subject of much attention in the newly constituted biochemical circles. This trend was not initiated by an interest in an aspect of physiological chemistry but rather, as we shall see, by prevailing interest in a clinical subject.

The main fatty acids undergoing catabolism in animal organisms contain even number of carbons and are straight-chain molecules, the number of carbons varying between 4 and 24. It was already shown by Wöhler[1] in 1824, and has been repeatedly confirmed since[2–5], that these fatty acids are completely oxidized to CO_2 and water. When Schotten[3] gave sodium salts of fatty acids to dogs he noted the presence of large amounts of sodium bicarbonate in their urine.

As noted in the introduction to this volume, one of the resources of biochemistry has been its access to biological material less complicated than a whole organism, allowing the use of simpler techniques. This was illustrated by experiments on yeast juice or macerated cells in fermentation studies. While glycolysis takes place in cytoplasm, the oxidation of fatty acids, like most respiration processes, occurs in mitochondria; this concept had to be formulated, and mitochondria characterised and isolated in bulk, before an active extract could be obtained from their contents. Furthermore, the enzymatic system of biological oxidations was believed to be linked to the integrity of the cell ("structure-bound"), a belief which in truth derived from the fact that cofactors were diluted and had to be added; this is not the case with highly concentrated yeast extracts or macerations obtained

References p. 317

from, say, dried yeast (Lebedev-juice) or by means of high pressure. In contrast, to obtain extracts from animal material it is necessary to add aqueous solutions.★

On the other hand, in the studies on preparations composed of unruptured cells, the chemical properties of fatty acids and of their salts, *e.g.* their slight solubility in water and their slow diffusion through cell walls occasionally made experimentation, *e.g.* with tissue slices, difficult and misleading. It was only half a century after its introduction by Knoop that the theory of β-oxidation of fatty acids could be formulated in terms of enzymology.

1. Acetoacetate considered as an obligatory intermediate in fatty acid catabolism

(a) Ketone bodies

Diabetics have been known since antiquity to produce large volumes of urine. Chemical aspects of the disease were described when Thomas Willis, in 1647, detected the sugary taste of the urine of diabetics; in 1857, Petters identified acetone as responsible for the smell of the breath of diabetics in the state of coma. (For these ancient aspects of the history of diabetes, the reader is referred to Lieben[6]). After Gerhardt[7] in 1865 found acetoacetic acid in the urine of diabetics and Arnold[8] in 1900 recognised it as the precursor of acetone, the problem of the biochemical origin of acetoacetic acid (and of its reduced form β-hydroxybutyric acid) came to the fore.

$$\underset{\text{acetone}}{CH_3COCH_3} \leftarrow \underset{\text{acetoacetic acid}}{CH_3COCH_2COOH} \rightleftharpoons \underset{\beta\text{-hydroxybutyric acid}}{CH_3CHOHCH_2COOH}$$

That ketone bodies (acetoacetic acid, β-hydroxybutyric acid, acetone) are derived in metabolism from fatty acids was first proposed by Geelmuyden (1897) in a paper entitled "Ueber Aceton als Stoffwechselprodukt" and confirmed in a number of studies[10–12].

★ It may also be remarked with respect to the concept of "structure bound" metabolism that when yeast extracts are diluted they lose their fermentation capacity very rapidly. Another point is that the metabolic activity of yeast cells is very high, compared to animal tissues. Muscle is exceptional in that respect and, as shown in a previous chapter, Meyerhof was able to obtain fermenting extracts from it.

(b) Knoop's theory of β-oxidation

As noted by Chaikoff and Brown[13], at the beginning of the century the merit of any proposed mechanism for the breakdown of fatty acids was judged in terms of the ability of that mechanism to explain the formation of ketone bodies. In his 1905 paper on β-oxidation, Knoop explicitly makes this point when, after formulating his theory, he writes[15]:

"Es mag hier darauf hingewiesen werden, dass bei dem Diabetiker beobachtete Vermehrung der ausgeschiedenen β-Oxybuttersäure, möglicherweise einen analogen Vorgang darstellt."

As Knoop himself states, another fact which inspired him is that fatty acids are contained in milk fat having 18, 16, 14, 12, 10, 8, 6 and 4 carbon atoms; this he considers "direkt einleuchtend". In contrast, no "ketone body" was known to derive from the 5-carbon fatty acid, levulinic acid $CH_3COCH_2CH_2COOH$.

TABLE III

Results of Knoop's[14,15] experiments

Acid fed	*Excreted product*	*Modification observed without reference to the existence of conjugates*
C_6H_5COOH (benzoic acid)	C_6H_5COOH	unmodified
$C_6H_5CH_2COOH$ (phenylacetic acid)	C_6H_5COOH	unmodified
$C_6H_5CHOHCOOH$	$C_6H_5CHOHCOOH$	unmodified
$C_6H_5CHNH_2COOH$	$C_6H_5CHOHCOOH$	desaminated
$C_6H_5CH_2CH_2COOH$ (phenylpropionic acid)	C_6H_5COOH	oxidation at the β carbon
$C_6H_5CHOHCH_2COOH$	C_6H_5COOH	
$C_6H_5COCH_2COOH$	C_6H_5COOH	
$C_6H_5CH{=}CHCOOH$	C_6H_5COOH	
$C_6H_5CH_2CHNH_2COOH$	O	apparently totally oxidized
$C_6H_5CH_2CHOHCOOH$		
$C_6H_5CH_2COCOOH$		
$C_6H_5CH{=}CNH_2COOH$		
$C_6H_5CH_2CH_2CH_2COOH$		
$C_6H_5CH_2CH_2CH_2COOH$ (phenylbutyric acid)	$C_6H_5CH_2COOH$	oxidation at the β carbon
$C_6H_5COCH_2CH_2COOH$		
$C_6H_5\cdot CH{=}CHCH_2COOH$		

References p. 317

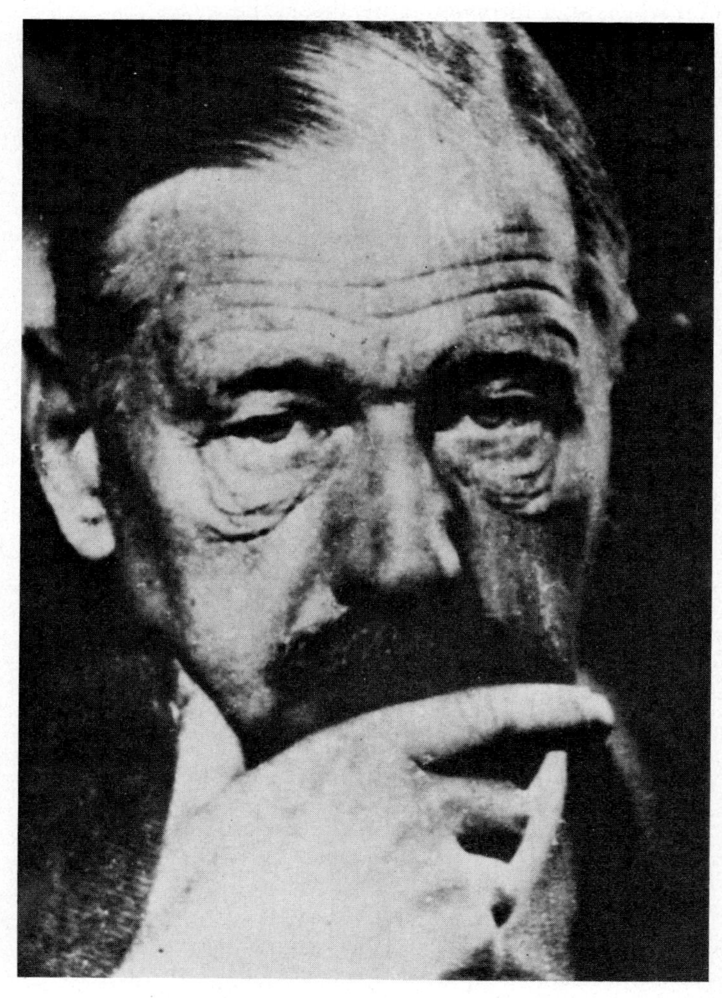

Plate 100. Franz Knoop.

Such facts led Knoop to conceive β-oxidation and to devise a method for testing it in the form of tracer experimentation.

To facilitate the identification of metabolites in the urine, Knoop[14,15] introduced into the terminal methyl of the fatty acid a radical which was not readily changed in the body when he administered to dogs the fatty acids containing from 1 to 5 carbons carrying a phenyl on their last methyl. He found that the derivatives of the odd carbon-number fatty acids broke down to benzoic acid and were excreted, coupled with glycine, in the form of hippuric acid. On the other hand, fatty acids with even numbers of carbon were broken down to phenylacetic acid, excreted as phenyl aceturic acid. Knoop's results are stated in Table III.

Thus (Fig. 19) phenylacetic acid does not undergo oxidation. In combination with glycine, it is excreted as phenaceturic acid

$$C_6H_5CH_2CONHCH_2COOH$$

β-phenylpropionic acid, on the other hand, is oxidized to benzoic acid which forms hippuric acid with glycine. From this Knoop concluded that the oxidation of the side chain of three carbons of phenylpropionic acid did not involve removal of one carbon since in that case phenaceturic acid would have been excreted. Two carbon atoms appeared to have been removed at a time from the lateral chain of phenylpropionic acid.

But γ-phenylbutyric acid was converted into phenaceturic acid, two carbon atoms having been removed from the side chains; and δ-phenylvaleric acid gave hippuric acid, *i.e.*, showed a loss of four carbon atoms.

On the basis of these observations Knoop formulated the theory of β-oxidation according to which the hydrogen attached to the β-carbon of the side chains of phenyl fatty acids is selected for oxidation. The side chain is thus shortened by the loss of two, or a multiple of two carbons at each step (Fig. 19).

The theory of β-oxidation was formulated by Knoop[14] in a thesis published in 1904. A more concise presentation[15] appeared in 1905.*

(c) Objections to the theory of β-oxidation

The theory of the degradation of fatty acids by loss of two carbon atoms

* In a book first published in 1912 Dakin[16] mentions the 1904 thesis (without dating it) and erroneously refers to the 1905 paper as dated 1904, a bibliographic mistake which has been repeated many times in the literature.

References p. 317

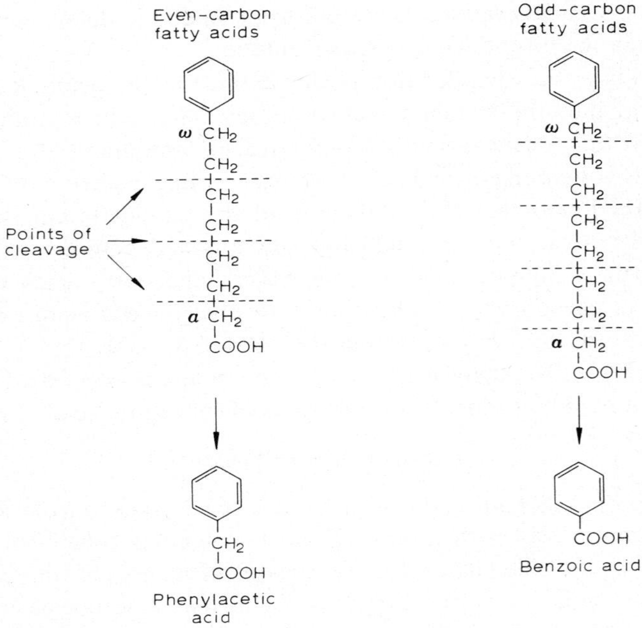

Even-carbon fatty acids

Odd-carbon fatty acids

Points of cleavage

Phenylacetic acid

Benzoic acid

Fig. 19. Metabolic products of an even-carbon and of an odd-carbon fatty acid.

at a time encountered opposition from chemists. Friedmann wrote:

"From the standpoint of pure chemistry the assumption of oxidation at the β-position is opposed to the facts that are known about the oxidation, substitution and condensation of fatty acids, since only the hydrogen atoms attached to the α-carbon have been found to be capable of reaction and no observation is known showing that the hydrogen atoms attached to the β-carbon atoms are capable of undergoing reaction" (ref. 17, translation by Dakin[16]).

This objection was discarded by Dakin[18] on the basis of oxidation experiments *in vitro* with H_2O_2 in which he showed that oxidation of the ammonium salts of the fatty acids by H_2O_2 at 90° resulted, admittedly on a small scale, in the formation of β-ketoacids. In the same year, by way of reply to Friedmann and others who wished to limit biochemistry to the establishment of analogies with known chemical facts, Knoop had formulated the philosophy of biochemistry in a text that deserves to be placed alongside the previously cited (Chapter 17) words of Buchner and of Hopkins:

"Ich habe nun nie behauptet, damit meine Befunde auf bekannte chemische Analogien zurückgeführt zu haben, sondern nur versucht, eine Anzahl physiologisch beobachtete

Tatsachen unter diesem einem Gesichtspunkte einheitlich aufzufassen. Mir war in Gegenteil sehr wohl bekannt, dass damit etwas prinzipiell Neues gefunden schien, dass den bekannten Oxydationsreaktionen der organischen Chemiker gegenüber in der Tat etwas ganz Unerwartetes darstellte. Aber der lebende Organismus gestattet sich gewiss noch manche Reaktion die dem organischem Chemiker noch unbekannt ist. Wenn sich deswegen in Fällen, wie hier, physiologische Analogien zu derartigen Beobachtungen finden lassen, die einen gewissen Grad von Wahrscheinlichkeit für sich haben, so halte ich die vorsichtige Diskussion einer solchen Hypothese nicht nur für erlaubt, sondern sogar für oft recht förderlich oder ganz unentbehrlich". (ref. 19)

(d) Experiments on perfused liver

Knoop's pioneering research on whole organisms was followed by experiments Embden and his collaborators performed on perfused liver. The authors[20–25] found that the small quantity of acetone formed in the perfusion of a freshly excised liver with defibrinated blood increased when fatty acids were added to the perfusion liquid. The most striking result was that only the fatty acids with even numbers of carbons increased acetone formation. As a result it was postulated that acetoacetate here was an intermediary product. This interpretation has found its way into current theories.

Experiments on diabetic animals gave wider scope to Embden's conclusions. Baer and Blum[26] showed that in these animals the administration of salts of butyric or isovaleric acids increased the excretion of β-hydroxybutyric acid, of acetoacetic acid and of acetone, while propionic acid or normal valeric acid did not.

(e) Dakin's views on the course of β-oxidation

Knoop's theory holds that the hydrogen carried on the β-carbon of a fatty acid chain is oxidized, so that the bond between the α and β carbon is cleaved.

Dakin[27] again investigated the fate of the phenylbutyric acid in the mammalian organism. As stated above, Knoop had administered this substance to dogs and had observed the excretion of phenaceturic acid in the urine. The conversion involves the intermediate formation of phenylacetic acid which is then coupled with glycine:

$$C_6H_5CH_2CH_2CH_2COOH \rightarrow C_6H_5CH_2COOH \rightarrow C_6H_5CH_2CONHCH_2COOH$$

The excretion of D-(−)-β-hydroxybutyric acid following the administration

References p. 317

of butyric acid to animals and the simultaneous presence of acetoacetic acid and acetone in the urine led to the belief that β-hydroxybutyric acid was the initial oxidation product and acetoacetic the second:

$$\underset{\text{butyric acid}}{CH_3CH_2CH_2COOH} \rightarrow \underset{\text{-hydroxybutyric acid}}{CH_3CHOHCH_2COOH} \rightarrow \underset{\text{acetoacetic acid}}{CH_3COCH_2COOH}$$

This was believed to be proven by the perfusion of surviving liver in which both butyric and β-hydroxybutyric acid were oxidized to acetoacetic acid (literature in Dakin[27]). A case analogous to the formation of β-hydroxybutyric acid was provided by Dakin's[27] observation of the excretion of L-β-hydroxyphenylpropionic acid following administration of phenylpropionic acid to animals.

It had long since been known[28] that β-phenylpropionic acid is converted to benzoic acid in the course of its passage through the animal body and excreted as hippuric acid.

$$\underset{\beta\text{-phenylpropionic acid}}{C_6H_5CH_2CH_2COOH} \rightarrow \underset{\text{benzoic acid}}{C_6H_5COOH} \rightarrow \underset{\text{hippuric acid}}{C_6H_5CONHCH_2COOH}$$

According to Knoop[14,15], evidence for the direct oxidation of the hydrogen carried on the β-carbon is that phenylacetic acid is not subject to the oxidation. Dakin represents the oxidation of phenylpropionic acid as follows:

$$\underset{\beta\text{-phenylpropionic acid}}{C_6H_5CH_2CH_2COOH} \rightarrow \underset{\beta\text{-phenyloxypropionic acid}}{C_6H_5CH(OH)CH_2COOH} \rightarrow \underset{\text{benzoylacetic acid}}{C_6H_5COCH_2COOH}$$

$$\rightarrow \underset{\text{acetophenone}}{C_6H_5COCH_3} + CO_2 \rightarrow \underset{\text{benzoic acid}}{C_6H_5COOH}$$

His evidence[16] for this scheme is that

"both β-phenyl-β-oxypropionic acid and acetophenone have been detected in the urine of dogs after subcutaneous injection of sodium β-phenylpropionate. The β,β-phenyloxypropionic acid contains an asymmetric carbon atom and the acid found in the urine proved to be levorotatory. Benzoylacetic acid was not detected but its formation must be inferred from the observed excretion of acetophenone which is readily formed from benzoylacetic acid through loss of carbon dioxide".

β,β-Phenyloxypropionic acid, benzoylacetic acid and acetophenone, when administered to an animal are oxidized and yield benzoic acid, the same end product as that obtained from phenylpropionic acid[14,15,29].

Dakin draws a scheme showing close resemblance between the oxidation of phenylpropionic acid and that of butyric acid:

$CH_3CH_2CH_2COOH$ butyric acid	$C_6H_5CH_2CH_2COOH$ β-phenylpropionic acid
↓	↓
levo $CH_3CH(OH)CH_2COOH$	*levo* $C_6H_5CH(OH)CH_2COOH$
	↓
CH_3COCH_2COOH	$C_6H_5COCH_2COOH$
↓	↓
CH_3COCH_3	$C_6H_5COCH_3$
↓	↓
CH_3COOH	C_6H_5COOH

To quote Dakin[27]:

"In each case a levorotatory β-oxyacid is first formed, this is oxidised to a β-ketonic acid which loses carbon dioxide passing into a ketone which then undergoes further oxidation with formation of lower acids. These results afford the most convincing evidence of the occurrence of β-oxidation in the animal body."

Dakin considered that α, β unsaturated acids may represent an intermediate stage in the β-oxidation of saturated fatty acids.

In Fig. 20, Dakin[16] shows the various changes involved in β-oxidations, including oxidation and reduction, hydration and dehydration.

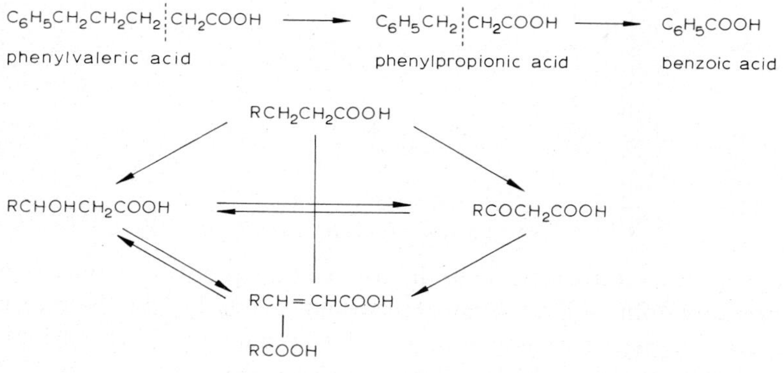

Fig. 20. β-Oxidation mechanism according to Dakin[16].

(*f*) *Dakin's theory of fatty acid oxidation ("Classical theory")*

The scheme proposed by Dakin[27] to account for the catabolism of fatty acids was a direct application of Knoop's β-oxidation concept along with the theory which holds that ketone bodies are obligatory intermediates in fatty acid catabolism.

According to this theory, a fatty acid is oxidized at the β-carbon and then loses both the α and the carboxyl carbons. The remainder of the chain is believed to undergo successive β-oxidations with removal of two carbons after each oxidation except the last (4C). In this scheme this last oxidized residue is believed to be acetoacetic acid.

The theory also holds that a molecule of fatty acid, whatever its length, should give rise to a single molecule of acetoacetic acid. In the case of octanoic acid the following diagram represents the proposed pathway:

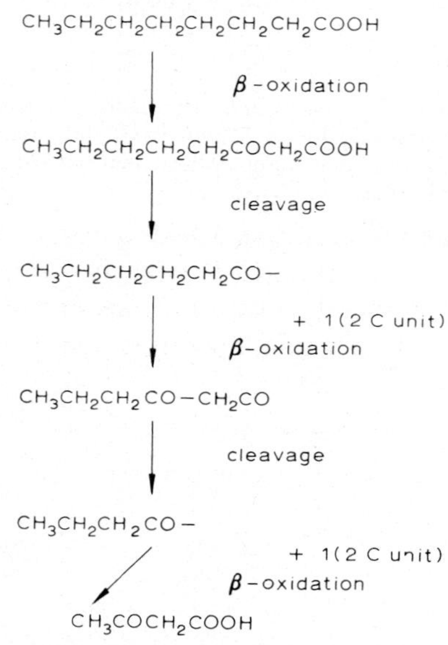

Total: 1 acetoacetate+2(C_2 units)

In this scheme acetoacetic acid, the β-oxidation product of butyric acid, derives only from the four terminal (ω) atoms of a fatty acid, the successive cleavages liberating carbon atoms and trimming the fatty acid chain, whatever its length, to the four-carbon acetoacetate.

The theory of fatty acid catabolism (often called the classical theory), which remained current until around 1945 and was expounded in many textbooks, distinguishes between a *primary phase* and a *secondary phase* (diphasic theory).

In the primary phase, the fatty acids are oxidized in the liver to ketone bodies; these enter the blood stream and the extrahepatic tissues, particularly kidney and muscle, and accomplish the oxidation of the ketone bodies to CO_2 and H_2O. The classical theory paid little heed to the carboxyl and α-carbons cleaved at each β-oxidation. In fact their fate was thought mysterious at the time. Acetate was not recognized as the split-off C_2 unit, because it had unsuccessfully been searched for in animal tissues. In 1912, Dakin[16] had suggested formic acid as an intermediate in the conversion of these two C atoms into CO_2:

$$RCH_2|CH_2COOH \rightarrow RCOOH + HCOOH + CO_2$$

but, Dakin[16] writes,

"there is no satisfactory evidence in favour of this belief"...
..."Attempts that have been made to demonstrate the production of formic acid as the result of the oxidation of higher fatty acids have not given decisive results. But Dakin and Wakeman have recently found that there is no increase of formic acid in the blood used for perfusing livers in which the oxidation of higher fatty acids has been in progress (unpublished results). Further information as to the fate of the carbon groupings set free from fatty acids by biochemical oxidation is much needed".

It must be pointed out that in the classical theory the carbon atoms cleaved off in β-oxidation were believed to contribute to the formation of CO_2 from fatty acids, but not to the formation of ketone bodies which, as the theory had it, were derived only from the last (ω) four carbons of the fatty acid chain. After the formulation of Thunberg's theory of carbohydrate metabolism (Fig. 14, Chapter 32) several authors considered that acetate deriving from fatty acids could follow the same path. But this view, like Thunberg's theory, came up against objections.

(g) Objections to Dakin's theory

Principal to this theory was the contention that one molecule of fatty acid provided, in catabolism, one single molecule of acetoacetate, corresponding to the four terminal carbon atoms at the ω end of the fatty acid. That such a process cannot be taken into account quantitatively for the acetoacetate formed has repeatedly been shown. Deuel *et al.*[30], Leloir and Munoz[31] and Witter, Newcomb and Stotz[32] found that, for instance, hexanoic acid yields more acetoacetate than butyric acid does. Another fact at variance with the theory was the formation of acetoacetate from valeric acid (5 C)[31,33–36], first shown by Quastel and Wheatley[33].

References p. 317

In accordance with the scheme of β-oxidation a fatty acid of the odd series would leave a 3-carbon residue CH_3CH_2CO- and not the 4-carbon residue CH_3COCH_2CO- from which acetocetate was supposed to arise.

(h) Alternate theories of acetoacetate formation

In 1915, Hurtley[37] had proposed that the whole of the fatty acid chain is oxidized at alternate carbons and that the whole oxidized chain is split up into 4 carbon blocks, the precursors of acetoacetate. This long-forgotten theory was revived in 1935 by Jowett and Quastel[38] as the "multiple-alternate-oxidation theory", which holds that fatty acids are not oxidized and broken down in stepwise fashion. Furthermore, from their experiments (with tissue slices) they concluded that butyric acid gives a lower yield of acetone bodies than do the acids containing 6 or 8 C, and that the acetic acid split-off would furnish very little acetoacetate.

(i) Acetate as precursor of acetoacetate

The theories which take ketone bodies to be intermediate products (either Dakin's theory or the multiple-alternate-oxidation theory), also take them to be derived from the residue that remains after the stepwise shortening of the fatty acid chains.

That the C_2 units split off from the chain could be a source of acetoacetate or other ketone bodies was first suggested by MacKay *et al.*[39] in 1940 ("theory of β-oxidation-condensation"). According to this view which agrees with Dakin's theory but not with that of multiple alternate oxidation, fatty acids are believed, as they were by Knoop, to be degraded by two carbons at a time. The C_2 fragments arising from fatty acids recondense to form acetoacetic acid. This theory was not in contradiction with the amounts of acetoacetate formed from fatty acids. That acetoacetate can be obtained upon perfusing liver with acetate was observed as early as 1912 by Loeb[40] but, as recalled above, attempts to isolate acetate from animal tissues had failed. Then the introduction of the isotopic method was taken advantage of by Weinhouse *et al.*[41] in experiments with liver slices. They used carboxyl-labeled octanoic acid and found that the ^{13}C isotope was incorporated only, and on an equal basis into the carboxyl and carbonyl carbons of acetoacetic acid. The incorporation as it should have taken place according to the theories is shown in Fig. 21. The results exclude Dakin's theory since

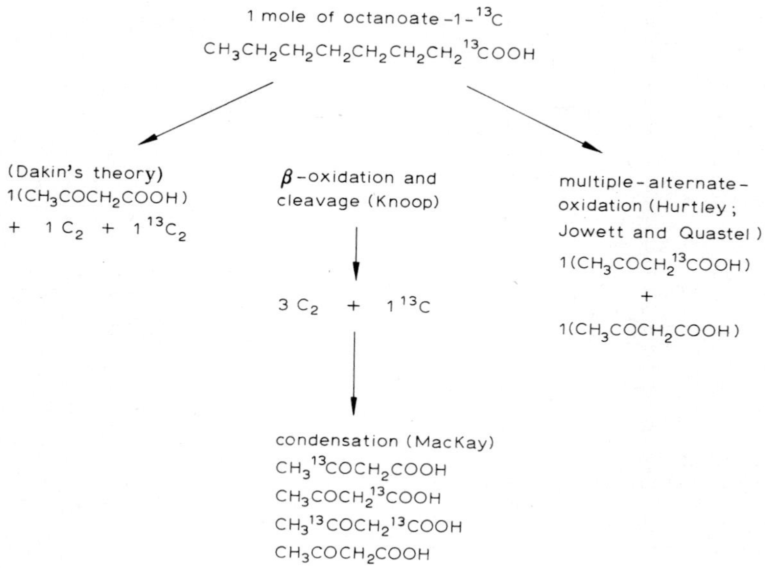

Fig. 21. Expected fate of [^{13}C]carboxyl according to several theories of fatty-acid oxidation.

according to it no isotope should have been found in the acetoacetate. They also exclude Hurtley's scheme according to which the isotope should have been found only in the carboxyl carbon of acetoacetic acid (Fig. 21). The observed incorporation both into the carbonyl and carboxyl carbons of acetoacetic acid was not in contradiction with MacKay's scheme, which had drawn support from a number of authors (literature in Chaikoff and Brown Jr.[13]).

"However, it is possible that acetoacetate labeled solely in the carboxyl position was first formed by four carbon fragmentations and that this substance was then rearranged into symmetrically labeled acetoacetate by reversible dissociation into two carbon units"[42]

as represented in Fig. 22. This possibility was suggested by Stadie[43].

The possibility of a rearrangement of singly labelled acetoacetate to give doubly labeled acetoacetate (by breaking down to C_2 fragments with subsequent random recondensation leading to symmetrically labeled acetoacetate) was tested by Buchanan *et al.*[44], who incubated carboxyl-labeled acetoacetate with liver preparations and observed that no isotope redistribution had taken place. Nonetheless, the results of other experiments disagreed with those of Weinhouse *et al.*[41]. For instance, isotope distribution in

References p. 317

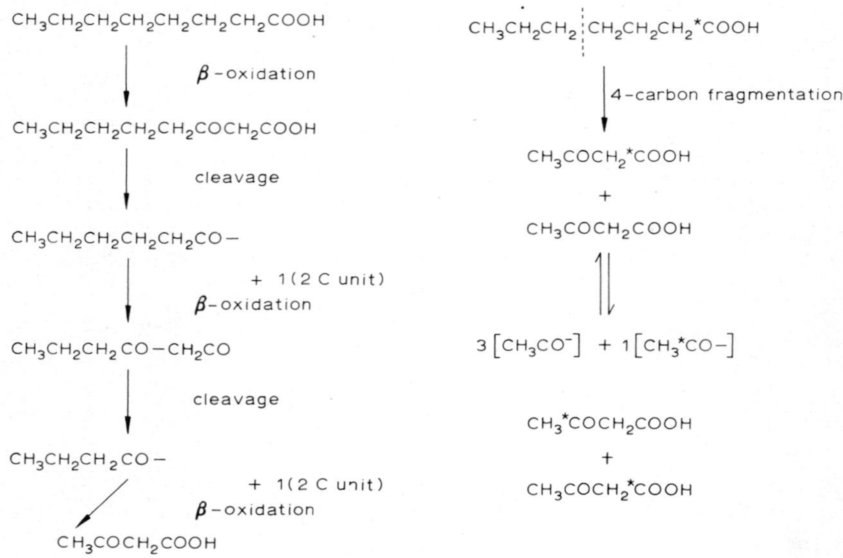

Fig. 22. (Gurin and Crandall[42]). The possibility of rearrangement of singly labeled acetoacetate to give doubly labeled acetoacetate.

acetoacetate formation from carboxyl-labeled fatty acids was found not to be symmetrical[44,45]. This led Crandall *et al.* to suggest that two different types of C_2 units derived from fatty acids: one (omega type) from the terminal 2 C atoms and the other (carboxyl type) from the other atoms (literature on this theory in Kennedy and Lehninger[46]).

This "two-species theory" was abandoned after a couple of years when it was shown that it did not agree with results obtained by Chaikoff *et al.*[47].

Taken together, these isotopic experiments seemed to show

"that two-carbon fragmentation *precedes* the formation of acetoacetate from fatty acids" (Gurin and Crandall[42]). This did not fail to reinforce the then prevailing belief that the pathway fatty acid→C_2→acetoacetate, was the essential fatty acid oxidation scheme.

(j) Detection, in vivo, of C_2 unit formation

Though the isotopic experiments, without undermining the view that an obligatory passage through acetoacetate exists, did bring into focus the fragments cleaved from the fatty acid chain; the C_2 fragment was not identified as such until its occurrence *in vivo* was detected by means of a method relying on the entrance of the C_2 unit in a compound which was excreted. The acetylation of foreign amines was assumed by Bernhard[48] who observed the excretion of deuterio-acetyl derivatives after administration to a dog of a foreign amine and of deuterio-acetic acid.

Bloch and Rittenberg[49] found the following deuterio fatty acids to be sources of deuterioacetyl groups: butyric acid, n-valeric acid, isovaleric acid and myristic acid. Propionic acid and undecylic acid failed to yield acetyl. Their results show that β-oxidation and removal of acetyl groups from the β-ketoacids was an important aspect in the metabolism of the fatty acids[50].

Though the theory of diphasic oxidation of fatty acid catabolism, as late as 1945, continued to hold onto an obligatory formation of acetoacetate in the liver and its subsequent metabolism in other tissues, attention (in the theory of acetoacetate formation) had nevertheless shifted from the residue left after β-oxidation to the C_2 units split off from the fatty acid chain.

2. New pathways for C_2 units split off from fatty acids

(a) Studies with mitochondria

The results obtained in the days of "physiological chemistry" were marked by puzzling discrepancies between the results of different sets of investigations (see Smedley-Maclean[51]). A new path of research was opened by Leloir and Munoz[31] in 1939 when they accomplished complete fatty acid oxidation in a cell-free system. They showed that washed liver particles oxidize fatty acids if the suspension is supplemented by an addition of inorganic phosphate, fumarate, adenylic acid, cytochrome *c* and Mg^{2+} or Mn^{2+}. After Claude (as shown in Chapter 16) developed the differential centrifugation method for the preparation of free mitochondria and showed that they were "the power plants of the cell", Schneider and Potter[52] in 1949 and Lehninger and Kennedy[53] in the same year identified as mitochondria the active particles in the preparation of Leloir and Munoz.

Lehninger[54] observed in 1945 that in the suspension prepared by Leloir and Munoz, fatty acid oxidation took place when there was a simultaneous oxidation of other metabolites, a coupling which became unnecessary in the presence of ATP. This suggested that the coupled oxidation produced ATP by oxidative phosphorylation.

References p. 317

Plate 101. Albert Lester Lehninger.

As observed by Knox *et al.*[55] a suspension of mitochondria supplemented with inorganic phosphate, Mg^{2+} and AMP, ADP or ATP, oxidizes any member of the Krebs cycle to CO_2 and water. On the other hand, it is without action on fatty acids except in the presence of a component of the Krebs cycle, the oxidation of which "sparks" the oxidation of the fatty acids.

Mitochondria prepared from kidney or from heart oxidize even-numbered fatty acids to CO_2 and water[56]. On the other hand, odd-numbered or branched fatty acids give rise to propionic acid[57]. A number of other experiments have also confirmed that in mitochondria the oxidation of fatty acids conforms to the pattern of β-oxidation (literature in Green[58]). In contrast with mitochondria of kidney and heart, liver mitochondria oxidize part of the fatty acid to CO_2 and water and another part to acetoacetate; this has been the subject of active investigation (literature in refs. 58, 59).

The study of mitochondria, though it produced no information regarding the mechanism at work, did re-establish on a firm footing Knoop's β-oxidation theory concerning mitochondria. Its results also refuted the "diphasic theory" and established the relation between the fatty-acid oxidizing system and the Krebs cycle. It should also be noted that the oxidation of acetate *via* the Krebs cycle in yeast was demonstrated in experiments by Sonderhoff and Thomas[60], by Virtanen and Sundmann[61] and by Lynen[62]. The same demonstration was afforded by Rittenberg and Bloch[63] for animal tissues. It should be kept in mind that this chapter concerns the oxidation of fatty acids to acetyl-CoA *in mitochondria* (including plant mitochondria), where it is known to occur *via* β-oxidation. In microsomal systems (derived from the endoplasmic reticulum), a conversion of the fatty acids to the ω-hydroxy compounds, further oxidized to dicarboxylic acids (ω-oxidation, Verkade[64]) is understood to take place. In microsomes, another enzymatic system (discovered by Fulco and Mead[65] in 1961) catalyses the 2-hydroxylation of long-chain fatty acids, in consequence of which decarboxylation of the α-hydroxy fatty acid results in odd-number fatty acids (α-oxidation). The mechanism of α-oxidation remains unknown and little is known of ω-oxidation. Also established by the mitochondrial studies was the equal rate of oxidation, at the level of these organelles, of saturated and unsaturated fatty acids[66]. Stoffel and Caesar[67] have reconstituted a sequence of reactions involving additional steps, by which acetyl-CoA derives from unsaturated fatty acids (see also Vol. 18 S of this Treatise).

(b) The concept of a common catabolic pathway involving the tricarboxylic cycle for pyruvate, acetate and acetoacetate

Once the metabolism of acetate (or rather an activated form of acetate) through the Krebs cycle (see Chapter 33) was understood, acetoacetate was no longer regarded as an obligatory intermediate in fatty-acid catabolism. So to regard it was, furthermore, at variance with the observations by Medes *et al.*[68] that none of the acetate utilized by the heart is converted to ketone bodies and that only part (± 50 per cent) of the acetate utilized by the kidney is so converted. These observations also showed that acetoacetate formation takes place in the kidney as well and thus is not a specific property of the liver as claimed by MacKay[69] and by Soskin and Levine[70]. It was known at the time that acetate and acetoacetate oxidation proceeds *via* the Krebs cycle. Yet, though CoA had not yet been discovered, there was a body of evidence showing that the substance which in carbohydrate oxidation reacts with oxaloacetate to form citrate was not pyruvate itself but an acetyl derivative formed by an oxidative decarboxylation of pyruvate (see Chapter 33, and also refs. 71, 72 and 73). These insights provided a basis

> "for the belief that acetate, pyruvate and acetoacetate are metabolized by a common pathway involving as intermediates components of the Krebs cycle. It is therefore conceivable that all three are convertible to a common intermediate..." providing "a common pathway for the complete oxidation of fatty acids and of carbohydrates, as well as certain amino acids, in so far as they can yield acetyl groups." (ref. 68)

(c) The activation of the β-carbon of fatty acids (formation of acyl-CoA's)

In Chapter 33, we explained the importance of CoA in the acetylation process (transacetylation) and of acetyl-CoA as the active 2-carbon intermediate. However, attempts to isolate acetyl-CoA were unsuccessful for several years until Lynen and Reichert[74] in 1951 isolated it from yeast and solved the problem of mechanism by showing the thiol group of CoA to be its active group, acetyl-CoA being recognized as a thiol ester of acetic acid and CoA. The activation of acetate to form acetyl-CoA was shown by Chou and Lipmann[75] to utilize the bond energy of ATP.

Berg[76] has formulated the reaction as follows:

$$\text{ATP} + \text{acetate} \rightleftharpoons \text{acetyl-AMP} + \text{PP}$$

$$\text{Acetyl-AMP} + \text{CoA} \rightleftharpoons \text{acetyl-CoA} + \text{AMP}$$

In such a scheme, the energy of ATP is preserved in the intermediate and subsequently transferred to acetyl-CoA. The activation of fatty acids takes place along the same lines, CoA being the active acyl acceptor[77]. In mammalian systems the activation of the fatty acids can be represented as follows:

$$RCH_2CH_2CH_2COOH + ATP + CoASH \rightleftharpoons RCH_2CH_2CH_2CH_2COSCoA + AMP + PP_i$$

while in bacteria acetyl CoA may directly activate the fatty acid without expending ATP. This is due to the presence in these bacteria of the propionate-CoA transferase (EN* 2.8.3.1), identified by Stadtman[78] in 1953:

$$\text{Acetyl-CoA} + RCOOH \rightleftharpoons \text{acetate} + RCOSCoA$$

β-Oxidation of fatty acids is a consequence of the action of CoA in the activation of the β-carbon of an acyl-thioester. This activation takes place in mammals in the presence of an enzyme and of ATP and Mg^{2+}:

$$RCOOH + CoASH + ATP \underset{Mg^{2+}}{\rightleftharpoons} RCOSCoA + AMP + PP_i$$

A number of laboratories have investigated the different enzymes involved (literature in Lynen[79]) and from this work the identification of three well-defined enzymes has emerged:

—Acetyl-CoA synthetase [acetate:CoA ligase (AMP-forming), EN 6.2.1.1]. This enzyme activates acetate, propionate and acrylate in decreasing order. It is also known as the acetyl-activating enzyme, and an acetate thiokinase; it was isolated by Rose *et al.*[80] from *S. haemolyticus* and *E. coli*. An acetyl-activating enzyme was also isolated by Jones *et al.*[81] from a variety of sources (animal and plant tissues, yeast, *Rhodospirillum rubrum*). The formation of acetyl-CoA from acetate occurs through pyrophosphate cleavage of ATP:

$$ATP + \text{acetate} + CoASH \rightleftharpoons \text{acetyl-CoA} + AMP + PPi$$

Only a single enzyme and ATP participate in the reaction, together with[81–83] Mg^{2+}. Webster[84] has found the molecular weight of the crystallized enzyme to be equal to 35200. He has shown that one mole of acetyladenylate is bound per mole of enzyme.

$$\text{Acetate} + ATP + E \rightleftharpoons \text{E-(acetyladenylate)} + PP_i$$

$$\text{E-(acetyladenylate)} + CoA \rightleftharpoons \text{acetyl-CoA} + AMP + E$$

* EN, Enzyme Nomenclature, 3rd ed., Amsterdam, 1972 (also 3rd ed. of Vol. 13 of this Treatise)

The enzyme has a multiple requirement monovalent (K^+) and divalent (Mg^{2+}) cations for the half-reaction leading to E-(acetyladenylate)[85,86].

—Butyryl-CoA synthetase [butyrate:CoA ligase (AMP-forming), EN 6.2.1.2] is the short-chain (C_4–C_{11}) fatty-acid activating enzyme. It was isolated from beef liver mitochondria by Mahler *et al.*[87].

—Acyl-CoA synthetase [acid:CoA ligase (AMP-forming), EN 6.2.1.3] is the long-chain fatty-acid (C_6–C_{20}) activating enzyme. In extracts of guinea-pig liver, Kornberg and Pricer[88] discovered an enzyme converting long-chain fatty acids into corresponding acyl-CoA's. This enzyme also converts unsaturated fatty acids (linolic, linolenic, oleic, chaulmoogric).

More recently Galzigna *et al.*[89] have found in mitochondrial membranes an enzyme which utilizes long-chain fatty acids with a phospholipid requirement, and uses GTP as substrate.

$$\text{RCOOH} + \text{GTP} + \text{CoASH} \rightleftharpoons \text{RCOSCoA} + \text{GDP} + \text{P}_i$$

Acyl-CoA can also be synthesized by a transacylation catalysed by a thiophorase, as shown by Stern *et al.*[90]:

$$\text{succinyl-CoA} + \underset{(\text{short chain};\ C_4,\ C_6)}{\text{RCOOH}} \rightleftharpoons \text{succinate} + \text{RCOSCoA}$$

(d) Oxidative degradation of fatty acids by complete soluble systems

Stadtman[78], using soluble enzymes from *Cl.kluyveri* obtained a degradation of butyrate to acetate or acetylphosphate by way of butyryl-CoA and acetoacetyl-CoA.

The first oxidation accomplished by a soluble system of animal origin was realized by Drysdale and Lardy[91] who found that clear extracts of rat liver mitochondria decolorize 2,6-dichlorphenol-indophenol in the presence of fatty acids (C_4 to C_{18}) and of ATP which was understood to be required for the activation of fatty acids. Depending upon whether or not oxaloacetate was added to the system, citrate or acetoacetate appeared as oxidation products. If the extract of mitochondria was previously treated with Dowex 1, the system required the addition of CoA.

Although fatty-acid oxidizing systems were successively studied at the level of the whole organism, of isolated liver, of liver slices, of liver homogenates and of particle preparations, none of these increasingly destructive methods allowed for an accumulation of an intermediate product (except acetoacetate) between fatty acid on one hand and CO_2 and H_2O on the other hand. Only with soluble enzymes was it possible to reconstitute parts of the sequence and to accumulate intermediates.

(e) Enzymes of β-oxidation

To search out the individual enzymes of β-oxidation it was necessary to know the coenzyme requirements of the overall process. This information had already been made available by Drysdale and Lardy[91], Green and Beinert[92] and Mahler[93]. The essential electron acceptor for fatty acid oxidation was NAD and it appeared that no other additional coenzymes were required other than CoA, Mg^{2+} and NADP (for citrate oxidation).

Assay systems for routine screening were designed by Mahler[93] who used triphenyl tetrazolium as final electron acceptor with catalytic amounts of methylene blue. Another assay system was designed by Drysdale and Lardy[91] who used 2,6-dichlorphenol indophenol as final electron acceptor. Both of these electron acceptors were satisfying as they did not oxidize CoASH, which would have prevented either the formation of acyl-CoA or the cleavage of acetoacetyl-CoA to acetyl-CoA.

Large-scale isolation of enzymes from mitochondria became feasible (literature in Green[58]). The stage was thus set for overall screening assays and it became possible for Lynen and Ochoa[94] and for Beinert *et al.*[95] simultaneously to identify a succession of enzymatic steps (see Fig. 23, steps 2–5): a dehydrogenation of fatty acyl-CoA, a hydration of unsaturated fatty acyl-CoA and a dehydrogenation of β-hydroxyacyl-CoA resulting in a β-ketoacyl-CoA. As the overall nature of the pathway was identified, one-step assays replaced overall screening and methods were devised, as had successfully been done in the case of glycolysis, to obtain preparations of the individual enzymes involved, which were successively isolated.

In Lynen's laboratory, which at the time lacked the activation enzyme for the preparation of acyl-CoAs, another method was adopted. The pathway, instead of being studied in the oxidative direction, was studied in the reducing direction, which is possible because all the steps are reversible. This called for organic chemistry to provide the synthetic substrates (for literature on the preparation of the thioesters of *N*-acetyl-cysteamine by chemical synthesis and of the CoA derivatives by enzymatic methods, see Lynen[79]. Acyl-CoAs have more recently been synthesized by the inter-

References p. 317

Plate 102. David Ezra Green.

1. Formation of the coenzyme A thioester (acyl-CoA)

$$RCOOH + CoASH + ATP \underset{Mg^{2+}}{\overset{\text{acyl-CoA synthetase}}{\rightleftharpoons}} RCOSCoA + AMP + PP_i \quad (1)$$

2. α,β-dehydrogenation of acyl-CoA

$$RCH_2CH_2COSCoA + \text{acceptor} \rightarrow \textit{trans}\text{-}RCH{=}CHCOSCoA + \text{reduced acceptor} \quad (2)$$

3. Hydration of α,β-unsaturated fatty acyl-CoA

$$\textit{trans}\text{-}RCH{=}CHCOSCoA + H_2O \rightleftharpoons \text{L-}RCHOHCH_2COSCoA \quad (3)$$

4. Oxidation of the β-hydroxyacyl-CoA ester to the corresponding β-ketoacyl-CoA derivative

$$\text{L-}RCHOHCH_2COSCoA + NAD^+ \rightleftharpoons R\overset{\displaystyle O}{\overset{\|}{C}}CH_2COSCoA + NADH \quad (4)$$

5. Thiolytic cleavage of the β-ketoacyl-CoA ester to acetyl-CoA and a new fatty acyl-CoA ester undergoing cleavage to acyl-CoA and acetyl-CoA.

$$R\overset{\displaystyle O}{\overset{\|}{C}}CH_2COSCoA + CoASH \rightleftharpoons \underset{\text{acyl-CoA}}{RCOSCoA} + \underset{\text{acetyl-CoA}}{CH_3COSCoA} \quad (5)$$

Fig. 23. Steps of β-oxidation of fatty acids preceded by acylation of CoA by fatty acids. Net result: an acetic acid is split off in the form of acetyl-CoA and a fatty acyl CoA ester with two carbons less is formed. The cycle is repeated until the chain is completely degraded to units of acetyl-CoA. The cycle involves CoA derivatives and not free acids as in Dakin's scheme (Fig. 20).

action of CoA either with thiol acids or with acid anhydrides).

Lynen and his group (see ref. 96), for instance, synthesized the *S*-acetoacetyl derivative of *N*-acetyl thioethanolamine as substrate for β-hydroxyacyl-CoA dehydrogenase in reverse and they prepared the crotonyl derivative for the study of the acyl-CoA dehydrogenase in reverse: the synthetic substrates were observed to take the place of the normal substrates. Fatty acyl-CoA is transformed to the corresponding β-ketoacyl derivative by the action of two dehydrogenases and a hydratase, and a cleaving enzyme insures a link between successive β-oxidation cycles as well as between fatty acid oxidation and the Krebs cycle.

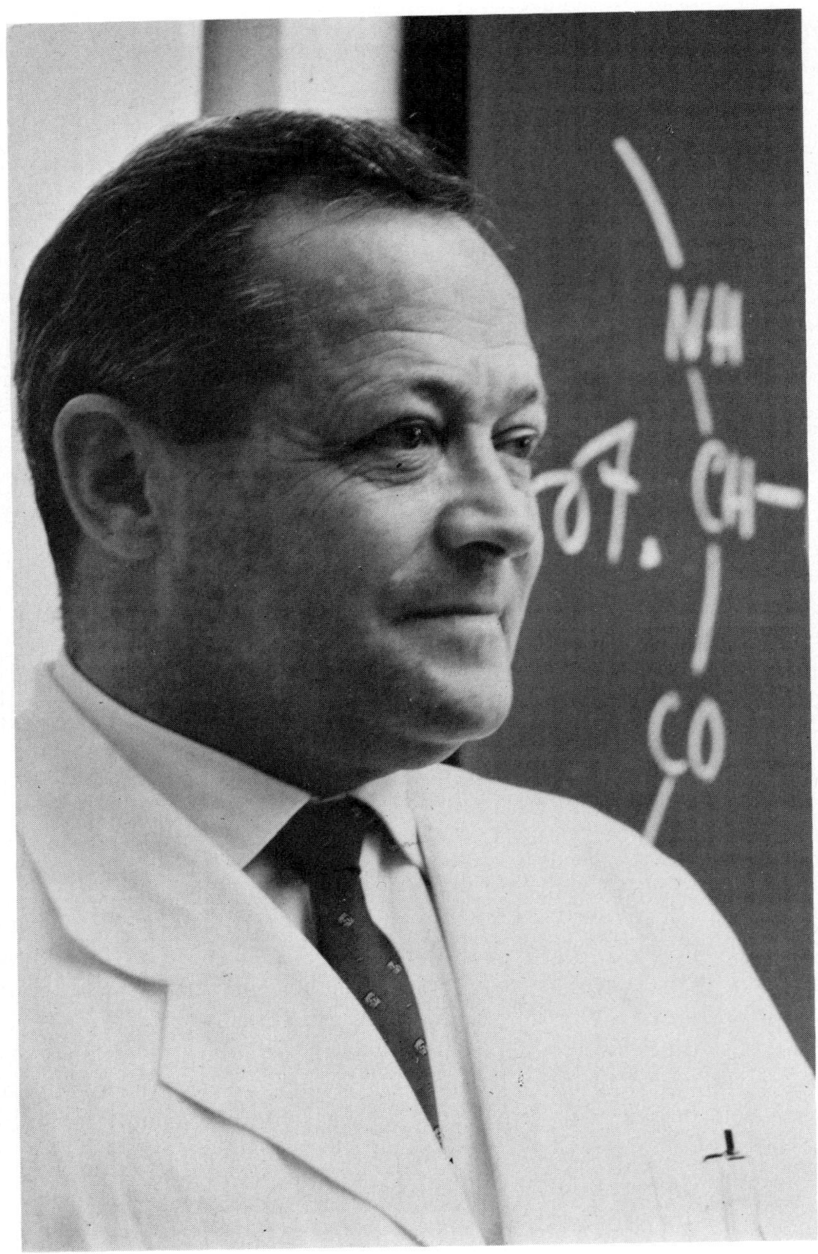

Plate 103. Feodor Lynen.

—Acyl-CoA dehydrogenases *e.g.* butyryl-CoA dehydrogenase [butyryl-CoA: (acceptor) oxidoreductase: EN 1.3.99.2] and two acyl-CoA dehydrogenases [acyl-CoA:(acceptor) oxidoreductase, EN 1.3.99.3].

The acyl-CoA dehydrogenases (step 2 in Fig. 23) have been shown to differ in chain-length specificity and together to bring about the degradation of acyl-CoA derivatives of C_4 to C_{20} chain length[97–99]. They are flavoproteins, two electrons being transferred to another FAD-flavoprotein which links to the cytochrome *b* of the respiratory chain most likely through CoQ, thence to cytochrome *b*[100–101]. An acyl-CoA dehydrogenase was first isolated by Green *et al.*[102] in the form of a green-colored preparation from beef liver mitochondria; it was the first metallo-flavoprotein (copper) to be recognized as such.

—Enoyl-CoA hydratase, sometimes called crotonase [L-3-hydroxyacyl-CoA hydro-lyase, EN 4.2.1.17]. This enzyme catalyses reaction 3 in Fig. 23. It was first isolated in crystallized form by Stern *et al.*[103] from ox liver, and by Wakil and Mahler[104]. Wakil[105] has shown that the enzyme has a broad specificity and can hydrate *cis-* as well as *trans-α,β*-unsaturated fatty acyl-CoA esters. (The latter is hydrated to the L(+)antipode of β-hydroxyacyl-CoA).

—3-Hydroxyacyl-CoA dehydrogenase, or β-hydroxyacyl dehydrogenase [L-3-hydroxyacyl-CoA:NAD oxidoreductase, EN 1.1.1.35]. As shown in step 4 of Fig. 23, the L-β-hydroxyacyl-CoA derived from step 3 is oxidized to the corresponding β-ketoacyl-CoA ester in the presence of both NAD^+ and the enzyme which was first purified from sheep liver by Lynen *et al.*[106], and from beef liver mitochondria by Wakil *et al.*[107]. The specificity of the enzyme for the L-isomer is strict but it can be used in a wide variety of chain lengths.

—3-ketoacyl-CoA thiolase or acetyl-CoA acyltransferase [acyl-CoA:acetyl-CoA *C* acyltransferase EN 2.3.1.16]. This enzyme is involved in step 5 of Fig. 23; it was independently discovered in three laboratories (Lynen and Ochoa[94] who called it thiolase; Goldman[108], who called it β-ketoacetyl-CoA cleavage enzyme, and Stern *et al.*[109] who named it 3-ketoacyl-CoA thiolase).

Given the fact that thiolase is inhibited by SH-poisons (iodoacetic acid, As_2O_3), Lynen[96] proposed the following mechanism of action:

$$RCOCH_2COSCoA + HS\text{–}E \rightarrow RCOS\text{–}E + CH_3COSCoA$$

$$RCOS\text{–}E + CoA\text{–}SH \rightarrow RCOS\text{–}CoA + HS\text{–}E$$

The thiolytic cleavage is in fact a displacement of the acetyl group by the

References p. 317

SH group. Lynen showed that the propionyl-CoA took up radioactivity on incubation with $H^{35}S$-CoA and purified thiolase, a fact in favour of the reaction:

$$\text{propionyl-SCoA} + \text{HS–E} \rightleftharpoons \text{propionyl–S–E} + \text{HSCoA}$$

Thiolase was crystallized from ox liver by Seubert, Lamberts, Kramer and Ohly[110] in 1968. Their preparation showed a broad substrate specificity.

For a more detailed account of recent progress in the study of the enzymes of β-oxidation, the reader is referred to reviews by Green and Altmann[59] and by Stumpf[111] as well as to the chapter on fatty-acid metabolism in volume 18 S of this Treatise.

As was noted by Green[58], the order in which the enzymatic steps from fatty acids to acetyl-CoA were discovered is the reverse of the order of the steps in the sequence.

(f) The site of β-oxidation and the role of carnitine

Acyl-CoAs are formed outside the mitochondrial inner membrane and their β-oxidation takes place in the matrix of the mitochondrion, whilst acyl-CoAs are unable to cross the inner membrane. This is only one of many aspects of the pools of CoA corresponding to the compartmentalization of mitochondria with respect to fatty-acid metabolism (see the chapter by R. Bressler in Vol. 18 of this Treatise). The problem of compartmentalization is solved by the utilization of carnitine, as acyl-carnitine esters, for instance, cross membranes that acyl-CoAs cannot penetrate. Carnitine (γ-trimethylammonium-β-hydroxybutyrate)

$$(CH_3)_3N^+CH_2\underset{\displaystyle \mathrm{OH}}{\underset{|}{C}}HCH_2COOH$$

has been known for a long time to be widely distributed in organisms though its role remained mysterious (for historical literature on carnitine, see Fraenkel and Friedman[112]). The compound is biosynthesized in the liver of mammals whence it is distributed to other parts[113]. The metabolic importance of carnitine was suggested by the observation that it is an obligatory component of the diet of insects, a phylum which has lost the capacity for its biosynthesis.

The role played by carnitine in fatty-acid metabolism consists of transferring acyl groups between two mitochondrial pools of CoA, the inner one being in contact with the enzymes of β-oxidation in the mitochondrial matrix and the outer one being accessible to fatty acids or acyl-CoAs unable to cross the inner mitochondrial membranes. Credit for the discovery of the activity of carnitine as carrier of acyl groups is due to Lundsgaard[114] who observed in 1950 that a factor obtained by perfusion of muscle increased the oxygen consumption of perfused liver. One of Lundsgaard's collaborators named Fritz[115], by scanning a number of muscle components showed, in 1955, that carnitine was the only compound to produce an increase in oxygen consumption by rat-liver slices, with palmitic acid as substrate. Fritz conceived the hypothesis that carnitine is necessary to allow the transfer of long fatty acids through mitochondrial membranes. Underlying this hypothesis was the fact that, if the oxidation of palmitate by washed particle preparations from liver muscle was increased by carnitine, this compound has no influence on the oxidation by extracts of liver particles[116].

Fritz and Yue[117] demonstrated the synthesis of palmitylcarnitine from carnitine and palmitate by heart muscle preparations in the presence of ATP and CoA or without any addition, from palmityl-CoA and carnitine.

$$\text{Palmityl-CoA} + \text{carnitine} \rightleftharpoons \text{palmitylcarnitine} + \text{CoA}$$

The enzyme catalyzing the reaction had been identified by Friedman and Fraenkel[118] as early as 1955. That carnitine is charged with transferring acyl groups between different pools of mitochondrial CoA appeared also from experiments on the effects of inhibitors on the oxidation of palmitate and palmitylcarnitine by isolated mitochondria[119].

3. Ketone bodies

(a) Formation of acetoacetyl-CoA

While in the framework of general biochemistry the pathway of fatty-acid oxidation leading to acetyl-CoA which enters the cycle of tricarboxylic acid (Lynen's *spiral*) came to be recognized as a feature of mitochondrial biochemistry, the case of fatty-acid metabolism in the whole mammalian organism remained a subject for discussion. The classical diphasic theory admitted that in the liver mitochondria, ketone bodies were formed and liberated into the blood stream to undergo degradation in other tissues. That acetoacetate can be obtained upon perfusing liver with acetate was known as early as 1912 (refs. 40, 120); this finding has been repeatedly confirmed

References p. 317

(literature in Bloch[50]). According to present theory, the accumulation of acetoacetate in *in vitro* experiments or in the urine of organisms in ketosis, is ascribed to circumstances interfering with the disposal of acetyl-CoA through the Krebs cycle. In isolated liver or in liver preparations this results from a lack of available carbohydrate.

Lehninger[54] found that in preparations of washed liver cells, acetoacetate is the sole product of octanoic acid oxidation whereas if fumaric acid is added the amount of acetoacetate produced diminishes.

After the discovery of acetyl-CoA, other pathways came to light. Acetoacetyl-CoA was identified by Lynen *et al.*[106] and by Stern *et al.*[103] and it was shown that it can be derived by oxidation from the terminal fragment of the β-oxidation of even-numbered fatty acids, butyryl-CoA, and also from the condensation of two molecules of acetyl-CoA in the presence of a thiolase

2 acetyl-CoA $\rightleftharpoons$ acetoacetyl-CoA + CoA (Green *et al.*[121])

In this reaction, as well as in the condensation of acetyl-CoA with oxaloacetate, CoA is regenerated to form citrate. In the absence of oxaloacetate, liver mitochondria convert fatty acids quantitatively into acetoacetate, *via* acetoacetyl-CoA. Acetyl-CoA formed by β-oxidation in liver mitochondria enters the Krebs cycle if oxaloacetate is available.

As shown by Lehninger[122], in the absence of oxaloacetate, liver mitochondria convert fatty acids almost quantitatively to acetoacetate. But even if the pathway through ketone bodies is followed, the coenzyme A derivatives are formed[94].

(b) Liberation of ketone bodies in the liver and their degradation in other tissues

Two types of metabolic processes have been proposed to account for the utilisation of acetoacetic acid in animal tissues: reduction to β-hydroxybutyric acid, or oxidation to CO_2 and H_2O. The reduction to β-hydroxybutyric acid was first observed and shown to be reversible, by Friedmann and Maase[123a] and by Wakeman and Dakin[124].

The occurrence of an oxidative breakdown of acetoacetate is part of the classical theory of fatty-acid catabolism through the intermediary of acetoacetate and it received additional support from experiments by Snapper and Grünbaum[125] in 1927.

Acetoacetate is derived from acetoacetyl-CoA and is carried to other regions of the organism as such or in its reduced form, β-hydroxybutyrate. In isolated liver mitochondria the end product of fatty-acid metabolism is mainly free acetoacetic acid[56,122,126].

This may be due to the presence of a deacylating system and to the lack of a specific thiokinase system for the synthesis of acetoacetyl-CoA from acetoacetate[59]. In different tissues the formation of free acetoacetate will be observed if the rate of deacylation exceeds the rate of conversion of acetoacetate to acetoacetyl-CoA. The mechanism of acetoacetate formation from acetoacetyl-CoA has been attributed to the interplay of two enzymes[127,128]. One, isolated from liver by Rudney[129], catalyzes the condensation of acetyl-CoA with acetoacetyl-CoA into β-hydroxy-β-methylglutaryl-CoA:

$$\mathrm{COOH{-}CH_2{-}\overset{\displaystyle OH}{\underset{\displaystyle CH_3}{C}}{-}CH_2{-}\underset{\displaystyle \|\atop O}{C}{-}S{-}CoA}$$

The other, isolated by Bachhawat *et al.*[130] cleaves this molecule into acetyl-CoA and acetoacetate.

(c) Catabolism of acetoacetate

Acetoacetate or its reduction product (β-hydroxybutyrate) carried to the tissues contributes, together with glucose liberated from liver glycogen, and fatty acids lipolyzed from adipose tissue, to the supply of energy-producing compounds.

Breusch[131] as well as Wieland and Rosenthal[132] suggested that the $-CH_2COOH$ grouping of acetoacetate is directly transferred as a whole to oxaloacetate with formation of citrate.

More recent work has shown that the pathway from acetoacetate to the tricarboxylic acid cycle takes place mainly through acetoacetyl-CoA.

Stern *et al.*[133] have presented evidence for the reversible enzymatic transfer of CoA from succinyl-CoA to acetoacetate according to the reaction:

$$\text{Succinyl-CoA} + \text{acetoacetate} \rightleftharpoons \text{acetoacetyl-CoA}^- + \text{succinate}^- + H^+$$

and Stern[134] has named the enzyme involved succinyl-β-ketoacyl-CoA transferase, or more simply CoA transferase.

The enzyme has been purified from pig heart by Stern *et al.*[135].

References p. 317

CoA transferase is involved in the entrance of most of the acetoacetate into the tricarboxylic acid cycle of peripheral tissues (particularly heart and kidney) after conversion to acetoacetyl-CoA. The Krebs cycle in turn generates the necessary succinyl-CoA by oxidation of α-ketoglutarate. That the heart metabolizes acetoacetate has been shown *in situ* both in man by Bing *et al.*[136] and in dog by Waters *et al.*[137], and by Barnes *et al.*[138].

Bing *et al.*[136] found that the human heart utilizes 0.50 to 6.1 μmoles of blood ketone per 100 g of heart, per minute, at 37°. From the content of heart muscle in CoA transferase and from the turnover number of the enzyme, Stern *et al.*[135] have calculated values compatible with the utlization rates observed by Bing *et al.*[136].

(d) The significance of physiological ketosis

In our present view, ketone body formation represents a lateral extension of the mitochondrial pathways of fatty-acid catabolism through β-carbon activation, splitting into C_2 units (acetyl-CoA), the condensation of C_2 units with oxaloacetate forming citrate which is converted to CO_2 and H_2O through the tricarboxylic acid cycle. Its identification in the course of studies on diabetes later led to a consideration of ketosis as a pathological feature. This is an oversimplified view.

Ketosis must be considered in the framework of the physiological concept of "caloric homeostasis", introduced by Fredrickson and Gordon[139] to describe the respective roles played by the circulatory energy-producing compound in providing the respiratory fuel to cells.

Heart muscle, for example, as shown by Williamson and Krebs[140], can use acetoacetate preferentially even when glucose, insulin and other substrates are available. Kidney cortex also tends to use acetoacetate in preference to glucose or lactate (Krebs and Speake[141]). The living sheep can derive as much as one-third of its total respiratory CO_2 from ketone bodies when these are supplied by intravenous infusion[142]. To quote Krebs[143]:

"Whenever the glucose level in the blood plasma is low, as in starvation or on a low carbohydrate diet, or when glucose is not available, as in diabetes, the concentration of free fatty acids in the plasma rises. This rise is roughly paralleled by an increase in the concentration of ketone bodies which provide a third source of energy. In other words, the moderate ketosis which occurs under a variety of circumstances is to be looked upon as a normal physiological process supplying the tissues with a readily utilizable fuel of respiration when glucose is short in supply (see also Mayes[144]). The production of oxaloacetate in this form of ketosis is normal and unconnected with the rise of the concentration of the ketone bodies."

According to Krebs, the production of ketone bodies is related to the rate of liberation of free fatty acids from adipose tissues and to the increase in concentration of the free fatty acids in the plasma.

(e) Pathological ketosis

As established by the classical observations of Lehninger, in mitochondria the production of ketone bodies is related to the availability of oxaloacetate which leads acetyl-CoA to the tricarboxylic acid cycle and diverts it from the formation of ketone bodies. According to Krebs[143], one of the causes of the pathological ketosis leading to diabetic acidosis results from the utilization of oxaloacetate for the formation of phosphopyruvate utilized in gluconeogenesis. To quote Krebs[143],

"The often quoted dictum of Rosenfeld[145] that fats and ketone bodies burn in the fire of carbohydrate is no longer to the point. It would be more correct to say that the synthesis of carbohydrate prevents the combustion of the ketone bodies because a catalyst needed for the combustion of acetyl-CoA and the ketone bodies-oxaloacetate- serves as the immediate precursor for carbohydrate. It is the excessive demand for carbohydrate, and not the failure to oxidize carbohydrate, which causes the accumulation of ketone bodies."

The liver compensates for the loss of energy resulting from the diminished rate of the tricarboxylic acid cycle by an increased rate of other forms of ATP-providing reactions.

More recent developments in the area of ketosis and its hormonal regulation are sketched in R. Bressler's chapter in Volume 18 of this Treatise.

4. Retrospect

The first studies on fatty acid catabolism were inspired by clinical observation of diabetic patients. This led to the brilliant discovery of β-oxidation by Knoop in 1904. Elaboration of this discovery through the methods of physiological chemists led to the assembly of a large amount of data resulting in Dakin's comprehensive classical theory according to which ketone bodies are obligatory intermediates resulting from the splitting off of C_2 units leaving a 4C oxidized product, acetoacetic acid. Other theories regarded the split-off C_2 units as sources of acetoacetate.

References p. 317

A second period started with the use of isolated mitochondria, which revealed the link between fatty acid catabolism and the tricarboxylic acid cycle.

A third period began with the realisation of complete fatty acid oxidation by clear extracts of rat liver mitochondria. This was accomplished by Drysdale and Lardy in 1952, half a century after Knoop's pioneer study. New metabolic pathways were opened for the C_2 units split off from fatty acids in the course of Knoop's β-oxidation, after Lipmann recognized the role of acetyl-CoA and after Lynen isolated this compound, and characterized the nature of its high-energy link and introduced the concept of acyl-CoA resulting from the activation of fatty acids by CoA. Energy-rich thioester compounds were recognized as intermediates in β-oxidation and the relevant enzymology was cleared up. While during the half century following Knoop's discovery attention had been focused on the residue left after β-oxidation, emphasis was now placed upon the C_2 units splitt off in the form of the energy-rich compound acetyl-CoA, which joins the acetyl-CoA derived from glucose in entering the tricarboxylic acid cycle.

In this case of fatty acids CoA was recognized as playing the activating role exerted by phosphates in carbohydrate metabolism.

"By conversion into the CoA derivative, the fatty acid is "activated" and thereupon yields acetyl-CoA and an acyl-CoA with two carbons less than the starting substance. This is subjected immediately to the same degradation sequence in a second cycle, again liberating acetyl-CoA and so on." (ref. 79)

The unravelling of the pathway of fatty-acid catabolism has followed a historical sequence which was characterized by the different preoccupations and methodologies which succeeded each other since the turn of the century. The experimentation on whole animals or on isolated organs has been followed by increasingly destructive methods finally reaching the purified enzymes of the system which can be reconstructed *in vitro*.

This historical development has been highlighted by Lynen's discovery of energy-rich thioester compounds, an acquisition which provided the key to the energetics of the process. Ketone bodies, though first regarded as obligatory intermediates in the process of fatty-acid oxidation, were eventually recognized as belonging to an ancillary system involved in the homeostasis of mammals, a lateral system which was mistakenly believed to be the main pathway, during the lengthy period in which experiments on mammals dominated the field of fatty-acid catabolism.

REFERENCES

1 F. Wöhler (cited after Dakin[16]), 1 (1824).
2 R. Buchheim, *Arch. Physiol. Heilk.*, 1 (1857) 122.
3 C. Schotten, *Z. Physiol. Chem.*, 7 (1882) 375.
4 J. Pohl, *Arch. Exptl. Pathol. Pharm.*, 37 (1895) 413.
5 P. Bergell, *Verhandl. Kongr. Inn. Med.*, 24 (1907) 236.
6 F. Lieben, *Geschichte der Physiologischen Chemie*, Leipzig and Vienna 1935 (Reprinted by Olms, Hildesheim, 1970).
7 C. Gerhardt, *Wien. Med. Presse*, 6 (1865) 672.
8 V. Arnold, *Zentr. Inn. Med.*, 21 (1900) 417.
9 H. C. Geelmuyden, *Z. Physiol. Chem.*, 23 (1897) 431.
10 A. Magnus-Levy. *Arch. Exptl. Path. Pharm.*, 42 (1899) 149.
11 R. Waldvogel, *Zentr. Inn. Med.*, 20 (1899) 729.
12 L. Schwarz, *Zentr. Stoffw. Verdauungskrankh.*, 1 (1900) 1.
13 I. L. Chaikoff and G. W. Brown Jr., in D. M. Greenberg (Ed.), *Chemical Pathways of Metabolism*, Vol. I, New York, 1954.
14 F. Knoop, *Der Abbau Aromatischer Fettsäure in Tierkörper*, Freiburg (Baden), 1904.
15 F. Knoop, *Beitr. Chem. Physiol. Pathol.*, 6 (1905) 150.
16 H. D. Dakin, *Oxidations and Reductions in the Animal Body*, London, 1912.
17 E. Friedmann, *Med. Klin.*, (1909) Nrs. 36 and 37.
18 H. D. Dakin, *J. Biol. Chem.*, 4 (1908) 77.
19 F. Knoop, *Beitr. Chem. Physiol. Pathol.*, 11 (1908) 411.
20 G. Embden and F. Kalberlah, *Beitr. Chem. Physiol. Pathol.*, 8 (1906) 121.
21 G. Embden, H. Salomon and F. Schmidt, *Beitr. Chem. Physiol. Pathol.*, 8 (1906) 129.
22 G. Embden and A. Marx, *Beitr. Chem. Physiol. Pathol.*, 11 (1908) 318.
23 G. Embden and H. Engel, *Beitr. Chem. Physiol. Pathol.*, 11 (1908) 323.
24 G. Embden and L. Michaud, *Beitr. Chem. Physiol. Pathol.*, 11 (1908) 332.
25 G. Embden and L. Michaud, *Biochem. Z.*, 13 (1908) 262.
26 J. Baer and L. Blum, *Arch. Exptl. Path. Pharm.*, 55 (1906) 89; 56 (1906) 92; 59 (1908) 321; 62 (1910) 129; (1911) 1.
27 H. D. Dakin, *J. Biol. Chem.*, 4 (1908) 419; 5 (1908–1909) 173, 303; 6 (1909) 203, 221.
28 E. and H. Salkowski, *Ber. deut. Chem. Ges.*, 12 (1979) 653.
29 M. Nencki, *J. Prakt. Chem.*, 18 (1878) 288.
30 H. J. Deuel, L. F. Hallman, J. S. Butts and S. Murray, *J. Biol. Chem.*, 116 (1936) 621.
31 L. F. Leloir and J. M. Munoz, *Biochem. J.*, 33 (1939) 734.
32 R. F. Witter, E. H. Newcomb and E. Stotz, *J. Biol. Chem.*, 185 (1950) 537; 195 (1952) 663.
33 J. H. Quastel and A. H. M. Wheatley, *Biochem. J.*, 27 (1933) 1753.
34 E. M. MacKay, A. N. Wick and C. P. Barnum, *J. Biol. Chem.*, 136 (1940) 503.
35 R. P. Geyer, M. Cunningham and J. Pendergast, *J. Biol. Chem.*, 185 (1950) 461.
36 R. P. Geyer, M. Cunningham and J. Pendergast, *J. Biol. Chem.*, 188 (1950) 185.
37 W. H. Hurtley, *Quart. J. Med.*, 9 (1915) 301.
38 M. Jowett and J. Quastel, *Biochem. J.*, 29 (1935) 2159.
39 E. M. MacKay, R. H. Barnes, H. O. Carne and A. N. Wick, *J. Biol. Chem.*, 135 (1940) 157.
40 A. Loeb, *Biochem. Z.*, 47 (1912) 118.
41 S. Weinhouse, G. Medes and N. F. Floyd, *J. Biol. Chem.*, 153 (1944) 689; 155 (1944) 143.
42 S. Gurin and D. I. Crandall, *Cold Spring Harbor Symp.*, 13 (1948) 118.

43 W. C. Stadie, *Physiol. Rev.*, 25 (1945) 395.
44 J. M. Buchanan, W. Sakami and S. Gurin, *J. Biol. Chem.*, 169 (1947) 411.
45 R. P. Geyer, M. Cunningham and J. Pendergast, *J. Biol. Chem.*, 185 (1950) 461.
46 E. P. Kennedy and A. L. Lehninger, in W. D. McElroy and B. Glass (Eds.), *A Symposium on Phosphorous Metabolism*, Vol. 2, Baltimore, 1952, p. 253.
47 I. L. Chaikoff, D. S. Goldman, G. W. Brown, W. G. Dauben and M. Gee, *J. Biol. Chem.*, 190 (1951) 229.
48 K. Bernhard, *Z. Physiol. Chem.*, 267 (1940) 91.
49 K. Bloch and D. Rittenberg, *J. Biol. Chem.*, 155 (1944) 243.
50 K. Bloch, *Physiol. Rev.*, 27 (1947) 574.
51 I. Smedley-Maclean, *The Metabolism of Fat*, London, 1943.
52 W. C. Schneider and V. R. Potter, *J. Biol. Chem.*, 177 (1949) 897.
53 A. L. Lehninger and E. P. Kennedy, *J. Biol. Chem.*, 179 (1949) 957.
54 A. L. Lehninger, *J. Biol. Chem.*, 161 (1945) 413.
55 W. E. Knox, B. N. Noyce and V. H. Auerbach, *J. Biol. Chem.*, 176 (1948) 117.
56 A. L. Grafflin and D. E. Green, *J. Biol. Chem.*, 176 (1948) 95.
57 W. A. Atchley, *J. Biol. Chem.*, 176 (1948) 123.
58 D. E. Green, *Biol. Rev.*, 29 (1954) 330.
59 D. E. Green and D. W. Altmann, in D. M. Greenberg (Ed.), *Metabolic Pathways*, 3rd ed., New York, 1968.
60 R. Sonderhoff and H. Thomas, *Ann. Chem.*, 530 (1937) 195.
61 A. I. Virtanen and J. Sundmann, *Biochem. Z.*, 313 (1942) 236.
62 F. Lynen, *Ann. Chem.*, 552 (1942) 270.
63 D. Rittenberg and K. Bloch, *J. Biol. Chem.*, 157 (1945) 749.
64 P. E. Verkade, *Chem. Ind. (London)*, 57 (1938) 704.
65 A. J. Fulco and J. F. Mead, *J. Biol. Chem.*, 236 (1961) 2416.
66 W. Stoffel and H. G. Scheefer, *Z. Physiol. Chem.*, 341 (1965) 84.
67 W. Stoffel and H. Caesar, *Z. Physiol. Chem.*, 341 (1965) 76.
68 G. Medes, N. F. Floyd and S. Weinhouse, *J. Biol. Chem.*, 162 (1946) 1.
69 E. A. MacKay, *J. Clin. Endocr.*, 3 (1943) 101.
70 S. Soskin and R. Levine, *Am. J. Dig. Dis.*, 11 (1944) 305.
71 H. A. Krebs, *Advan. Enzymol.*, 3 (1943) 191.
72 F. Lipmann, *Advan. Enzymol.*, 1 (1941) 99.
73 C. Martius, *Z. Physiol. Chem.*, 279 (1943) 96.
74 F. Lynen and E. Reichert, *Z. Angew. Chem.*, 63 (1951) 47.
75 T. Chou and F. Lipmann, *J. Biol. Chem.*, 196 (1952) 89.
76 P. Berg, *J. Biol. Chem.*, 222 (1956) 991.
77 M. B. Hoagland, M. L. Stephenson, J. F. Scott, L. I. Hecht and P. C. Zameknik, *J. Biol. Chem.*, 231 (1958) 241.
78 E. R. Stadtman, *J. Biol. Chem.*, 203 (1953) 501.
79 F. Lynen, *Ann. Rev. Biochem.*, 24 (1955) 653.
80 I. Rose, M. Grunberg-Manago, S. Korey and S. Ochoa, *Federation Proc.*, 13 (1954) 283.
81 M. G. Jones, S. Black, R. M. Flynn and F. Lipmann, *Biochim. Biophys. Acta*, 12 (1953) 141.
82 P. Hele, *J. Biol. Chem.*, 206 (1954) 671.
83 H. Hilz, *Doctoral Thesis*, Munich 1953 (cited after Lynen[79])
84 L. T. Webster Jr., *J. Biol. Chem.*, 240 (1965) 4158.
85 L. T. Webster Jr., *J. Biol. Chem.*, 240 (1965) 4164.
86 L. T. Webster Jr., *J. Biol. Chem.*, 241 (1966) 5504.
87 H. R. Mahler, S. J. Walker and R. M. Bock, *J. Biol. Chem.*, 204 (1953) 453.

88 A. Kornberg and W. E. Pricer, *J. Biol. Chem.*, 204 (1953) 329.
89 L. Galzigna, C. R. Rossi, L. Sartorelli and D. M. Gibson, *J. Biol. Chem.*, 242 (1967) 2111.
90 J. R. Stern, M. J. Coon, A. Del Campillo and M. C. Schneider, *J. Biol. Chem.*, 221 (1956) 15.
91 G. R. Drysdale and H. A. Lardy, in W. D. McElroy and B. Glass (Eds.), *Sympposium on Phosphorous Metabolism*, Vol. 2, Baltimore, 1952, p. 281.
92 D. E. Green and H. Beinert, *ibid.*, 1 (1951) 330.
93 H. R. Mahler, *ibid.*, 2 (1952) 286.
94 F. Lynen and S. Ochoa, *Biochim. Biophys. Acta*, 12 (1953) 299.
95 H. Beinert, R. M. Bock, D. S. Goldmann, D. E. Green, H. R. Mahler, S. Mii, P. G. Stansly and S. J. Wakil, *J. Am. Chem. Soc.*, 75 (1953) 4111.
96 F. Lynen, *Federation Proc.*, 12 (1953) 683.
97 F. L. Crane, S. Mii, J. G. Hauge, D. E. Green and H. Beinert, *J. Biol. Chem.*, 218 (1956) 701.
98 F. L. Crane and H. Beinert, *J. Biol. Chem.*, 218 (1956) 717.
99 J. G. Hauge, F. L. Crane and H. Beinert, *J. Biol. Chem.*, 219 (1956) 727.
100 F. L. Crane, J. G. Hauge and H. Beinert, *Biochim. Biophys. Acta*, 17 (1955) 293.
101 P. B. Garland, B. Chance, L. Ernster, C. Lee and D. Wong, *Proc. Natl. Acad. Sci. (U.S.)*, 58 (1967) 1696.
102 D. E. Green, S. Mii, H. R. Mahler and R. M. Bock, *J. Biol. Chem.*, 206 (1954) 1.
103 J. R. Stern, M. J. Coon and A. Del Campillo, *J. Am. Chem. Soc.*, 75 (1953) 2377.
104 S. J. Wakil and H. R. Mahler, *J. Biol. Chem.*, 207 (1954) 125.
105 S. J. Wakil, *Biochim. Biophys. Acta*, 19 (1956) 497.
106 F. Lynen, L. Wessely, O. Wieland and L. Rueff, *Angew. Chem.*, 64 (1952) 687.
107 S. J. Wakil, D. E. Green, S. Mii and H. R. Mahler, *J. Biol. Chem.*, 207 (1954) 631.
108 D. S. Goldman, *J. Biol. Chem.*, 208 (1954) 345.
109 J. R. Stern, A. Del Campillo and I. Raw, *J. Biol. Chem.*, 218 (1956) 971.
110 W. Seubert, I. Lamberts, R. Kramer and B. Ohly, *Biochim. Biophys. Acta*, 164 (1968) 498.
111 P. K. Stumpf, *Ann. Rev. Biochem.*, 38 (1969) 159.
112 G. Fraenkel and S. Friedman, *Vitam. and Horm.*, 15 (1957) 73.
113 C. Corredor, C. Mansbach and R. Bressler, *Biochim. Biophys. Acta*, 114 (1967) 366.
114 E. Lundsgaard, *Biochim. Biophys. Acta*, 4 (1950) 322.
115 I. B. Fritz, *Acta Physiol. Scand.*, 34 (1955) 367.
116 I. B. Fritz, *Am. J. Physiol.*, 197 (1959) 297.
117 I. B. Fritz and K. T. N. Yue, *J. Lipid Res.*, 4 (1963) 279.
118 S. Friedman and G. Fraenkel, *Arch. Biochem. Biophys.*, 59 (1955) 491.
119 J. B. Chappell and A. R. Crofts, *Biochem. J.*, 95 (1965) 393.
120 E. Friedmann, *Biochem. Z.*, 55 (1913) 436.
121 D. E. Green, D. S. Goldman, S. Mii and H. Beinert, *J. Biol. Chem.*, 202 (1953) 137.
122 A. L. Lehninger, *J. Biol. Chem.*, 164 (1946) 291.
123 A. L. Lehninger and G. D. Greville, *Biochim. Biophys. Acta*, 12 (1953) 188.
123a E. Friedmann and C. Maase, *Biochem. Z.*, 27 (1910) 474.
124 A. J. Wakeman and H. D. Dakin; *J. Biol. Chem.*, 6 (1910) 373; 8 (1910) 108.
125 I. Snapper and A. Grünbaum, *Biochem. Z.*, 181 (1927) 418.
126 A. L. Lehninger, *J. Biol. Chem.*, 165 (1946) 131.
127 F. Lynen, N. Henning, C. Bublitz, B. Sörbo and L. Kröplin-Rueff, *Biochem. Z.*, 330 (1958) 269.
128 I. C. Caldwell and G. I. Drummond, *J. Biol. Chem.*, 238 (1963) 64.
129 H. Rudney, *J. Biol. Chem.*, 227 (1957) 363.

130 B. K. Bachhawat, W. G. Robinson and M. J. Coon, *J. Biol. Chem.*, 216 (1955) 727.
131 F. L. Breusch, *Science*, 97 (1943) 490.
132 H. Wieland and C. Rosenthal, *Ann. Chem.*, 554 (1943) 241.
133 J. R. Stern, M. J. Coon and A. del Campillo, *J. Biol. Chem.*, 221 (1956) 1.
134 J. R. Stern, in S. P. Colowick and N. O. Kaplan (Eds.), *Methods in Enzymology*, Vol. 1, New York, 1955, p. 573.
135 J. R. Stern, M. J. Coon, A del Campillo and M. C. Schneider, *J. Biol. Chem.*, 221 (1956) 15.
136 R. J. Bing, A. Siegel, I. Ungar and M. Gilbert, *Am. J. Med.*, 16 (1954) 504.
137 E. T. Waters, J. P. Fletcher and I. A. Mirsky, *Am. J. Physiol.*, 122 (1938) 542.
138 R. H. Barnes, E. M. MacKay, G. K. Moe and M. B. Visscher, *Am. J. Physiol.*, 123 (1938) 272.
139 D. S. Fredrickson and R. S. Gordon Jr., *Physiol. Rev.*, 38 (1958) 585.
140 J. R. Williamson and H. A. Krebs, *Biochem. J.*, 80 (1961) 540.
141 H. A. Krebs and R. N. Speake (quoted after Krebs[143]).
142 E. N. Bergman, K. Kon and M. L. Katz, *Am. J. Physiol.*, 205 (1963) 658.
143 H. A. Krebs, *Advan. Enzyme Regul.*, 4 (1966) 339.
144 P. A. Mayes, *Metabolism*, 11 (1962) 781.
145 G. Rosenfeld, *Dtsch. Med. Wochenschr.*, 11 (1885) 683.

Chapter 35

From Amino Acids to the Tricarboxylic Acid Cycle

1. Introduction

In animal organisms, fatty acids and glucose are oxidized in pathways joining the tricarboxylic acid cycle and reaching the respiratory chain derived from it. The tricarboxylic acid cycle is also the pathway for the dissimilation of the carbon skeletons of amino acids which in higher animal forms for instance, are a not negligible as part of the energy-providing compounds. In a man weighing 70 kg, about 100 g of amino acids are oxidized daily[1]. Whether exogenous or endogenous, these are the 20 amino acids commonly found in proteins; in higher plants many other amino acids are catabolized as well. The present chapter is limited to the oxidative degradation through the tricarboxylic acid cycle of the carbon structure of the 20 amino acids. They are catabolized through pathways leading to the tricarboxylic acid cycle but entering it at different points along its course, as shown in Fig. 24.

The pathways described from amino acids to the portals of the tricarboxylic acid cycle do not represent the total number followed by these amino acids. Elsewhere in this History we shall consider other important metabolic pathways.

The pathways of amino acid catabolism constituting sources of free energy have been analysed by means of techniques involving either the whole organisms, liver perfusion, surviving slices or homogenates; however, such analyses have also relied heavily on the isotopic method and on the isolation and purification of the enzymes involved in the pathways, allowing their reconstruction *in vitro*. Occasionally studies on inborn errors of metabolism and on mutants of *Neurospora* have afforded important data.

All in all, these techniques have led to the reconstruction of a series of pathways only a few sections of which remain incompletely understood, and to the identification of a large number of the biocatalysts involved.

References p. 362

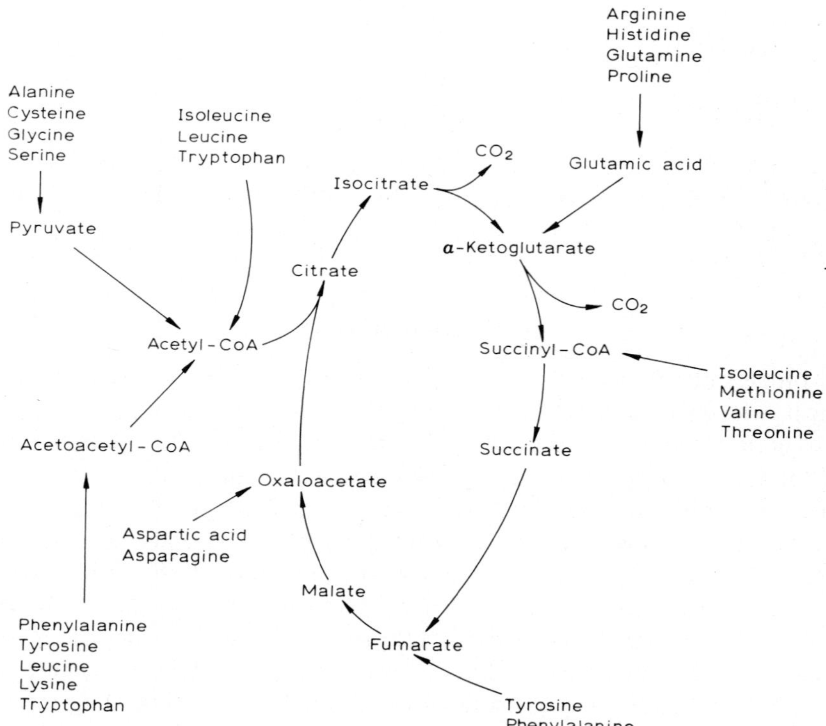

Fig. 24. Entry of the carbon skeletons of amino acids in the tricarboxylic acid cycle.

2. Amino acid deamination

(a) Oxidative deamination

The first step of amino acid catabolism is the removal of the α-amino group by an enzymatic process which is quite similar in different cases. Historically, this process was first studied in terms of oxidative deamination.

Embden[2] used the method of isolated liver perfusion to study amino acid catabolism. He observed that the fate of an α-amino acid was similar to that of the corresponding fatty acid minus one carbon atoms. For instance, leucine as well as isovaleric acid give acetoacetate, the oxidation being accomplished at the α-position in both cases.

$$\begin{matrix} CH_3 \\ CH_3 \end{matrix}\!\!>\!CHCH_2CHNH_2COOH + O_2 \longrightarrow \begin{matrix} CH_3 \\ CH_3 \end{matrix}\!\!>\!CHCH_2COOH + NH_3^+ + CO_2$$

leucine isovaleric acid

The first milestone in this analysis was the concept introduced by O. Neubauer[3,4] of the formation of an α-ketonic acid in the oxidation of an amino acid in the animal body. The train of thought which inspired Neubauer was continued by Schotten[5] who observed that, if phenylaminoacetic acid is fed in large amounts to a dog, it is partly excreted unchanged and partly converted into mandelic acid. Mandelic acid itself is highly resistant to further oxidation.

Repeating these experiments, Neubauer[6] found, after injecting a dog with DL-phenylaminoacetic acid, phenylaminoacetic acid in the urine with an excess of the laevo-component, *laevo*-mandelic acid, phenylglyoxylic acid and benzoic acid in the form of its derivative, hippuric acid (see Fig. 25). A logical conclusion was that the *laevo* component of phenylaminoacetic acid is less readily oxidized than the *dextro* compound. This was confirmed by the observation that the *laevo*-compound if administered was mainly excreted unchanged while on feeding the *dextro*-compound, *laevo*-mandelic acid was found in the urine in addition to phenylglyoxylic acid, suggesting that an optical inversion had occurred.

DL-$C_6H_5CHNH_2COOH$
DL-phenylaminoacetic acid

- L-$C_6H_5CHNH_2COOH$ — L-phenylaminoacetic acid
- L-$C_6H_5CHOHCOOH$ — L-mandelic acid
- $C_6H_5COCOOH$ — phenylglyoxylic acid
- C_6H_5COOH — benzoic acid

Fig. 25. Substances excreted after the introduction of DL-phenylaminoacetic acid.

These remarkable results were explained by the fact that the *laevo*-mandelic acid was a secondary product of the asymmetric reduction of phenylglyoxylic acid which is itself optically inactive, containing no asymmetric carbon.

$$C_6H_5CHNH_2COOH \underset{\text{oxidation}}{\rightarrow} C_6H_5COCOOH \underset{\text{reduction}}{\rightarrow} C_6H_5CHOH\,COOH$$

References p. 362

Plate 104. Otto Neubauer.

In conclusion, the oxidation of DL-phenylaminoacetic acid may be considered according to Neubauer as taking place as indicated in Fig. 26. It is important to emphasize that, in this scheme, α-ketonic acid is the intermediate in the series of oxidations while the hydroxyacid mandelic acid is a secondary reduction product. This concept was confirmed by perfusion experiments with liver, by Neubauer and Fischer[7].

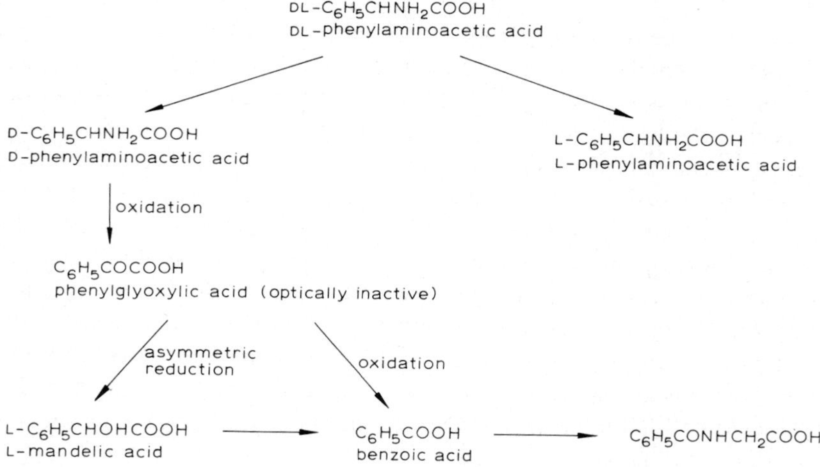

Fig. 26. The process of oxidation of DL-phenylaminoacetic acid according to Neubauer.

It may be recalled here that prior to the experiments done by Neubauer, the theory according to which the formation of α-hydroxyacids represents the first step in the catabolism of amino acids was considered probable. This was based on considerable experimental data. After administering alanine to starving rabbits, Neuberg and Langstein[8] had obtained lactic acid from the urine.

Hydroxyacids were generally recognized as secondary reduction products after Neubauer reported new experiments relating to inborn errors of metabolism. He observed that both tyrosine and *p*-hydroxyphenyl pyruvic acid when given to an alcaptonuric give homogentisic acid while *p*-hydroxyphenyl lactic acid does not. This shows that the α-ketonic acids are not derivatives of hydroxyacids, a view which as reinformed by a number of subsequent observations (literature in Dakin[9]). For instance, Flatow[10] gave to rabbits *meta*-hydroxyphenylalanine (*meta*-tyrosine), *meta*-chloro-

References p. 362

phenylalanine and furylalanine and observed an excretion of ketonic acid in all cases.

From these developments the view emerged that α-ketonic acids are obligatory products of the direct oxidation of amino acids whereas the hydroxy acids are not directly derived from them. As shown above, this view was the outcome of experiments on whole organisms (normal or diseased) or on isolated surviving organs.

In 1933 Krebs[11] studied amino acid oxidation with animal tissues (surviving slices of kidney and liver) and with homogenates. It had previously been thought that in mammals the main seat of amino acid deamination was the liver. Krebs showed that besides the liver, the kidney is a principal seat of deamination. Concerning the mechanism of deamination, an enzyme capable of liberating ammonia had already been detected by Jacoby[12] and Loewi[13]. Krebs showed the existence in the kidney of an enzyme which in presence of molecular oxygen transforms α-amino acids into the corresponding α-ketonic acids (alanine → pyruvic acid; α-aminobutyric acid → α-ketobutyric acid; phenylalanine → phenylpyruvic acid; leucine → α-ketoisocaproic acid; norleucine → α-keto-*n*-caproic acid). In addition, he inferred an oxidative deamination of other amino acids (aminodicarboxylic acids, diamino acids, glycine) from the fact that their deamination took place in the presence of oxygen and was inhibited in anaerobic conditions.

Though London *et al.*[14] had concluded from experiments performed on dogs by means of the physiological technique of "angiostomia" that the intestinal wall was the main, and the kidney the secondary site of oxidative deamination, Borsook and Jeffreys[15] confirmed by studying surviving slices of rat tissues that the deamination of amino acids takes place in the kidney and liver and that, due to the size of the latter, the main bulk of it probably results from deamination in the liver.

In his work of 1933, Krebs[11] had observed that both optical isomers of the amino acids were deaminated and even that the amino acids of the D-series were deaminated more rapidly than the natural ones. In 1935, Krebs[16] showed that kidney and liver contain two different enzymatic systems, one of which deaminates the non-natural D-amino acids without being destroyed by drying while the other system, which deaminates natural L-amino acids, was.

Later (1938), the D-amino acid oxidase was shown by Warburg and Christian[17] to be a flavoprotein, with isoalloxazin-adeninedinucleotide as prosthetic group. The enzyme is registered in the scientific nomenclature as [D-amino acid: oxygen oxidoreductase (deaminating), EN 1.4.3.3] A large amount of experimental work has been done to establish its physical properties, its specificity and its inhibitors (literature in Sallach and Fahien[18]). The L-amino acid oxidase [L-amino acid: oxygen oxidoreductase (deaminating), EN 1.4.3.2] was isolated from kidney and liver by Blanchard *et al.*[19,20]. Such enzymes have been isolated from different sources *e.g.* snake venom, rat kidney, turkey liver and molds, and their properties have been studied under many aspects (literature in Sallach and Fahien[18]). The enzyme is a flavoprotein and its prostethic group is FMN[20].

(b) Transamination

Though considered for some time the essential mechanism of deamination, oxidative deamination was later understood to play only a minor part in the course of mammalian amino acid metabolism. L-amino acid oxidase is practically absent in many tissues and its turnover number is low. On the other hand, it has been impossible so far to define the role of D-amino acid oxidases in metabolism.

The discovery of transaminases opened a new and efficient pathway of amino acid deamination. The first suggestion of the concept was formulated in 1930 by D. M. Needham[21]. It is related by herself in her book *Machina carnis*[22] as follows:

> "In 1930 Needham[21] had found that glutamic or aspartic acid added to minced muscle under anaerobic conditions gave rise to succinic, malic and fumaric acids without formation of ammonia or any change in the soluble nitrogen fractions. Aerobically these two amino acids increased the oxygen uptake, again with no change in the soluble N fractions, although it was shown that the aspartic acid actually disappeared. She suggested that the amino group entered into combination with some reactive carbohydrate residue and was retained in a new amino acid."

Several years later, Braunstein and Kritzmann[23] formulated the concept of transamination* (transfer of an amino group from an amino compound to a keto compound, or aldehyde, without an intermediate formation of ammonia). They observed that in muscle tissue glutamic acid disappears in anaerobiosis while alanine is produced. The addition of pyruvate quickened the disappearance of glutamate. This was shown to be a reaction between glutamate and α-keto acids. Several of these were observed to be

---

* The term "transamination" was introduced by Schaeffer and Le Breton[23a] in 1938.

Plate 105. Alexander Evseyevich Braunstein.

able to replace pyruvate and it was also shown that a number of amino acids were able to transfer their nitrogen to α-ketoglutarate.

An extract of muscle containing a transaminase specific for glutamic acid and inactive for aspartic acid was prepared from muscle by Kritzmann[24]. She later showed the presence of aspartic transaminase in such extracts and detected the presence of a thermostable coenzyme[25].

It was not until P. P. Cohen[25], working in the laboratory of Krebs, introduced a method for the quantitative microdetermination of glutamic acid that the necessary quantitative measurements of the activity of various transaminases could be made. These experiments showed that aspartate, and much less so alanine, react readily with α-ketoglutarate and that all other amino acids react at rates of a different, very much lower, order of magnitude.

Braunstein[26] has gathered data showing that the oxidative deamination of L-amino acids goes through the coupled action of L-amino acid–α-ketoglutarate transaminases and glutamate dehydrogenase (transdeamination). The amino acid is involved in a transamination with α-ketoglutarate, resulting in glutamate and the corresponding α-keto acid. The glutamate is oxidatively deaminated by glutamic dehydrogenase with a formation of ammonia and with a regeneration of α-ketoglutarate.

L-Glutamate dehydrogenase [L-glutamate-NAD[P⁺] oxidoreductase, EN 1.4.1.3] catalyzes the reaction

$$H_2O + \begin{array}{c} COO^- \\ | \\ CH_2 \\ | \\ CH_2 \\ | \\ CHNH_3^+ \\ | \\ COO^- \end{array} + NAD^+ \text{ (or } NADP^+) \rightleftharpoons \begin{array}{c} COO^- \\ | \\ CH_2 \\ | \\ CH_2 \\ | \\ C{=}O \\ | \\ COO^- \end{array} + NADH \text{ (or } NADPH) + H^+ + NH_4^+$$

and has recently been the subject of extensive studies related by Sallach and Fahien[18].

The primary importance of transdeamination in the oxidative deamination of L-amino acids has been substantiated by a number of experimental results. For example, in tissue slices the rate of transamination of L-α-amino acids with α-ketoglutarate is comparable to the rates of ammonia or urea formation[27].

It is therefore now believed that the amino groups originating by transamination from the various amino acids appear ultimately in the α-amino group of glutamic acid which collects the amino groups of other

Plate 106. Esmond Emerson Snell

amino acids. It appears that glutamate dehydrogenase, rather than amino acid oxidases, plays the key role in amino acid deamination.

Another aspect of transaminases is their common coenzyme, pyridoxal phosphate (Fig. 27).

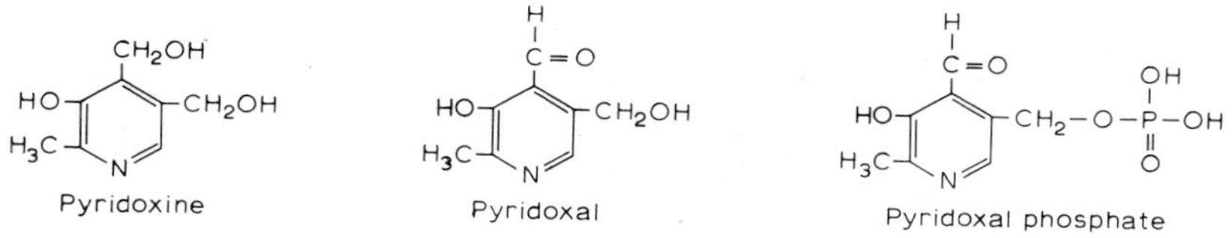

Fig. 27. Pyridoxine, pyridoxal, pyridoxal phosphate.

This major development, which made possible the determination of the mechanism of the transaminase reaction, originated in an observation by Snell[28] on the reversible interconversion reactions. Snell, who was working on B_6 vitamins observed that at raised temperature pyridoxal reacts with glutamic acid to form pyridoxamine and α-ketoglutaric acid. He had the good sense to relate this observation to the transaminase studies in Braunstein's laboratory and in collaboration with Schlenk, Snell[29] showed that in rats deficient in vitamin B_6 the tissues had a lowered transaminase activity. The following year, Kritzmann and Samarina[30] showed a reactivation of glutamic acid transaminase by pyridoxal phosphate.

In Chapter 4 of Volume 11 of this Treatise, Snell has outlined recent developments in pyridoxal biochemistry.

3. Pathways from amino acids to portals of the tricarboxylic acid cycle

These pathways, constituting only one of many aspects of amino acid metabolism (that constituting a source of free energy) have been the subject of reviews by Greenberg[31] and by Rodwell[32], to which the following presentation is greatly indebted.

(a) Alanine, aspartic acid, glutamic acid

Deamination brings these amino acids directly to the portals of pyruvate, oxaloacetate and α-ketoglutarate respectively.

References p. 362

(b) Glutamine

Glutamine is a natural precursor of glutamic acid, the deamination of glutamine being catalyzed by a widely distributed glutaminase [L-glutamine amidohydrolase, EN 3.5.1.2]. That the deamination of glutamine was catalyzed by preparations of a number of animal tissues was shown by Lang[33] in 1904. The enzyme has been purified about 300-fold from homogenates of hog and dog kidney by Klingman and Handler[34], Shepherd and Kalnitsky[35] and Sayre and Roberts[36].

(c) Asparagine

Asparagine is a natural precursor of aspartic acid. Asparaginase [L-asparagine amidohydrolase, EN 3.5.1.1] catalyzes the reaction. The enzyme is widely distributed in nature (literature in Cohen and Sallach[37] and in Varner[38]). From guinea pig serum, Yellin and Wriston[39] have isolated a homogeneous enzyme with a molecular weight of $1.4 \cdot 10^5$. The high-substrate specificity of the enzyme was demonstrated by Meister *et al.*[40].

(d) Glycine

After a number of confusing reports, related by Greenberg[41], the main pathway of glycine catabolism has been recognized as initiated by a conversion to serine by enzymatic addition of a hydroxymethyl group donated by the coenzyme N^{10}-hydroxymethyl-tetrahydrofolate.

This tetrahydrofolate-dependent pathway was recognized when it appeared that the α-carbon of glycine is converted to methylene-folate-H_4 and its carboxyl in CO_2 without the obligatory passage through glycolate, glyoxylate or formaldehyde, as several other theories had postulated (literature in Greenberg[31]).

Important information concerning intermediary steps has come from the study of the fermentation of glycine to acetate, CO_2 and NH_3 by the bacterium *Pepticoccus glycinophilus*, in which the acetate derives from the C-2 and the CO_2 from the C-1 of glycine[42,43].

$$NH_2CH_2COOH + \text{folate} \cdot H_4 \rightarrow CH_2\ \text{folate} \cdot H_4 + CO_2 + NH_3 + 2\ H$$

$$CH_2\ \text{folate} \cdot H_4 + \text{glycine} \rightarrow \text{serine} + \text{folate} \cdot H_4$$

$$\text{serine} \rightarrow \text{pyruvate} + NH_3$$

$$\text{Pyruvate} \rightarrow\rightarrow\rightarrow \text{acetate}$$

The reversible conversion glycine ⇆ serine is catalyzed by the enzyme serine hydroxymethyltransferase [5,10-methylene tetrahydrofolate:glycine hydroxymethyltransferase; EN 2.1.2.1], which has been the subject of many studies (reviews by Huennekens and Osborn[44] and by Rabinowitz[45]).

(e) Serine

An important pathway of serine degradation is constituted by its conversion to pyruvate in the presence of serine dehydrogenase, a pyridoxal-P-dependent enzyme. Among many arguments in favour of this concept is the observation by Lien and Greenberg[46] that [3-^{14}C]serine is mainly transformed by mitochondrial preparations into isotopic alanine which is formed, by transamination, from pyruvate.

(f) Threonine

Several pathways of metabolism have been described for threonine, one of which leads to acetyl-CoA through acetaldehyde. This pathway begins with a scission of threonine into glycine and acetaldehyde. This reaction was first reported by Braunstein and Vilenkina[47] and confirmed by Meltzer and Sprinson[48].

The main pathway of threonine degradation (Fig. 28) is now recognized to enter the tricarboxylic acid cycle by the portal of succinyl-CoA, threonine being first converted into α-ketobutyric acid and ammonia in the presence of threonine dehydrase. Highly purified preparations of threonine dehydrase have been obtained from mammalian liver by Sayre and Greenberg[49] and by Nishimura[50]. The α-ketobutyric acid is transformed into succinyl-CoA by the pathway described in the sections on isoleucine and on methionine.

(g) Histidine

Among the many pathways followed by histidine in metabolism, one leads to the tricarboxylic acid cycle, entering it at the α-ketoglutarate level after passing through glutamate.

Edlbacher[51] and Györgi and Rothler[52] independently showed the decomposition of histidine by liver *brei*. On the other hand, urocanic acid was detected by Darby and Lewis[53] in the urine of animals fed L-

References p. 362

Plate 107. David M. Greenberg

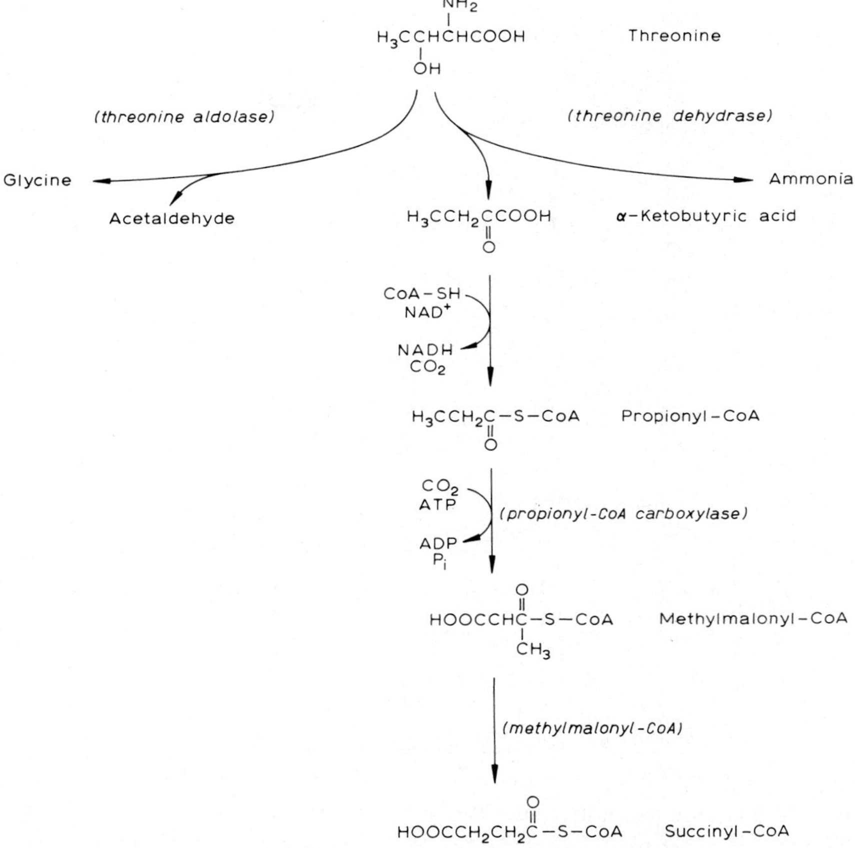

Fig. 28. From threonine to succinyl-CoA.

histidine. As urocanic acid when fed to animals shows a very sluggish metabolism, it was not considered important in histidine metabolism until, in 1939, Sera and Yada[54] observed a hydrolytic splitting of urocanic acid by liver extracts and proposed that histidine is first deaminated to urocanic acid in the presence of histidase [L-histidine ammonia-lyase; EN 4.3.1.3], the urocanic acid being converted to a product yielding glutamate and formate. Many data have since been gathered concerning histidase (literature in Greenberg[31]).

Borek and Waelsch[55], in 1953, showed that urocanic acid is transformed

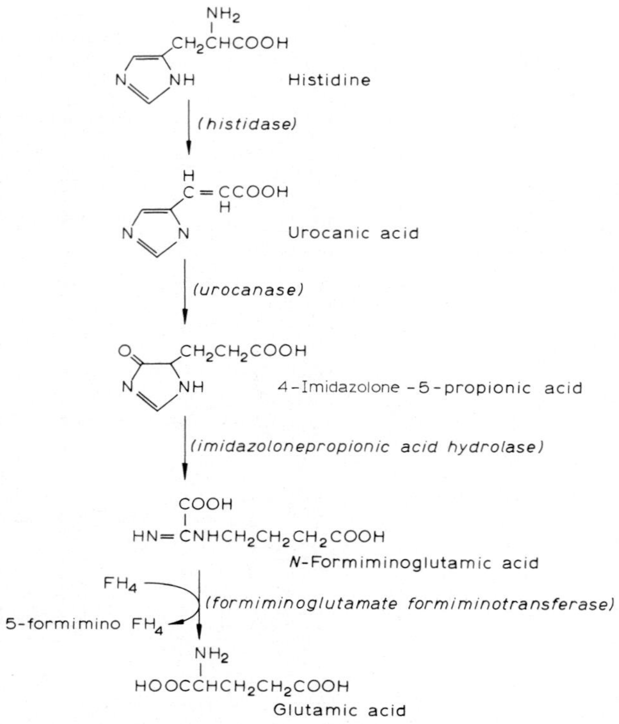

NH_2
$CH_2CHCOOH$
N NH
Histidine

(histidase)

H
$C=CCOOH$
H
N N
Urocanic acid

(urocanase)

O CH_2CH_2COOH
N NH
4-Imidazolone-5-propionic acid

(imidazolonepropionic acid hydrolase)

COOH
$HN=CNHCH_2CH_2CH_2COOH$
N-Formiminoglutamic acid

FH_4
(formiminoglutamate formiminotransferase)
5-formimino FH_4

NH_2
$HOOCCHCH_2CH_2COOH$
Glutamic acid

Fig. 29. From histidine to glutamic acid.

into *N*-formimino-L-glutamate by liver extracts or bacterial preparations. Two enzymes are involved in this process, as was demonstrated by Feinberg and Greenberg[56,57] and by Revel and Magasanik[58]. There is convincing if indirect evidence showing that the first enzyme, urocanase, produces 4(5)-imidazolone-5(4) propionic acid while the second, imidazolone propionic acid hydrolase, hydrolyzes this product into formiminoglutamate (Fig. 29). Up to this point the scheme of Fig. 29 corresponds to that tentatively suggested by Suda and his group[59]. Another enzyme, isolated by Miller and Waelsch[60] in 1957, converts formiminoglutamic acid to glutamic acid. It has been named formiminoglutamate formiminotransferase.

(h) Tyrosine

The pathway of the carbon structure of this amino acid reaches the tricarboxylic acid cycle at the portal of acetyl-CoA, passing through acetoacetate and acetoacetyl-CoA before reaching this point. The pathway reaches the stages of acetoacetic acid and fumaric acid through the intermediary stage of homogentisic acid. This pathway through C-2 and C-5 hydroxylation of the phenyl ring goes to homogentisic acid which is cleaved at the level of the phenyl ring. The pathway was identified by means of isotopic tracer experiments, consideration of inborn errors of metabolism, purification of the enzymes involved, etc.

In the first years of this century it was suggested that a migration of the side chain of tyrosine takes place, the side chain being replaced by a second hydroxyl group as it appears in homogentisic acid[61,62].

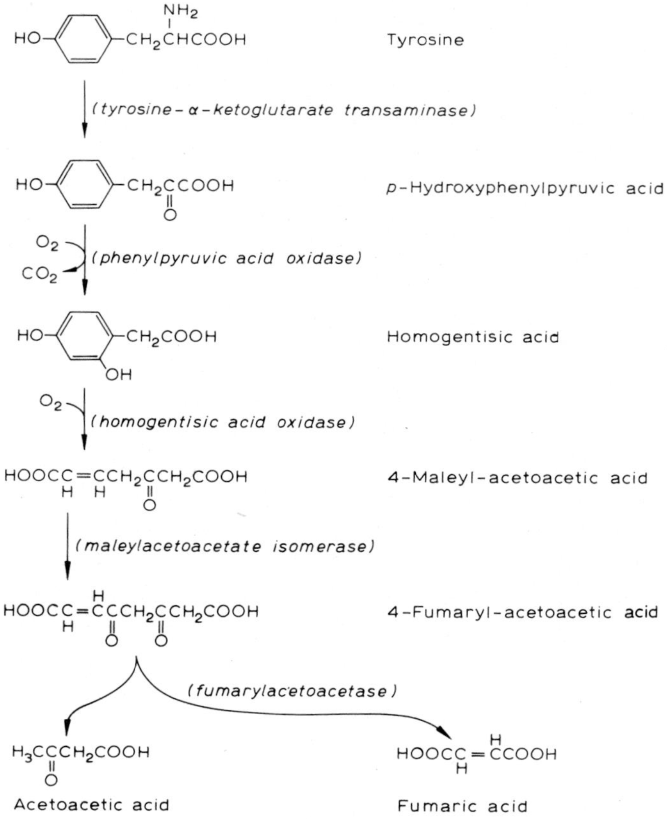

HO–C_6H_4–$CH_2CH(NH_2)COOH$ Tyrosine

↓ (tyrosine-α-ketoglutarate transaminase)

HO–C_6H_4–$CH_2C(=O)COOH$ *p*-Hydroxyphenylpyruvic acid

↓ O_2 → CO_2 (phenylpyruvic acid oxidase)

HO–C_6H_3(OH)–CH_2COOH Homogentisic acid

↓ O_2 (homogentisic acid oxidase)

$HOOCCH{=}CHCH_2C(=O)CH_2COOH$ 4-Maleyl-acetoacetic acid

↓ (maleylacetoacetate isomerase)

$HOOCCH{=}CHC(=O)CH_2C(=O)CH_2COOH$ 4-Fumaryl-acetoacetic acid

↓ (fumarylacetoacetase)

$H_3CC(=O)CH_2COOH$ Acetoacetic acid + $HOOCCH{=}CHCOOH$ Fumaric acid

Fig. 30. From tyrosine to acetoacetic acid and fumaric acid.

References p. 362

While there is evidence that the human organism is able to completely oxidize tyrosine, this property is lost in the inborn error of metabolism known as "alcaptonuria", in which the metabolic product is homogentisic acid (Fig. 30). Its origin from tyrosine and probably from phenylalanine was proposed by Wolkow and Baumann[63] who determined its constitution. Its origin from phenylalanine was definitely proved by Falta and Langstein[64]. From these studies resulted the theory of the nature of alcaptonuria according to which homogentisic acid is a normal intermediate of tyrosine and phenylalanine metabolism which is not oxidized further in alcaptonurics while in normal individuals it is completely oxidized. This view was strengthened by observations on perfusion of surviving liver in which phenylalanine, tyrosine and homogentisic acid all gave large amounts of acetoacetic acid which was believed to undergo complete oxidation in whole organisms.

In his book on oxidations and reductions in the animal body, Dakin[9] opposed this theory and claimed that the pathway of phenylalanine and tyrosine catabolism "does not lie by way of homogentisic acid". The arguments in favour of his position, which the reader will find in the book mentioned above, deserve attention. By analogy with the contributions of Neubauer and Knoop, Dakin, though this was demonstrated only later, surmised that the first products of tyrosine and phenylalanine catabolism must be the corresponding α-keto acids, phenylpyruvic acid and *p*-hydroxyphenylpyruvic acid

$$C_6H_5\text{–}CH_2\text{–}CO\text{–}COOH \quad \text{phenylpyruvic acid}$$

$$C_6H_4OH\text{–}CH_2\text{–}CO\text{–}COOH \quad p\text{-hydroxyphenylpyruvic acid}$$

a view for which he provides some indirect evidence. Relying on the claim that phenylacetic acid and *p*-hydroxyphenylpyruvic acid do not yield homogentisic acid when given to alcaptonurics[65], Dakin concludes that homogentisic acid has to be excluded as an intermediate. According to him, the acetoacetic acid derived from phenylpyruvic acid results from the separation of two carbon atoms of the nucleus and two of the side chains. In his opinion, alcaptonuria

> is a condition in which there is not only an abnormal formation of homogentisic acid but also an abnormal failure to catabolize it when formed"[9].

Enzymes of tyrosine catabolism. The identification and purification of the enzymes involved in tyrosine catabolism were made difficult by the instability of a number of the intermediary steps of the pathway. After 1950, enzymes were detected which specifically took part in the oxidation respectively of tyrosine, of *p*-hydroxyphenyl pyruvate and of homogentisic acid to acetoacetic acid. The enzymes were separated and purified and the pathway could be reconstructed and its steps defined.

(a) Tyrosine-α-ketoglutarate transaminase. The oxidation of tyrosine to acetoacetate involves only four oxygen atoms. The reaction is stimulated by α-ketoglutarate. When the transaminase coenzyme is removed, the oxidation is decreased. Addition of pyridoxal-P increases it, however.

When the substrate is *p*-hydroxyphenyl-pyruvate, keto acids do not increase the oxidation rate[66,67].

(b) Phenylpyruvic acid oxidase, p-hydroxyphenylpyruvic acid hydroxylase [4-hydroxyphenylpyruvate dioxygenase EN 1.13.11.27]. The enzyme has been purified and studied by LaDu and Zannoni[66] and by Knox and Le May-Knox[68]. It is the enzyme lacking in alcaptonuria as was shown on autopsy material of liver and kidney (LaDu, Zannoni, Laster and Seegmiller[69]).

(c) Homogentisate oxygenase [homogentisate:oxygen 1,2-oxidoreductase (decyclizing); EN 1.13.11.5] (formerly EC 1.99.2.5 and EC 1.13.1.5). In 1951, Ravdin and Crandall[70], isolated from rat liver homogenate two enzyme fractions, one catalyzing the oxidation of an open-chain acid. Their second enzyme cleaved this component to fumarate and acetoacetate. From rabbit liver a preparation catalyzing the opening of the benzene ring was obtained by Suda and Takeda[71]. Homogentisate oxygenase has been purified from extracts of acetone powders of beef liver[72] and from calf liver[73]. Fe^{2+} is a component of homogentisate oxygenase[74].

Whereas Ravdin and Crandall had believed the product of the enzyme action to be 4-fumarylacetoacetic acid, it was shown that in fact this compound results from the action of a second enzyme, an isomerase, the initial product being 4-maleyl-acetoacetic acid[75].

(d) Maleylacetoacetate isomerase [4-maleylacetoacetate-cis-trans-isomerase, EN 5.2.1.2]. As stated above, the existence of this enzyme was detected by Knox and Edwards[75] who were able to isolate it from rat liver homogenate by ethanol fractionation. Glutathione is a coenzyme required for the action of the isomerase.

(e) Fumarylacetoacetase [4-fumarylacetoacetate fumaryl hydrolase, EN 3.7.1.2]. This enzyme is identical with enzymes catalyzing the

References p. 362

Plate 108. Alton Meister

hydrolysis of triacetic acids, formerly described by Connors and Stotz[76]. The products of the action are acetoacetate and fumarate.

(i) Phenylalanine

This amino acid is converted to tyrosine, as shown by the feeding experiments of Womack and Rose[77] as early as 1934. When phenylalanine is available in sufficient amounts, tyrosine is not essential. The conversion of phenylalanine to tyrosine was clearly established by the isotopic experiments of Moss and Schoenheimer[78] who fed deutero-DL-phenylalanine to rats and isolated the corresponding labeled L-tyrosine.

Udenfriend and Cooper[79] obtained from liver, in a very unstable form, an enzyme catalyzing the hydroxylation of phenylalanine to tyrosine.

"Phenylalanine hydroxylase" consists of two enzymes, respectively requiring a biopterin and NADPH as coenzymes[80,81]. It belongs to the category of "mixed-function oxygenases". One oxygen atom of the oxygen

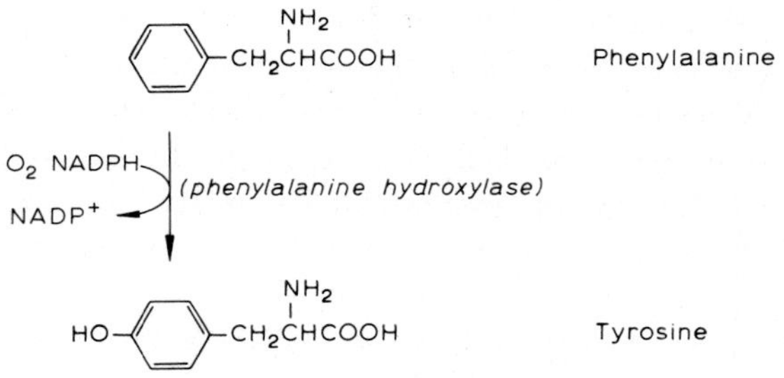

Fig. 31. Hydroxylation of phenylalanine to tyrosine.

molecule is incorporated in the –OH on the benzene ring[82], the other is reduced to water by NADPH. The scheme of Fig. 32 represents the hydroxylation mechanism.

In the disease phenylketonuria the trouble has been traced back to a lack of Enzyme I (refs. 83, 84).

(j) Tryptophan

Many of the pathways followed by tryptophan in metabolism have been reviewed by Dalgliesh[85] and by Greenberg[31]. When tryptophan labeled

References p. 362

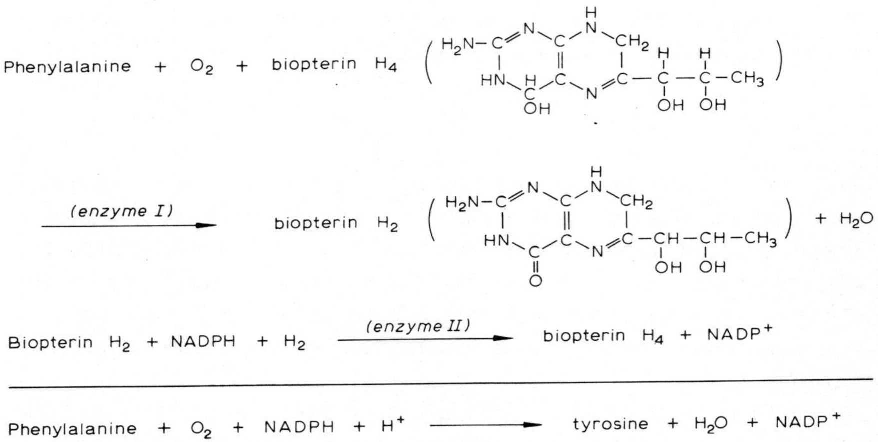

Phenylalanine + O_2 + biopterin H_4 (biopterin H_4 structure)

(enzyme I) → biopterin H_2 (biopterin H_2 structure) + H_2O

Biopterin H_2 + NADPH + H_2 —(enzyme II)→ biopterin H_4 + $NADP^+$

Phenylalanine + O_2 + NADPH + H^+ → tyrosine + H_2O + $NADP^+$

Fig. 32. Mechanism of phenylalanine hydroxylation.

in the benzene ring became available, Henderson and Hankes[86], in experiments on rats, showed that a substantial portion of the carbon in position 3α, 7 and 7α is rapidly oxidized to CO_2. Similar experiments by Dalgliesh and Tabechian[87] with tryptophan uniformly labeled by biosynthesis also showed that tryptophan was converted to CO_2. Gholson *et al.*[88], used tryptophan labeled with ^{14}C in position 7α on the benzene ring, and after 12 h found approximately 36 per cent of the ^{14}C in expired CO_2, 7 to 15 per cent in the urine and 40 to 50 per cent in the tissues.

The first compound proposed as the intermediate on the pathway of tryptophan catabolism to CO_2 was kynurenic acid, isolated by Ellinger[89] in 1904 from the urine of dogs fed tryptophan. The case of kynurenic acid is typical of the extreme difficulty involved in recognizing intermediary steps of metabolic pathways which existed before the advent of the isotopic method and of the enzymological techniques which made possible the isolation and purification of enzymes. It also exemplifies the importance of a knowledge of the correct organic chemistry of biomolecules. Liebig[90] had isolated kynurenic acid from dog's urine in 1853, before tryptophan was known. The organic chemistry of kynurenic acid remained confused for some time (literature in Robson[91]). The current information concerning the assumed excretion of kynurenic acid by some animal species and not by others reinforced the prevailing notion that kynurenic acid is not an intermediate in the degradative pathway of tryptophan in animals. Though not metabolized or possibly metabolized to a limited degree in animals by dehydroxylation to quinaldic acid which is excreted in the urine[92], kynurenic acid is in certain bacteria on the path of the quinolinic pathway of tryptophan dissimilation, starting with a transamination of kynurenine, the result of which cyclyzes to form kynurenic acid.

The key to the problem was afforded by Matsuoka and Yoshimatsu[93] and by Kotake and Iwao[94]. After the administration of large amounts of tryptophan into rabbits, the Japanese authors observed the excretion not only of kynurenic acid but also of an unknown product which they isolated, purified and named kynurenine, and which they found convertible to kynurenic acid, *in vivo* and *in vitro*. Unfortunately an incorrect structural formula was assigned to kynurenine by these authors. The correct structure was established by Butenandt *et al.*[95] only in 1942.

In experiments on pyridoxine-deficient rats, 3-hydroxy-kynurenine was first shown to be a tryptophan metabolite in animals[96,97].

That 3-hydroxyanthranilic acid is an obligatory intermediate in the so-called kynurenine-anthranilic acid pathway leading to a crossroad whence the pathway goes either to niacin formation or to complete dissimilation, was suggested by Gholson *et al.*[98] on the basis of their observation that the rate and extent of conversion of [1-^{14}C] 3-hydroxyanthranilate was what would be expected of an intermediate. The pathway from 3-hydroxyanthranilic acid to glutaryl-CoA (see Fig. 33) was proposed for mammalian liver by Nishizuka *et al.*[99] who identified glutaryl-CoA as the primary reaction product of oxidative decarboxylation of α-ketoadipic acid.

These authors have reported experiments with [5-^{14}C] 3-hydroxyanthranilic acid showing that carbon 5 of the acid is converted to carbon atom 2 or 4 of glutaric acid. Their results strongly suggest that carbon atoms 2, 1, 6, 5 and 4 of 3-hydroxyanthranilic acid directly provide carbon atoms 1, 2, 3, 4 and 5 of glutaryl-CoA, respectively.

The path followed by glutaryl-CoA was originally inferred from the knowledge of the β-oxidation of fatty acids. Tustanoff and Stern[100] showed in 1960 that glutaconyl-CoA can be enzymatically formed by carboxylation of crotonyl-CoA.

(i) Enzymes of tryptophan catabolism. Kotake and Masayama[101] proposed the name tryptophan pyrrolase for the enzyme which, in mammalian liver, catalyzes the formation of kynurenine from tryptophan. This designation was abandoned since Knox and Mehler[102] showed, in

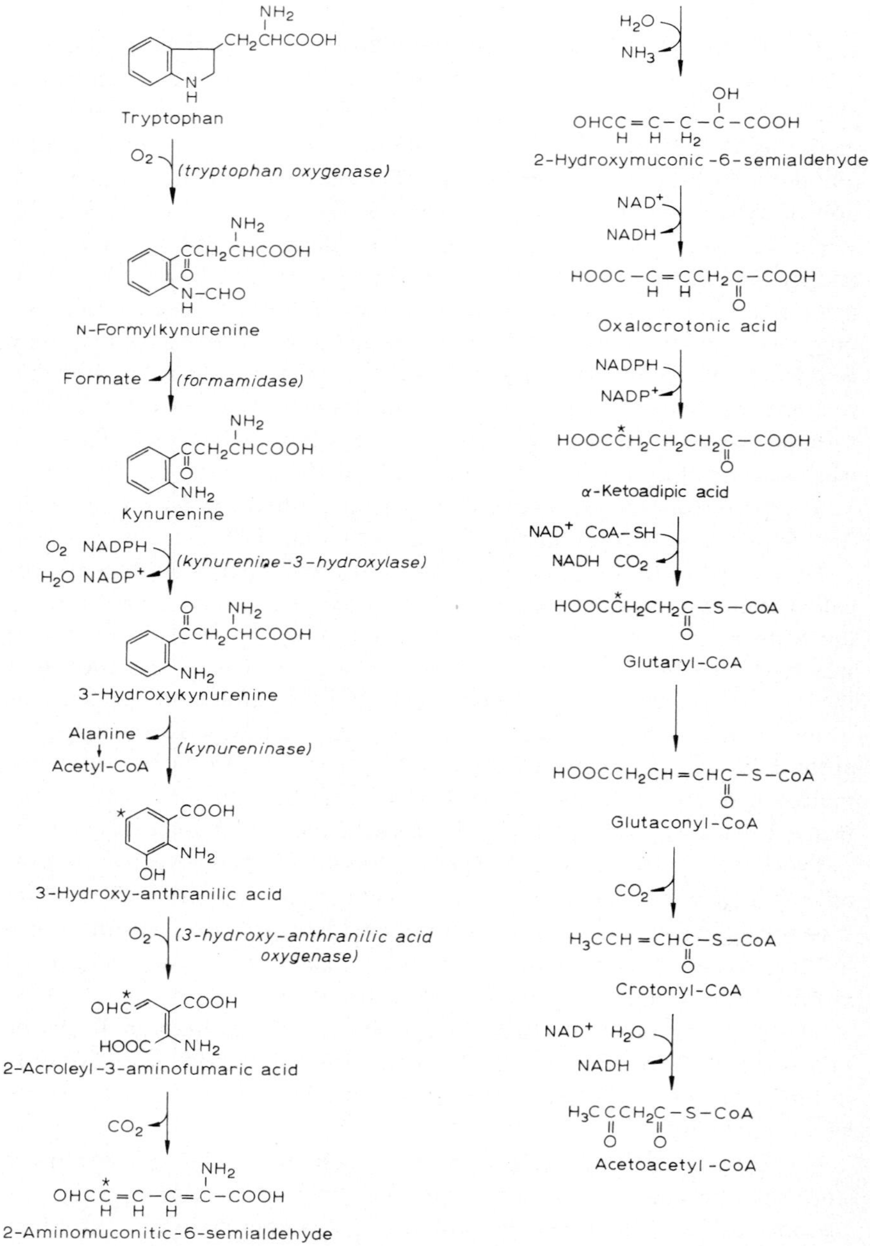

Tryptophan

O_2 (tryptophan oxygenase)

N-Formylkynurenine

Formate (formamidase)

Kynurenine

O_2 NADPH, H_2O $NADP^+$ (kynurenine-3-hydroxylase)

3-Hydroxykynurenine

Alanine → Acetyl-CoA (kynureninase)

3-Hydroxy-anthranilic acid

O_2 (3-hydroxy-anthranilic acid oxygenase)

2-Acroleyl-3-aminofumaric acid

CO_2

OHCC*H=CH–CH=C(NH₂)–COOH

2-Aminomuconitic-6-semialdehyde

H_2O, NH_3

OHCCH=CH–CH₂–CH(OH)–COOH

2-Hydroxymuconic-6-semialdehyde

NAD^+, NADH

HOOC–CH=CHCH₂C(=O)–COOH

Oxalocrotonic acid

NADPH, $NADP^+$

HOOCC*H₂CH₂CH₂C(=O)–COOH

α-Ketoadipic acid

NAD^+ CoA-SH, NADH CO_2

HOOCC*H₂CH₂C(=O)–S–CoA

Glutaryl-CoA

HOOCCH₂CH=CHC(=O)–S–CoA

Glutaconyl-CoA

CO_2

H₃CCH=CHC(=O)–S–CoA

Crotonyl-CoA

NAD^+ H_2O, NADH

H₃CC(=O)CH₂C(=O)–S–CoA

Acetoacetyl-CoA

Fig. 33. From tryptophan to acetoacetyl-CoA and acetyl-CoA.

1950, that two enzymatic steps take place between tryptophan and kynurenine, first an oxidation to formylkynurenine (on this theory see Dalgliesh[85]). However it was shown in experiments with ^{18}O by Hayashi *et al.*[103] that if it is true that H_2O_2 activates the reaction, and if it is true that molecular O_2 is incorporated in *N*-formylkynurenine (Fig. 33), it is not the ^{18}O from $H_2{}^{18}O$.

Tryptophan oxygenase [L-tryptophan:oxygen 2,3-oxidoreductase, EN 1.13.11.11] (formerly EC 1.11.1.4 and EC 1.13.1.12) has been the subject of many studies. It is a widely distributed iron porphyrin protein (literature in Greenberg[31]). The conversion of kynurenine to 3-hydroxykynurenine was shown in 1956 by de Castro, Price and Brown[104] to be catalyzed by liver and kidney mitochondria. The enzyme kynurenine 3-hydroxylase [L-kynurenine, reduced-NADP:oxygen oxidoreductase (3-hydroxylating) EN 1.14.13.9] (formerly EC 1.14.1.2) was partially purified by Saito *et al.*[105] who recognized that O_2 and NADPH are required in the reaction

$$\text{L-kynurenine} + \text{NADPH} + \text{H}^+ + \text{O}_2 \rightarrow \text{3-hydroxykynurenine} + \text{NADP}^+ + \text{H}_2\text{O}$$

The step from 3-hydroxykynurenine to 3-hydroxyanthranilic acid and alanine is catalyzed by kynureninase [L-kynurenine hydrolase, EN 3.7.1.3], an enzyme which has been identified in a variety of sources. Alanine is produced by the scission of the side chain and pyridoxal-P is required as coenzyme[106].

In contrast with this concept, Sanadi and Greenberg[107] proposed for the side chain a scheme involving the conversion of carboxyl carbon to CO_2 and the two remaining carbons to acetate. Gholson *et al.*[98] have used [α-^{14}C]tryptophan and relied on the labeling pattern of several non-essential amino acids to indicate the fate of the tryptophan side chain. Their results clearly indicated that the side chain of [α-^{14}C]-tryptophan is removed as [2-^{14}C]pyruvate or a metabolic equivalent. In the scheme of Fig. 33 [α-^{14}C]tryptophan should yield [2-^{14}C]alanine which was observed by the authors, while if metabolized as suggested by Sanadi and Greenberg it would give rise to [1-^{14}C] acetate.

The labeling patterns observed for serine, alanine, aspartic acid and glutamic acid by Gholson *et al.*[98] indicate that the side chain is removed as pyruvate or its metabolic equivalent. The results are therefore in agreement with the kynurenine-3-hydroxyanthranilic acid pathway of Fig. 33 and exclude any significant participation of the route proposed by Sanadi and Greenberg.

That the kynurenine-3-hydroxyanthranilic acid pathway is the major route for the oxidation of the benzene ring of tryptophan was placed in doubt by Rothstein and Greenberg[108] who reported that under certain conditions tryptophan is oxidized to CO_2 by rat liver homogenates to a greater extent than is kynurenine. Gholson *et al.*[109] have concluded in favor of the pathway of Fig. 33 on the basis of experiments on rats with [1-^{14}C]-3-hydroxyanthranilate, showing that the conversion to $^{14}CO_2$ is sufficient to account for the observed oxidation of tryptophan. They observed that [7α-^{14}C]tryptophan and [^{3}H]3-hydroxyanthranilate were converted to radioactive glutamic acid by the rat.

The cleavage of the benzene ring is catalyzed by 3-hydroxyanthranilate oxygenase [3-hydroxyanthranilate: oxygen 3,4-oxidoreductase; EN 1.13.11.6] (formerly EC 1.13.1.6), which has been highly purified in a number of laboratories (literature in Greenberg[31]). It is admitted on the basis of indirect evidence that the resulting oxidation product corresponds to 2-acroleyl-3-aminofumaric acid, though the product has not been isolated nor the enzymology unraveled. The enzymatic decarboxylation product of 2-acroleyl-3-amino fumarate is 2-amino-muconic-6-semialdehyde which is transformed non-enzymatically into 2-hydroxymuconic-6-semialdehyde.

(k) Leucine

From experiments on perfused organs, Embden *et al.*[110] concluded that isovaleric acid is a product of leucine dissimilation, a conclusion which was confirmed after the introduction of the isotopic method, by whole-animal studies with deuteroisoleucine and later [^{14}C]leucine and [^{14}C]isovalerate (see Rodwell's review[32]).

Leucine has long been known as a ketogenic amino acid; it was accounted for as such when isotopic experiments demonstrated that leucine and isovaleric acid formed acetoacetate *via* an acetic-like intermediary product derived from C-2 and C-3 of leucine (C-1 and C-2 of isovaleric acid), with C-3 of leucine (C-2 of isovalerate) forming the methyl acetoacetic acid[111–113].

That the isopropyl group also yielded acetoacetate was shown later[113,114]. The three carbons of the isopropyl group were shown to be incorporated as a unit into acetoacetate, its fourth carbon being supplied by an incorporation of CO_2 (refs. 114, 115).

The nature of the enzymatic steps in leucine dissimilation was elucidated by Coon and his associates[116,117] and by Lynen and his associates[118] (Fig. 34). The first step is a transamination.

That the oxidative decarboxylation leads to isovaleryl-CoA and not to isovalerate was demonstrated by Coon *et al.*[119], who also observed that mammalian extracts contain activating enzymes for the esterification of isovalerate to isovaleryl-CoA. In the bacterium *Proteus vulgaris* isovaleraldehyde is first formed and isovaleryl-CoA derived from it[120]. But to the author's knowledge there is only indirect evidence for the implication of two oxidative decarboxylases in mammalian organisms, and the subject remains controversial.

The dehydrogenation step from isovaleryl-CoA to 3-methylcrotonyl-CoA was hypothesized by Coon *et al.*[119] in analogy with the dehydrogenation of straight-chain acyl-CoA thioesters. Leucine was converted into β-hydroxyisovaleryl-CoA[116]. That a specific oxidoreductase is implied in the step from isovaleryl-CoA to 3-methylcrotonyl-CoA is suggested by the existence of a genetic defect, "isovaleric-acid acidemia" in which isovalerate accumulates, probably *via* deacetylation of accumulated isovaleryl-CoA[121].

As shown by Bacchawat *et al.*[116] liver and heart extracts convert 3-methylcrotonyl-CoA plus CO_2 to HMG-CoA and cleave this compound into acetoacetate and acetyl-CoA. Similar observations were made by Knappe and Lynen[122] with regard to mycobacterial extracts. This fixation of CO_2 made sense of the observation that the conversion of the isopropyl group of leucine to acetoacetate, which explains its ketogenic effect, is accompanied by a fixation of stoichiometric quantities of CO_2 as observed by Coon[114] and by Plaut and Lardy[115]. The formation of 3-methylglutaconyl-CoA as an intermediate between 3-methylcrotonyl-CoA and HMG-CoA was suggested by Lynen *et al.*[123] and experimentally demonstrated by Himes *et al.*[118] using purified Achromobacter 3-methylcrotonoyl-CoA carboxylase [3-methylcrotonoyl-CoA: carbon dioxide ligase (ADP-forming); EN 6.4.1.4]. That 3-methylglutaconyl-CoA is also an intermediate in liver was demonstrated by Del Campillo-Campbell *et al.*[124].

As stated in Chapter 33, HMG-CoA is an intermediate in the acetoacetate formation from acetoacetyl-CoA and from acetyl-CoA. The enzyme involved in the cleavage into acetoacetate and acetyl-CoA is HMG-CoA lyase [3-hydroxy-3-methylglutaryl-CoA acetoacetate lyase; EN 4.1.3.4].

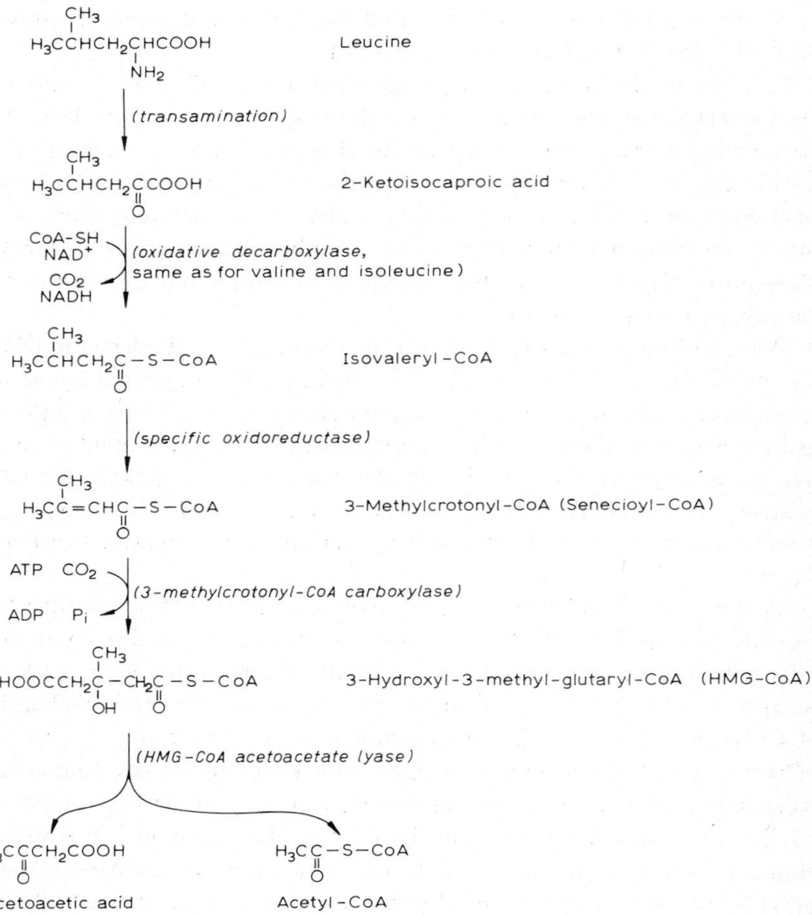

Fig. 34. From leucine to acetoacetic acid and acetyl-CoA.

(l) Valine

It has long been known that valine is a glycogenic amino acid. This was documented by nutrition studies done by Butts and Sinnhuber[125] and Rose *et al.*[126]. When isotopic markers became available, the conversion of valine into glucose and glycogen was confirmed in 1951 by Fones *et al.*[127] who used [^{13}C]valine. These first isotopic experiments pointed to a three-carbon intermediate but subsequent studies by Kinnory *et al.*[128] with [^{14}C]valine implicated isobutyric acid and β-hydroxyisobutyric acid as intermediates.

The conversion of isobutyryl-CoA to metacrylyl-CoA is believed to be catalyzed by the general acyl-CoA dehydrogenase. The metabolic step was demonstrated with dialyzed liver extract by Coon's group[129] and the enzyme itself, acyl-CoA dehydrogenase [acyl-CoA:(acceptor) oxidoreductase; EN 1.3.99.3] was purified by Crane *et al.*[130]. The hydration of methylacrylyl-CoA to 3-hydroxyisobutyryl-CoA can occur spontaneously without enzymatic catalysis[119,129].

That the acyl-CoA thioesters were involved as intermediates was demonstrated by Coon, *et al.*[119], and in 1957 Coon and his colleagues[129,131] documented the pathway from valine to methylmalonate semialdehyde as shown in Fig. 35.

The initial transamination of valine to α-ketoisovalerate and its oxidative decarboxylation have been studied in experiments on isobutyric acid oxidation by kidney homogenates[132,133] and of [^{14}C]valine dissimilation by liver homogenates[128].

It can also be catalyzed by an enzyme of broad specificity for L-β-hydroxyacyl-CoA, the hydro-lyase crotonase[134]. As the CoA ester cannot serve as a substrate for the dehydrogenation leading to methylmalonic semialdehyde, Coon and his collaborators[129] have postulated the deacylation which has been shown to result from the action of a deacylase isolated from pig's heart by Rendina and Coon[135] in 1957 [3-hydroxyisobutyryl-CoA hydrolase; EN 3.1.2.4]. The only other known substrate is β-hydroxypropionyl-CoA[135].

Deacylation precedes oxidation; with fatty acids (see Chapter 34) the reverse is the case.

The oxidation to methylmalonic semialdehyde was demonstrated[129] by incubating β-hydroxyisobutyrate with a fraction of pig's heart in the presence of NAD.

The enzyme, 3-hydroxyisobutyrate dehydrogenase [3-hydroxyisobutyrate: NAD$^+$ oxidoreductase, EN 1.1.1.31] has been shown by Den *et al.*[136] to differ from β-hydroxypropionate dehydrogenase.

Methylmalonic semialdehyde can follow different pathways. One of them leads to succinyl-CoA by oxidation of the semialdehyde to methylmalonate and the conversion of the latter to methylmalonyl-CoA which is isomerized to succinyl-CoA. This isomerization is catalyzed by the cobamide coenzyme-dependent methylmalonyl-CoA mutase [methylmalonyl-CoA:CoA car-

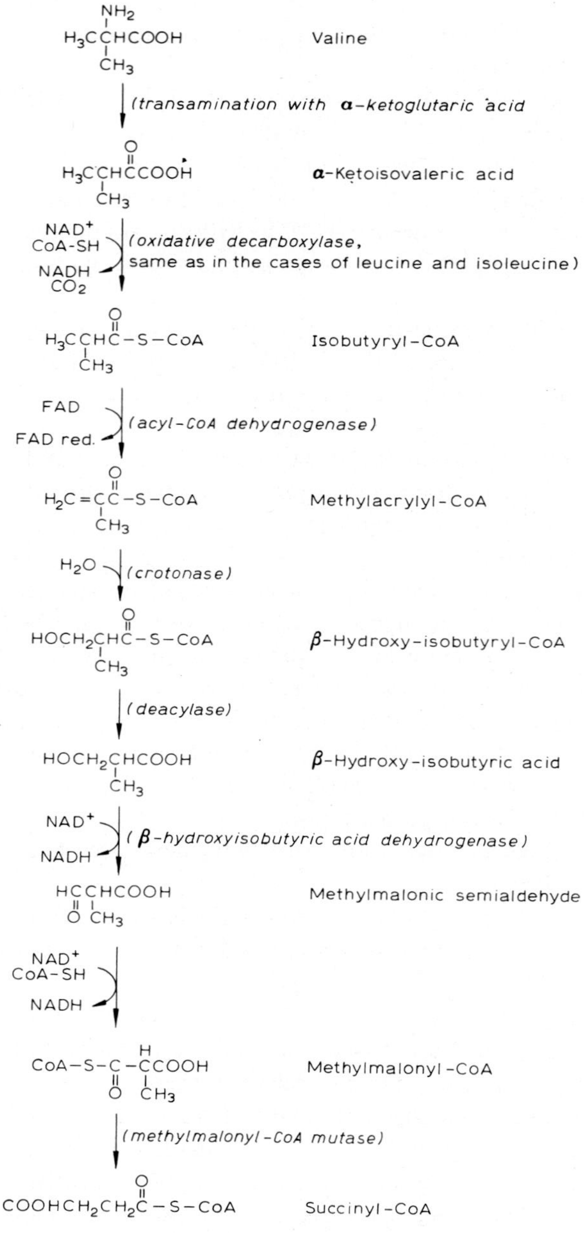

Fig. 35. From valine to succinyl-CoA.

bonyl mutase; EN 5.4.99.2] also important in propionate catabolism[137].

(m) Isoleucine

Nutrition studies on whole animals have revealed that, contrary to leucine, isoleucine is glycogenic[138,139]. Glycogen synthesis from isoleucine was confirmed as soon as isotopic tracers became available, with the help of D_2O (ref. 140).

Coon and Abrahamsen[141] showed, with the help of labeled [α-^{14}C]-methylbutyrate, that the isoleucine carbon skeleton was cleaved into a two-carbon atom fragment (C-3 and C-4 of α-methylbutyrate, β- and γ-carbons of isoleucine) and CO_2 (α–C of isoleucine).

The two-carbon fragment was identified as acetyl-CoA by Robinson *et al.*[142]. These authors also identified propionyl-CoA as an intermediate. Coon *et al.*[119] have proposed the other steps of the pathway of Fig. 36. Meister[151] has demonstrated the reversible transamination of L-isoleucine to α-keto-β-methylvaleric acid in the presence of hog heart tissue, of *E.coli* and of lactic bacteria.

The oxidative decarboxylation to α-methylbutyryl-CoA is catalyzed by the same enzyme as in the case of valine and leucine (oxidative decarboxylase). Tiglyl-CoA, the CoA ester of *cis*-2-methyl-2-butenoic acid, results from a dehydrogenation conceived by analogy with fatty-acid metabolism by Coon *et al.*[119].

Hydration of tiglyl-CoA has been observed to be catalyzed by a dialyzed pig heart preparation or by crystalline crotonase[142]. The same authors have shown presumably, by indirect ways, the conversion of tiglyl-CoA to α-methyl-β-hydroxybutyryl-CoA, as reaction mixtures treated with hydroxylamine contained α-methyl-β-hydroxybutyryl hydroxamate. The same authors also showed the conversion of tiglyl CoA to propionyl-CoA by isolation of the latter.

That propionyl-CoA is carboxylated to methylmalonyl-CoA (a) at the expense of 1 mole of ATP was shown by Flavin and Ochoa[143] in 1955. The enzyme propionyl-CoA carboxylase [propionyl-CoA:CO_2 ligase (ADP-forming), EN 6.4.1.3], has been highly purified and shown to contain biotin as a prosthetic group[144,145]. The racemization of methylmalonyl-CoA (a) to its optical enanthiomorph, methylmalonyl-CoA (b) is catalyzed by methylmalonyl-CoA racemase, discovered by Mazumder *et al.*[146]. Methylmalonyl-CoA (b) is isomerized in the presence of methylmalonyl-

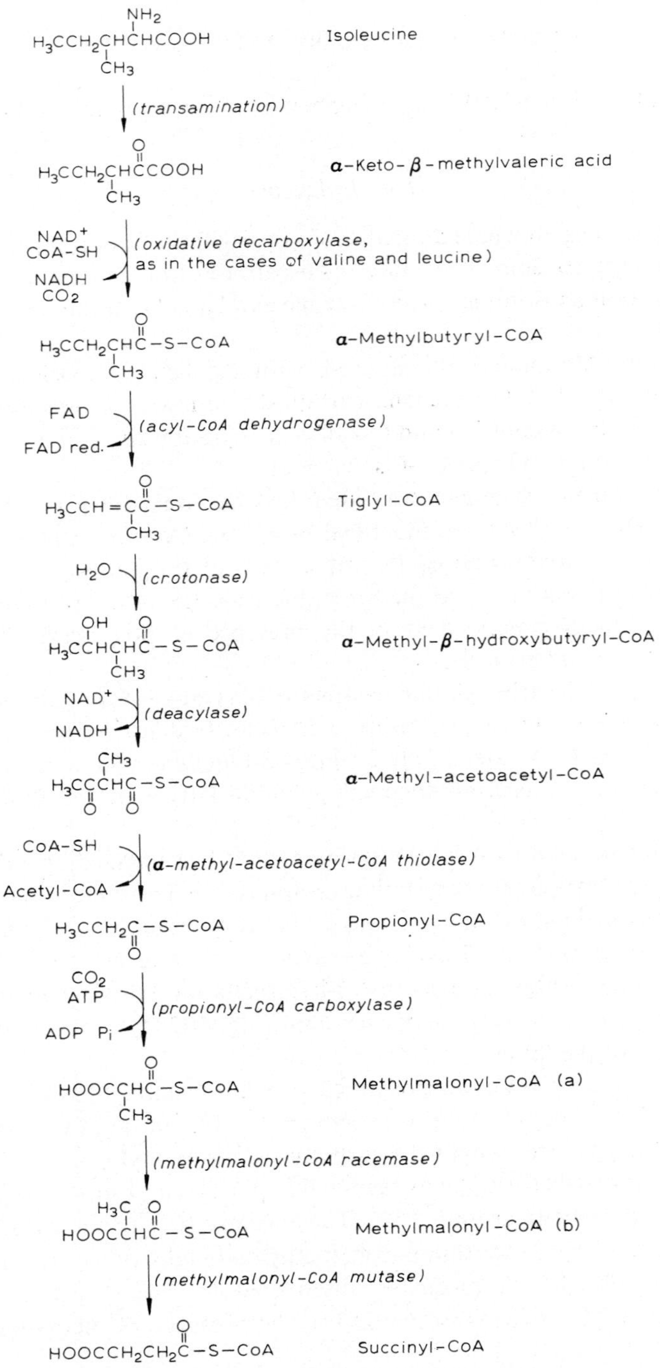

$H_3CCH_2CH(CH_3)CH(NH_2)COOH$ — Isoleucine

↓ *(transamination)*

$H_3CCH_2CH(CH_3)C(=O)COOH$ — **α**-Keto-**β**-methylvaleric acid

↓ NAD^+, CoA-SH → NADH, CO_2 *(oxidative decarboxylase, as in the cases of valine and leucine)*

$H_3CCH_2CH(CH_3)C(=O)-S-CoA$ — **α**-Methylbutyryl-CoA

↓ FAD → FAD red. *(acyl-CoA dehydrogenase)*

$H_3CCH=C(CH_3)C(=O)-S-CoA$ — Tiglyl-CoA

↓ H_2O *(crotonase)*

$H_3CCH(OH)CH(CH_3)C(=O)-S-CoA$ — **α**-Methyl-**β**-hydroxybutyryl-CoA

↓ NAD^+ → NADH *(deacylase)*

$H_3CC(=O)CH(CH_3)C(=O)-S-CoA$ — **α**-Methyl-acetoacetyl-CoA

↓ CoA-SH → Acetyl-CoA *(**α**-methyl-acetoacetyl-CoA thiolase)*

$H_3CCH_2C(=O)-S-CoA$ — Propionyl-CoA

↓ CO_2, ATP → ADP P_i *(propionyl-CoA carboxylase)*

$HOOCCH(CH_3)C(=O)-S-CoA$ — Methylmalonyl-CoA (a)

↓ *(methylmalonyl-CoA racemase)*

$HOOCCH(CH_3)C(=O)-S-CoA$ — Methylmalonyl-CoA (b)

↓ *(methylmalonyl-CoA mutase)*

$HOOCCH_2CH_2C(=O)-S-CoA$ — Succinyl-CoA

Fig. 36. From isoleucine to succinyl-CoA and acetyl-CoA.

CoA mutase [methylmalonyl-CoA:CoA carbonylmutase; EN 5.4.99.2], a cobamide enzyme which has also been highly purified[147].

(n) Arginine

The discovery of arginase by Kossel and Dakin[148] led to the suggestion that arginine first undergoes hydrolysis, converting arginine into urea and ornithine. In fact it was observed by W. H. Thompson[149] that when arginine is administered to a dog an amount of urea which corresponds to the urea derived from arginine by the action of arginase was rapidly excreted in the urine and an additional quantity of urea believed to be derived from ornithine was excreted more slowly.

Arginase [L-arginine amidinohydrolase; EN 3.5.3.1] catalyzes the transfer of the formamidine group to water. It has been highly purified from mammalian liver (see Greenberg[150]). That ornithine is transaminated by liver extracts to glutamate γ-semialdehyde has been shown by Meister[151]. The enzyme involved, ornithine transaminase [L-ornithine:2-oxoacid aminotransferase; EN 2.6.1.13], is widely distributed[152]. Glutamate γ-semialdehyde is oxidized into glutamate in the presence of an oxidoreductase (see Fig. 8 of ref. 32).

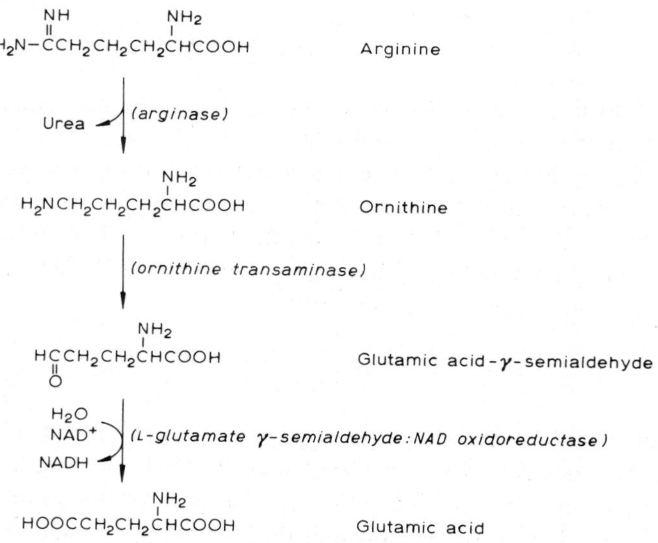

Fig. 37. From arginine to glutamic acid.

(o) Proline

That both L- and D-proline are oxidized by kidney or liver homogenates has been shown by several experimenters[153–156].

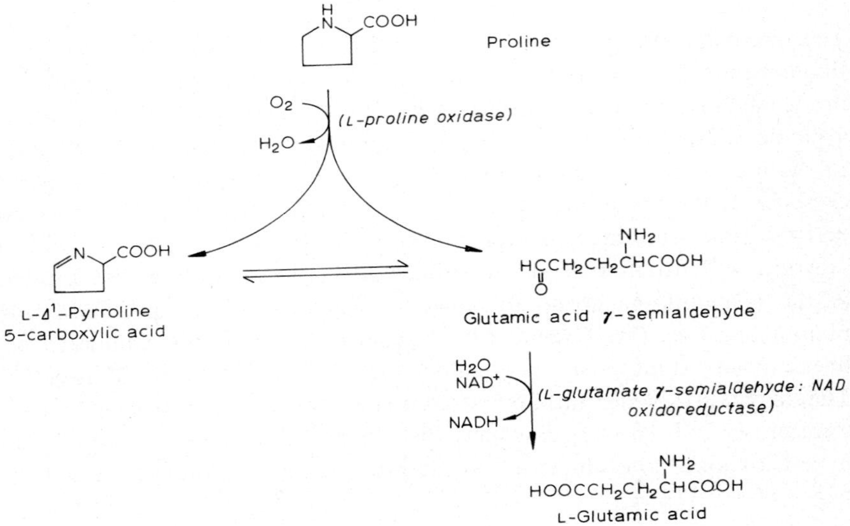

Proline

O_2 → H_2O (L-proline oxidase)

L-Δ^1-Pyrroline 5-carboxylic acid ⇌ Glutamic acid γ-semialdehyde ($HCCH_2CH_2CH(NH_2)COOH$)

H_2O, NAD^+ → NADH (L-glutamate γ-semialdehyde: NAD oxidoreductase)

$HOOCCH_2CH_2CH(NH_2)COOH$

L-Glutamic acid

Fig. 38. From proline to glutamic acid.

The initial attack at C-5 is catalyzed by the L-proline oxidase which has been found in liver[155,156] and in *E. coli*[157,158].

As shown by Strecker[158], an equilibrium mixture of L-Δ^1-pyrroline-5-carboxylic acid and of L-glutamate γ-semialdehyde is obtained. In liver[159,160], in *E. coli*[158] and in *B. subtilis*[161] this equilibrium mixture is oxidized to L-glutamate by L-glutamate γ-semialdehyde: NAD oxidoreductase isolated by Frank and Ranhand[162].

(p) L-*Lysine*

In mammals, four of the six carbon atoms of L-lysine are introduced into the tricarboxylic-acid cycle at the level of acetyl-CoA after passing through acetoacetyl-CoA. Until the late 1960s, the pathway of the dissimilation of L-lysine was believed to pass through the intermediary stage of pipecolic acid (literature in Meister[163] and in Rodwell[32]).

Schoenheimer[164,165], having observed little incorporation of the isotope into the amino groups of lysine after the administration of [^{15}N]ammonia or [^{15}N]amino acids to rats, concluded that lysine did not participate in reversible transamination. This, he suggested, could be explained if the degradation of lysine went through a cylic intermediate such as pipecolic acid

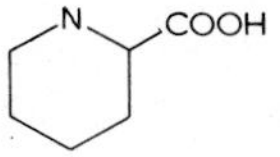

Pipecolic acid

The first intermediate product identified in the mammalian pathway had been α-aminoadipic acid. This was the result of the use of the isotopic technique by Borsook and his group[166] who synthesized [^{14}C]lysine and showed in 1948 its conversion to α-aminoadipic acid by homogenates of guinea pig liver and by slices of rat liver. Not only was α-aminoadipic acid later isolated by Boulanger and Biserte[167] from the urine of guinea pig and rat fed lysine, but α-ketoadipic acid was also isolated from the urine by Cavallini and Mondovi[168]. When pipecolic acid was recognized in 1952 as a natural compound and isolated from plants[169], Rothstein and Miller[170] gave [^{14}C]lysine to rats along with large amounts of pipecolic acid serving as a "trap". From the urine they isolated pipecolate with a relatively high specific activity. Similar experiments were performed by Hulme and Arthington[171] in 1952. These authors gave to rats [α-^{15}N]lysine and [ε-^{15}N]-lysine and they isolated labeled pipecolate only after the administration of [ε-^{15}N]lysine.

From these data, a pathway was derived going from lysine to α-amino-adipic acid through pipecolic acid.

In the late 1960s, new data brought new viewpoints. The experiments by Rothstein and Miller[170] and by Hulme and Arthington[171] reported above were accomplished with DL-lysine and interpreted on the basis of the assumed metabolic inertness of D-lysine, which was believed to be excreted unchanged when fed to rats[172]. As a result D-lysine was supposed to be inactive following administration of DL-lysine. In 1968, however, Grove and Henderson[173] discovered that D-lysine contributes 96 per cent of the pipecolate recovered from the urine whereas L-lysine contributes very little.

Grove *et al.*[174] confirmed the conversion of D-lysine to L-pipecolate after removal of the L-amino group and showed that it is the major metabolite of

D-lysine appearing in rat urine but constitutes a very minor part of the products of L-lysine dissimilation.

Rothstein and Miller[170] had concluded that L-lysine was converted to L-amino adipate following removal of the α-amino group, a conclusion they had drawn from experiments in which DL-[α-^{15}N]- or DL-[ε-^{15}N]-lysine was administered together with an overloading dose of L-α-aminoadipate. The results obtained by Rothstein and Miller had suggested that the initial step involved removal of α-amino group, thus leading to α-keto-ε-amino-caproic acid. Paik *et al.*[175] had suggested that α-deamination of lysine would be made easier by previous acetylation of the ε-amino group and they demonstrated that the mammalian liver was able to accomplish the conversion. They suggested that lysine was first converted to an α-acyl derivative, then to an ε-acylated form of α-keto-ε-aminocaproic acid and finally deacylated to α-keto-ε-aminocaproic acid.

In 1964, Boulanger *et al.*[176] administered [^{14}C]lysine, ^{14}C-labeled Δ′-piperidine-2-carboxylic acid and ^{14}C-labeled pipecolic acid to the sterile intact rat and from the results obtained they questioned whether pipecolic acid can be considered a major lysine metabolite.

Higashino *et al.*[177] demonstrated in 1965 that rat liver catalyses the reaction of lysine to saccharopine [ε-*N*-(L-glutaryl-2)-L-lysine]. This is the reverse of one of the steps of biosynthesis of lysine in *Neurospora* in which, as demonstrated by Broquist[178–180] saccharopine is an essential intermediate. Higashino and his colleagues[177] consider the saccharopine pathway as the main pathway of lysine catabolism in mammals. In 1968, Hutzler and Dancis[181] reported that they had observed in two children a genetic disease with an accumulation of lysine in the blood. The children were recognized as being unable to degrade lysine, the block apparently having to do with an early stage of the catabolic pathway. Hutzler and Dancis[181] showed, by isolating radioactive saccharopine, the conversion of [^{14}C]lysine to saccharopine by human liver *in vitro*.

In the same year (1968), Carson *et al.*[182] found highly concentrated saccharopine in the urine and in the cerebrospinal fluid of a mentally retarded patient whose urine also contained large quantities of lysine.

The scheme of Fig. 39 is in agreement with the view that L-pipecolate is relatively inert and is not on the major pathway of L-lysine degradation but probably represents a dead-end. Fig. 39 represents what appears today as the main pathway of L-lysine dissimilation leading to the tricarboxylic

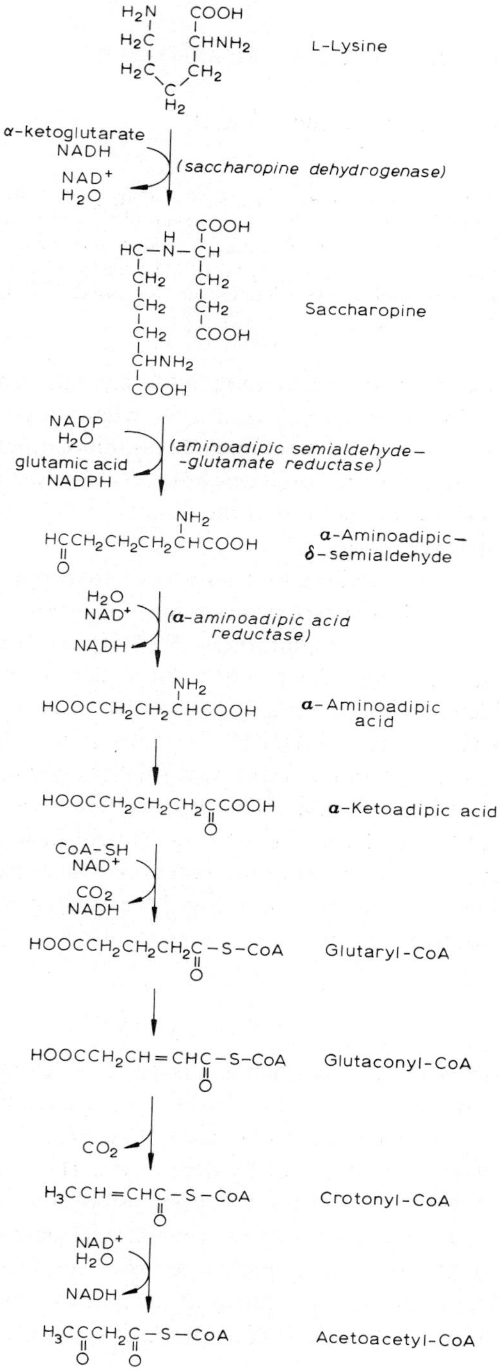

H_2N COOH

H_2C $CHNH_2$ L-Lysine

H_2C CH_2

C

H_2

α-ketoglutarate, NADH → NAD⁺, H_2O (saccharopine dehydrogenase)

COOH

H

HC—N—CH

CH_2 CH_2

CH_2 CH_2 Saccharopine

CH_2 COOH

$CHNH_2$

COOH

NADP, H_2O → glutamic acid, NADPH (aminoadipic semialdehyde--glutamate reductase)

NH_2

$HCCH_2CH_2CH_2CHCOOH$ (=O) α-Aminoadipic-δ-semialdehyde

H_2O, NAD^+ → NADH (α-aminoadipic acid reductase)

NH_2

$HOOCCH_2CH_2CHCOOH$ α-Aminoadipic acid

↓

$HOOCCH_2CH_2CH_2CCOOH$ (=O) α-Ketoadipic acid

CoA-SH, NAD^+ → CO_2, NADH

$HOOCCH_2CH_2CH_2C$(=O)$-S-CoA$ Glutaryl-CoA

↓

$HOOCCH_2CH{=}CHC$(=O)$-S-CoA$ Glutaconyl-CoA

→ CO_2

$H_3CCH{=}CHC$(=O)$-S-CoA$ Crotonyl-CoA

NAD^+, H_2O → NADH

H_3CC(=O)CH_2C(=O)$-S-CoA$ Acetoacetyl-CoA

Fig. 39. From L-lysine to acetoacetyl-CoA.

acid cycle. To quote Hutzler and Dancis[181]:

"The formation of saccharopine, followed by cleavage to α-aminoadipic semialdehyde and glutamic acid, results in the transfer of the ε-NH_2 group of lysine to the common acceptor, α-ketoglutarate, with saccharopine representing a stable intermediate. Saccharopine is analogous in this respect to arginosuccinic acid in the transfer of the amino group from aspartic acid to citrulline, and to cystathionine in the transfer of the SH from homocysteine to serine."

Saccharopine is converted to α-aminoadipate and glutamate by rat liver mitochondria[177]. In this sequence (contrary to what happens with D-lysine) the α-amino group of L-lysine is retained as α-aminoadipate. The pathway which had previously been proposed was not consistent with the inclusion of saccharopine, as this route permitted the removal of the α-amino group of lysine as the initial reaction.

As stated above, the reaction of L-lysine to saccharopine is the reverse of one of the steps of L-lysine biosynthesis in Neurospora (saccharopine + $NAD^+ + H_2O \rightleftharpoons$ lysine + α-ketoglutarate + NADH) catalyzed by saccharopine dehydrogenase[178], and the reaction from saccharopine to α-aminoadipic-δ-semialdehyde is the reverse of another step (α-aminoadipic-δ-semialdehyde + glutamate + NADPH $\rightleftharpoons$ saccharopine + $NADP^+$ + H_2O) catalyzed by the enzyme aminoadipic semialdehyde-glutamate reductase.

Saccharopine dehydrogenase has been isolated from rat liver by Higashino[177], and from human tissues by Hutzler and Dancis[181], who call the enzyme lysine-ketoglutarate reductase. The pathway from α-ketoadipic acid to acetoacetyl-CoA, in Fig. 39, was already discussed in the section on tryptophan.

(q) Cystine and cysteine

The degradation of these amino acids has been reviewed by Kun[183,184].

These two amino acids can be transformed to pyruvate by loss of sulphur resulting from desulphhydrase activity, but, unlike the oxidation pathway, of the sulphur, through sulphinic acid, by which the carbon atoms are also led to pyruvate, this pathway is of no metabolic importance.

That cysteine sulphinic acid is an intermediate in the metabolic pathway was established through experiments *in vivo* in Fromageot's laboratory[185,186]. The pathway between cysteine and cysteine sulphinic acid remains to be unraveled. In animal and microbial systems, cysteine sulphinic

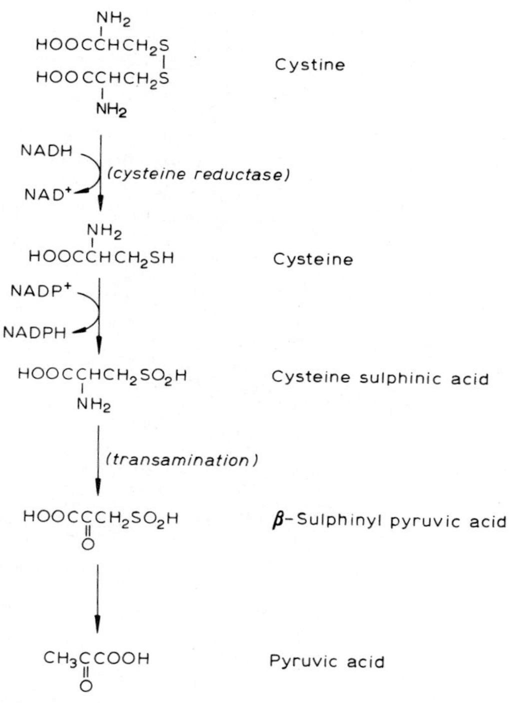

Fig. 40. Main pathway from cystine to pyruvic acid.

acid is transformed by transamination into β-sulphinylpyruvate by glutamate-aspartate transaminase as shown by Cohen[187] and by Kearney and Singer[188].

(r) Methionine

The dissimilation pathway of this amino acid enters the Krebs cycle at the level of succinyl-CoA.

The recognition of homoserine as an intermediate in the dissimilation of the carbon skeleton of methionine is based on the view that homocysteine combines with serine to form cystathionine, which is cleaved into cysteine and homoserine. The fact that homoserine can furnish the carbon skeleton for the synthesis of methionine in *Neurospora*[189] emphasizes relations between methionine and homoserine. In 1949, Carroll *et al.*[190] showed that

References p. 362

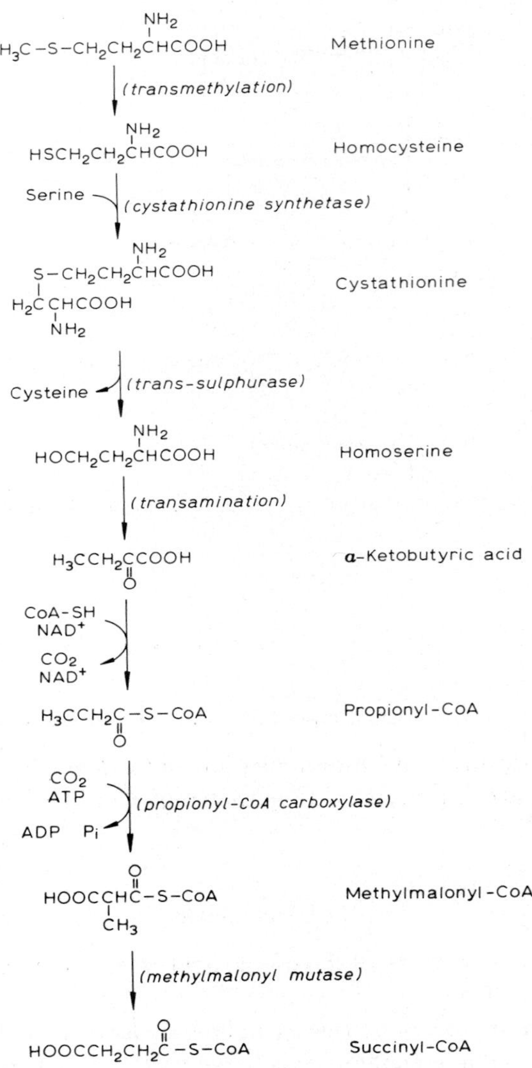

Fig. 41. From methionine to succinyl-CoA.

homoserine can be converted into α-ketobutyric acid by liver extracts under anaerobic conditions.

The pathway from α-ketobutyric acid is completed by oxidative decarboxylation followed by the path described for propionyl-CoA in the section on isoleucine.

References p. 362

REFERENCES

1 A. Lehninger, *Biochemistry. The Molecular Basis of Cell Structure and Function*, New York, 1970.
2 G. Embden, *Beitr. Chem. Physiol. Pathol.*, 11 (1908) 438.
3 O. Neubauer, *Verh. d. Kongr. f. inn. Med.*, (1910).
4 O. Neubauer, in A. Bethe, G. V. Bergmann, G. Embden and A. Ellinger (Eds.), *Handbuch der Normalen und Pathologischen Physiologie*, Vol. 5, 1928, p. 775.
5 C. Schotten, *Z. Physiol. Chem.*, 7 (1882) 375.
6 O. Neubauer, *Deut. Arch. Klin. Med.*, 95 (1909) 211.
7 O. Neubauer and H. Fischer, *Z. Physiol. Chem.*, 67 (1910) 230.
8 C. Neuberg and L. Langstein, *Verhandl. d. Physiol. Gesellsch. His-Engelmanns Arch., Physiol. Abt.*, Suppl. Band (1903) 514.
9 H. D. Dakin, *Oxidations and Reductions in the Animal Body*, London, 1912.
10 L. Flatow, *Z. Physiol. Chem.*, 64 (1910) 367.
11 H. A. Krebs, *Z. Physiol. Chem.*, 217 (1933) 191.
12 M. Jacoby, *Z. Physiol. Chem.*, 30 (1900) 149.
13 O. Loewi, *Z. Physiol. Chem.*, 25 (1898) 511.
14 E. S. London, A. M. Dubinsky, N. L. Wassilewskaja and M. J. Prochorona, *Z. Physiol. Chem.*, 227 (1934) 223.
15 H. Borsook and C. E. P. Jeffreys, *J. Biol. Chem.*, 110 (1935) 495.
16 H. A. Krebs, *Biochem. J.*, 29 (1935) 1620.
17 O. Warburg and W. Christian, *Biochem. Z.*, 298 (1938) 150.
18 H. J. Sallach and L. A. Fahien, in D. M. Greenberg (Ed.), *Metabolic Pathways*, 3rd ed., Vol. 3, New York and London, 1969.
19 M. Blanchard, D. E. Green, V. Nocito and S. Ratner, *J. Biol. Chem.*, 155 (1944) 421.
20 M. Blanchard, D. E. Green, V. Nocito and S. Ratner, *J. Biol. Chem.*, 161 (1945) 583.
21 D. M. Needham, *Biochem. J.*, 24 (1930) 208.
22 D. M. Needham, *Machina carnis. The Biochemistry of Muscular Contraction in its Historical Development*, Cambridge, 1971, p. 404.
23 A. E. Braunstein and M. G. Kritzmann, *Enzymologia*, 2 (1937–1938) 129.
23a G. Schaeffer and E. Le Breton, *L'Action Dynamique Spécifique des Protides*, Paris, 1938.
24 M. G. Kritzmann, *Biochimia*, 3 (1938) 603.
25 M. G. Kritzmann, *Nature*, 143 (1939) 603.
25a P. P. Cohen, *Biochem. J.*, 33 (1939) 551.
26 A. E. Braunstein, *Advan. Enzymol.*, 19 (1957) 335.
27 E. V. Rowsell, *Biochem. J.*, 64 (1956) 235.
28 E. E. Snell, *J. Am. Chem. Soc.*, 67 (1945) 194.
29 F. Schlenk and E. E. Snell, *J. Biol. Chem.*, 157 (1945) 425.
30 M. Kritzmann and O. Samarina, *Nature*, 158 (1946) 104.
31 D. M. Greenberg, in D. M. Greenberg (Ed.), *Metabolic Pathways*, 3rd ed., Vol. 3, New York and London, 1969.
32 V. W. Rodwell, in D. M. Greenberg (Ed.), *Metabolic Pathways*, 3rd ed., Vol. 3, New York and London, 1969.
33 S. Lang, *Beitr. Chem. Physiol. Pathol.*, 5 (1904) 321.
34 J. D. Klingman and P. Handler, *J. Biol. Chem.*, 232 (1958) 369.
35 J. A. Shepherd and G. Kalnitsky, *J. Biol. Chem.*, 192 (1951) 1.
36 F. W. Sayre and E. Roberts, *J. Biol. Chem.*, 233 (1958) 1128.

37 P. P. Cohen and H. J. Sallach, in D. M. Greenberg (Ed.), *Metabolic Pathways*, 2nd ed., Vol. 2, New York, 1961.

38 J. E. Varner, in P. D. Boyer, H. Lardy and K. Myrbäck (Eds.), *The Enzymes*, 2nd ed., Vol. 4, New York, 1960.

39 T. O. Yellin and J. C. Wriston, Jr., *Biochemistry*, 5 (1966) 1605.

40 A. Meister, L. Levintow, R. E. Greenfield and P. A. Abendschein, *J. Biol. Chem.*, 215 (1955) 441.

41 D. M. Greenberg, in D. M. Greenberg (Ed.), *Chemical Pathways of Metabolism*, Vol. 2, New York, 1954.

42 R. D. Sagers and I. C. Gunsalus, *J. Bacteriol.*, 81 (1961) 541.

43 S. M. Klein and R. D. Sagers, *J. Bacteriol.*, 83 (1962) 121.

44 F. M. Huennekens and M. J. Osborn, *Advan. Enzymol.*, 21 (1959) 369.

45 J. C. Rabinowitz, in P. D. Boyer, H. Lardy and K. Myrbäck (Eds.), *The Enzymes*, Vol. 2, New York and London, 1960.

46 O. G. Lien and D. M. Greenberg, *J. Biol. Chem.*, 195 (1952) 637.

47 A. E. Braunstein and G. Y. Vilenkina, *Dokl. Acad. Nauk SSSR*, 66 (1949) 243.

48 H. L. Meltzer and D. B. Sprinson, *J. Biol. Chem.*, 197 (1952) 461.

49 F. W. Sayre and D. M. Greenberg, *J. Biol. Chem.*, 220 (1956) 787.

50 J. Nishimura, *Ph.D. Thesis, Univ. of California, 1959* (cited by D. M. Greenberg, in P. D. Boyer, H. Lardy and K. Myrbäck (Eds.), *The Enzymes*, 2nd ed., Vol. 5, New York and London, 1961.)

51 S. Edlbacher, *Z. Physiol. Chem.*, 157 (1926) 106.

52 P. Györgi and H. Rothler, *Biochem. Z.*, 173 (1926) 334.

53 W. J. Darby and H. B. Lewis, *J. Biol. Chem.*, 146 (1942) 225.

54 Y. Sera and S. Yada, *J. Osaka City Med. Soc.*, 38 (1939) 1107.

55 B. A. Borek and H. Waelsch, *J. Biol. Chem.*, 205 (1953) 459.

56 R. H. Feinberg and D. M. Greenberg, *Nature*, 181 (1958) 897.

57 R. H. Feinberg and D. M. Greenberg, *J. Biol. Chem.*, 234 (1959) 2670.

58 H. R. B. Revel and B. Magasanik, *J. Biol. Chem.*, 233 (1958) 930.

59 M. Suda, I. Miyahara, K. Tomihata and A. Kato, *Med. J. Osaka Univ.*, 3 (1952) 115.

60 A. Miller and H. Waelsch, *J. Biol. Chem.*, 228 (1957) 383,397.

61 E. Mayer, *Deut. Arch. Klin. Med.*, 70 (1901) 1443.

62 E. Friedmann, *Beitr. Chem. Physiol. Pathol.*, 11 (1908) 304.

63 M. Wolkow and E. Baumann, *Z. Physiol. Chem.*, 15 (1891) 228.

64 W. Falta and L. Langstein, *Z. Physiol. Chem.*, 37 (1903) 513.

65 E. and H. Salkowski, *Ber. deut. Chem. Ges.*, 12 (1879) 653.

66 B. N. LaDu and V. G. Zannoni, *Ann. N.Y. Acad. Sci.*, 92 (1961) 175.

67 B. N. LaDu and D. M. Greenberg, *J. Biol. Chem.*, 190 (1951) 245.

68 W. E. Knox and M. LeMay-Knox, *Biochem. J.*, 49 (1951) 686.

69 B. N. LaDu, V. G. Zannoni, L. Laster and J. E. Seegmiller, *J. Biol. Chem.*, 230 (1958) 251.

70 R. G. Ravdin and D. I. Crandall, *J. Biol. Chem.*, 189 (1951) 137.

71 M. Suda and Y. Takeda, *J. Biochem. (Jap.)*, 37 (1950) 375.

72 K. Tokuyama, *J. Biochem. (Jap.)*, 46 (1959) 1379, 1453.

73 W. G. Flamm and D. I. Crandall, *J. Biol. Chem.*, 238 (1963) 389.

74 D. I. Crandall, *J. Biol. Chem.*, 212 (1955) 565.

75 W. E. Knox and S. W. Edwards, *J. Biol. Chem.*, 216 (1955) 479, 489.

76 W. M. Connors and E. Stotz, *J. Biol. Chem.*, 178 (1949) 881.

77 M. Womack and W. C. Rose, *J. Biol. Chem.*, 107 (1934) 449.

78 A. R. Moss and R. Schoenheimer, *J. Biol. Chem.*, 135 (1940) 415.

79 S. Udenfriend and J. R. Cooper, *J. Biol. Chem.*, 194 (1952) 503.

80 S. Kaufman and B. Levenberg, *J. Biol. Chem.*, 234 (1959) 2683.
81 S. Kaufman, *J. Biol. Chem.*, 230 (1958) 931.
82 C. Mitoma, *Arch. Biochem. Biophys.*, 60 (1956) 476.
83 H. W. Wallace, K. Moldave and A. Meister, *Proc. Soc. Exptl. Biol. Med.*, 94 (1957) 632.
84 C. Mitoma, R. M. Auld and S. Udenfriend, *Proc. Soc. Exptl. Biol. Med.*, 94 (1957) 634.
85 C. E. Dalgliesh, *Advan. Protein Chem.*, 10 (1955) 31.
86 L. M. Henderson and L. V. Hankes, *J. Biol. Chem.*, 222 (1956) 1069.
87 C. E. Dalgliesh and H. Tabechian, *Biochem. J.*, 62 (1956) 625.
88 R. K. Gholson, D. R. Rao, L. M. Henderson, R. J. Hill and R. E. Koeppe, *J. Biol. Chem.*, 1958 (230) 179.
89 A. Ellinger, *Z. Physiol. Chem.*, 43 (1904) 325.
90 J. von Liebig, *Ann. Chem.*, 86 (1953) 125.
91 W. Robson, *J. Biol. Chem.*, 62 (1924–1925) 495.
92 J. M. Price and L. W. Dodge, *J. Biol. Chem.*, 223 (1956) 699.
93 Z. Matsuoka and N. Yoshimatsu, *Z. Physiol. Chem.*, 143 (1925) 206.
94 Y. Kotake and J. Iwao, *Z. Physiol. Chem.*, 195 (1931) 139.
95 A. Butenandt, W. Weidel and W. von Derjugin, *Naturwissenschaften*, 30 (1942) 51.
96 C. E. Dalgliesh, W. E. Knox and A. Neuberger, *Nature*, 168 (1951) 20.
97 C. E. Dalgliesh, *Biochem. J.*, 52 (1952) 3.
98 R. K. Gholson, L. M. Henderson, G. A. Mourkides, R. J. Hill and R. E. Koeppe, *J. Biol. Chem.*, 234 (1959) 96.
99 Y. Nishizuka, A. Ichiyama, R. K. Gholson and O. Hayashi, *J. Biol. Chem.*, 240 (1965) 733.
100 E. R. Tustanoff and J. R. Stern, *Biochem. Biophys. Res. Commun.*, 3(1960) 81.
101 Y. Kotake and T. Masayama, *Z. Physiol. Chem.*, 243 (1936) 237.
102 W. E. Knox and A. H. Mehler, *J. Biol. Chem.*, 187 (1950) 419.
103 O. Hayashi, S. Rothberg, A. H. Mehler and Y. Saito, *J. Biol. Chem.*, 229 (1957) 889.
104 F. T. de Castro, J. M. Price and R. R. Brown, *J. Am. Chem. Soc.*, 78 (1956) 2904.
105 Y. Saito, O. Hayashi and S. Rothberg, *J. Biol. Chem.*, 229 (1957) 921.
106 A. E. Braunstein, E. V. Goryachenkova and T. S. Pashkina, *Biokhymia*, 14 (1949) 163.
107 D. R. Sanadi and D. M. Greenberg, *Arch. Biochem.*, 25 (1950) 323.
108 M. Rothstein and D. M. Greenberg, *Biochim. Biophys. Acta*, 34 (1959) 598.
109 R. K. Gholson, L. V. Hankes and L. M. Henderson, *J. Biol. Chem.*, 235 (1960) 132.
110 G. Embden, M. Salmon and F. Schmidt, *Beitr. Chem. Physiol. Pathol.*, 8 (1906) 129.
111 K. Bloch, *J. Biol. Chem.*, 155 (1944) 255.
112 M. J. Coon and S. Gurin, *J. Biol. Chem.*, 180 (1949) 1159.
113 I. Zabin and K. Bloch, *J. Biol. Chem.*, 185 (1950) 117.
114 M. J. Coon, *J. Biol. Chem.*, 187 (1950) 71.
115 G. W. E. Plaut and H. A. Lardy, *J. Biol. Chem.*, 192 (1951) 435.
116 B. K. Bachhawat, W. G. Robinson and M. J. Coon, *J. Biol. Chem.*, 216 (1955) 727.
117 B. K. Bachhawat, W. G. Robinson and M. J. Coon, *J. Biol. Chem.*, 219 (1956) 539.
118 R. H. Himes, D. L. Young, E. Ringelmann and F. Lynen, *Biochem. Z.*, 337 (1963) 48.
119 M. J. Coon, W. G. Robinson and B. K. Bachhawat, in W. D. McElroy and H. B. Glass (Eds.), *A Symposium on Amino Acid Metabolism*, Baltimore, 1955.
120 S. Sasaki, *Nature*, 189 (1961) 400.
121 K. Tanaka, M. A. Budd, M. L. Efron and K. J. Isselbacher, *Proc. Natl. Acad. Sci. (U.S.)*, 56 (1966) 236.
122 J. Knappe and F. Lynen, *Proc. 4th Intern. Congr. Biochem.; Abstr. Comm., Vienna, Sept. 1958* (1959–1960) 49.
123 F. Lynen, J. Knappe, E. Lorch, G. Jütting, E. Ringelmann and J. P. La Chance, *Biochem. Z.*, 335 (1961) 123.

124 A. del Campillo-Campbell, E. E. Dekker and M. J. Coon, *Biochim. Biophys. Acta*, 31 (1959) 290.
125 J. S. Butts and R. O. Sinnhuber, *J. Biol. Chem.*, 139 (1941) 963.
126 W. C. Rose, J. E. Johnson and W. J. Haines, *J. Biol. Chem.*, 145 (1942) 679.
127 W. S. Fones, T. P. Waalkes and J. White, *Arch. Biochem. Biophys.*, 32 (1951) 89.
128 D. S. Kinnory, Y. Takeda and D. M. Greenberg, *J. Biol. Chem.*, 212 (1955) 385.
129 W. G. Robinson, R. Nagle, B. K. Bachhawat, F. P. Kupiecki and M. J. Coon, *J. Biol. Chem.*, 224 (1957) 1.
130 F. L. Crane, S. Mii, J. G. Hauge, D. E. Green and H. Beinert, *J. Biol. Chem.*, 218 (1956) 701.
131 F. P. Kupiecki and M. J. Coon, *J. Biol. Chem.*, 229 (1957) 743.
132 W. A. Atchley, *J. Biol. Chem.*, 176 (1948) 123.
133 I. Gray, P. Adams and H. Hauptmann, *Experientia*, 6 (1950) 430.
134 J. R. Stern, A. del Campillo and I. Raw, *J. Biol. Chem.*, 218 (1956) 971.
135 G. Rendina and M. J. Coon, *J. Biol. Chem.*, 225 (1957) 523.
136 H. Den, W. G. Robinson and M. J. Coon, *J. Biol. Chem.*, 234 (1959) 1666.
137 Y. Kaziro and S. Ochoa, *Advan. Enzymol.*, 26 (1964) 283.
138 J. S. Butts, H. Blunden and M. S. Dunn, *J. Biol. Chem.*, 120 (1937) 289.
139 A. N. Wick, *J. Biol. Chem.*, 141 (1941) 897.
140 L. C. Terriere and J. S. Butts, *J. Biol. Chem.*, 190 (1951) 1.
141 M. J. Coon and N. S. B. Abrahamsen, *J. Biol. Chem.*, 195 (1952) 805.
142 W. G. Robinson, B. K. Bachhawat and M. J. Coon, *J. Biol. Chem.*, 218 (1956) 391.
143 M. Flavin and S. Ochoa, *J. Biol. Chem.*, 229 (1957) 965.
144 Y. Kaziro, E. Leone and S. Ochoa, *Proc. Natl. Acad. Sci. (U.S.)*, 46 (1950) 1319.
145 Y. Kaziro, S. Ochoa, R. C. Warner and J. Y. Chen, *J. Biol. Chem.*, 236 (1961) 1917.
146 R. Mazumder, T. Sasakawa, Y. Kaziro and S. Ochoa, *J. Biol. Chem.*, 236 (1961) Pc 53; 237 (1962) 3065.
147 R. Stjernholm and H. G. Wood, *Proc. Natl. Acad. Sci. (U.S.)*, 47 (1961) 303.
148 A. Kossel and H. D. Dakin, *Z. Physiol. Chem.*, 41 (1904) 321.
149 W. H. Thompson, *J. Physiol.*, 33 (1905) 106.
150 D. M. Greenberg, in P. D. Boyer, H. Lardy and K. Myrbäck (Eds.), *The Enzymes*, 2nd ed., Vol. 4, New York, 1960.
151 A. Meister, *J. Biol. Chem.*, 206 (1954) 587.
152 W. I. Sher and H. J. Vogel, *Proc. Natl. Acad. Sci. (U.S.)*, 43 (1957) 796.
153 H. Weil-Malherbe and H. A. Krebs, *Biochem. J.*, 29 (1935) 2077.
154 J. V. Taggart and R. B. Krakaur, *J. Biol. Chem.*, 177 (1949) 641.
155 K. Lang and G. Schmid, *Biochem. Z.*, 322 (1951) 1.
156 A. B. Johnson and H. J. Strecker, *J. Biol. Chem.*, 237 (1962) 1876.
157 L. Frank and P. Rybicki, *Arch. Biochem. Biophys.*, 95 (1961) 441.
158 H. J. Strecker, *J. Biol. Chem.*, 225 (1957) 825; 235 (1960) 2045.
159 H. J. Strecker and P. Mela, *Biochim. Biophys. Acta*, 17 (1955) 580.
160 H. J. Strecker, *J. Biol. Chem.*, 235 (1960) 3218.
161 G. de Hauwer, R. Lavalle and J. M. Wiame, *Biochim. Biophys. Acta*, 81 (1964) 257.
162 L. Frank and B. Ranhand, *Arch. Biochem. Biophys.*, 107 (1964) 325.
163 A. Meister, *Biochemistry of the Amino Acids*, 2nd ed., New York, 1965.
164 N. Weissman and R. Schoenheimer, *J. Biol. Chem.*, 140 (1941) 779.
165 R. Schoenheimer and D. Rittenberg, *Physiol. Rev.*, 20 (1940) 218.
166 J. W. Dubnoff and H. Borsook, *J. Biol. Chem.*, 173 (1948) 425.
167 P. Boulanger and G. Biserte, *Compt. Rend.*, 232 (1951) 1451.
168 D. Cavallini and B. Mondovi, *Arch. Sci. Biol.*, 36 (1952) 468.
169 R. M. Zacharius, J. F. Thomson and F. C. Steward, *J. Am. Chem. Soc.*, 74 (1952) 2949.

170 M. Rothstein and L. L. Miller, *J. Am. Chem. Soc.*, 75 (1953) 4371.
171 A. C. Hulme and W. Arthington, *Nature*, 170 (1952) 659.
172 M. Rothstein, C. G. Bly and L. L. Miller, *Arch. Biochem. Biophys.*, 50 (1954) 252.
173 J. Grove and L. M. Henderson, *Biochim. Biophys. Acta*, 165 (1968) 113.
174 J. A. Grove, T. J. Gilbertson, R. H. Hammerstedt and L. M. Henderson, *Biochim. Biophys. Acta*, 184 (1969) 329.
175 W. K. Paik, L. Bloch-Frankenthal, S. M. Birnbaum, M. Winitz and J. P. Greenstein, *Arch. Biochem. Biophys.*, 69 (1957) 56.
176 P. Boulanger, E. Sacquet, R. Osteux and H. Charlier, *Compt. Rend.*, 259 (1964) 932.
177 K. Higashino, K. Tsukada and L. Lieberman, *Biochem. Biophys. Res. Commun.*, 20 (1965) 285.
178 J. S. Trupin and H. P. Broquist, *J. Biol. Chem.*, 240 (1965) 2524.
179 E. E. Jones and H. P. Broquist, *J. Biol. Chem.*, 240 (1965) 2531.
180 P. P. Saunders and H. P. Broquist, *J. Biol. Chem.*, 241 (1966) 3435.
181 J. Hutzler and J. Dancis, *Biochim. Biophys. Acta*, 158 (1968) 62.
182 N. A. J. Carson, B. G. Scally, D. W. Neill and I. J. Carré, *Nature*, 218 (1968) 679.
183 E. Kun, in D. M. Greenberg (Ed.), *Metabolic Pathways*, 2nd ed., Vol. 2, New York and London, 1961.
184 E. Kun, in D. M. Greenberg (Ed.), *Metabolic Pathways*, 3rd ed., Vol. 3, New York and London, 1969.
185 B. Bergeret and F. Chatagner, *Biochim. Biophys. Acta*, 14 (1954) 297.
186 F. Chapeville and P. Fromageot, *Biochim. Biophys. Acta*, 17 (1955) 275.
187 P. P. Cohen, *J. Biol. Chem.*, 136 (1940) 565.
188 E. B. Kearney and T. P. Singer, *Biochim. Biophys. Acta*, 11 (1953) 270; 11 (1953) 276; 14 (1954) 570.
189 H. J. Teas, N. H. Horowitz and M. Fling, *J. Biol. Chem.*, 172 (1948) 651.
190 W. R. Carroll, G. W. Stacey and V. du Vigneaud, *J. Biol. Chem.*, 180 (1949) 375.

Chapter 36

The Respiratory Chain

1. Introduction

The anaerobic phosphorylations which accompany glycolysis release a modest amount of energy by the conversion of organic compounds to other organic compounds with lower energy but with the same oxidation level. The aerobic phosphorylation of "respiration" transforms organic compounds into inorganic oxides such as CO_2 and H_2O. The recognition that the tricarboxylic acid cycle insures the degradation of carbon chains of different natures (carbohydrates, fatty acids, amino acids) has greatly modified prevailing views on biological oxidations. Specifically, it has introduced a clear distinction between the scheme of carbon chain degradation on the one hand and the electron transfer chains derived from it on the other.

In each of these chains, starting either from NADH (or NADPH in the case of isocitrate) or from succinate, two electrons flow through the chain of an enzyme system and reach molecular oxygen, the final electron acceptor in cell respiration. This chain is located in the cell membrane of prokaryotes and in the inner membrane of mitochondria in eukaryotes. The energy of the flowing electrons is partially conserved in the form of the energy-rich pyrophosphate bond of ATP, the remaining part being dissipated as heat.

Another important highlight was the recognition by Claude, in 1940, of mitochondria as the "power plants" of cells and the seat of biological oxidation (see Chapter 16).

These two developments focused attention on a highly restricted field; this contrasts with the previous view that a large number of potential systems might be included in the process. Furthermore, the application of the term "catalysts" to what now came to be called "carriers" had also been a source of confusion.

When it was recognized that every oxidation is coupled with a reduction,

References p. 403

the overall process consisting mostly in a mere redistribution of electrons in the system, it was realized also that a thermodynamically weaker acceptor cannot bring about the hydrogenation of a "stronger" donator. In other words, in order to function as an acceptor of hydrogen, a system must have a higher (less negative) potential than the system dehydrogenated. General acceptance of the view that the hydrogen is transferred to the activated oxygen led to the novel insight that all kinds of nutrients abandon hydrogen to a chain of carriers leading to oxygen.

As we saw in Chapter 29, Keilin had tried to reconstruct such a chain of carriers.

When it was generally recognized that cytochrome oxidase is the only autoxidizable heme compound, the others were understood to be carriers and the concept arose that the respiratory system, far from having a single hydrogen carrier to link the activated substrate with the activated oxygen, in fact used a chain of several carriers.

Analysis of the respiratory chain resulted from the introduction of a number of leading concepts: the activation of the hydrogen of the substrate (Wieland), the activation of the oxygen (Warburg), the participation of cytochrome (the significance of which was shown by Keilin), the participation of pyridine-linked dehydrogenases and of flavin-linked dehydrogenases (Warburg) and the recognition of the binding of the enzymes of the chain to the membranes, either cellular or mitochondrial, in the form of organized assemblies.

In the first years of this century it was considered that biological oxidations occurred in the reaction

$$\tfrac{1}{2}\,O_2 + H_2A \rightarrow H_2O + A$$

Then a period followed in which two opposite theories prevailed.

According to Wieland, the activation of the hydrogen of the substrate was all that was required. Around 1925, the general view was that probably the activation of both oxygen and hydrogen were required.

2. Respiratory chain considered as an electron transfer chain

In 1927 Warburg[1] and Keilin[2] suggested a preliminary scheme, a tentative outline of the sequence based on the discovery of the involvement of hemochromogens in the respiratory process.

Keilin was influenced by Wieland's theories and thus conceived a chain of compounds involved in the transfer to oxygen of hydrogens from the substrate activated by a dehydrogenase. According to Keilin[2]

"In the living cell, the cytochrome is in intimate relation on the one hand, with the thermolabile reducing systems (the dehydrogenase) and, on the other hand, with the oxidase or the thermolabile indophenolase. They all constitute a part of the complex system of oxidizing enzymes."

According to Warburg[1], who at the time of writing his 1927 book, claimed never to have seen a dehydrogenase at work, the "oxygen-transferring enzyme" takes up oxygen, activates it and, *via* MacMunn's histohematin, passes the activated oxygen to the organic molecule. As he saw it, the respiratory chain was essentially a device for the transfer of activated oxygen.

We have retraced the history of cytochrome oxidase in Chapter 29. We may recall that following Warburg's brilliant indirect photospectroscopic study, the "oxygen-transferring enzyme" or "respiratory ferment" was recognized as an iron compound of the nature of a hematin derivative. We also pointed out that Keilin proposed to replace the term "respiratory ferment" by "indophenoloxidase". This hurt Warburg's feelings and caused some bitterness. For several years Warburg avoided the terms "dehydrogenase" and "cytochrome". When he had shown that his "respiratory ferment" was inhibited by its combination with CO, the question naturally arose whether one of the components of the cytochrome was responsible for that combination. As we have pointed out, it was in 1938 that Keilin and Hartree attributed the activity of cytochrome oxidase to what they called cytochrome a_3.

During the thirties, the concept of the rôle of oxygen activation won general agreement and for some time Wieland's views on this point were considered unfounded. In contrast, the participation in the chain of the group of compounds which Keilin called "cytochrome" (as against Warburg's term "MacMunn's histohematin") was also recognized.

By the end of the twenties, Keilin thought in terms of hydrogens, which Wieland believed to be transferred to oxygen to form water. In the early thirties, the transfer of electrons was considered an essential feature of the respiratory chain. It was Warburg who stated in 1931 that biological oxidations were based on changes of iron valency, *i.e.* on oxidation-reduction.

"Es ist den Valenzwechsel einer Eisenverbindung — des Sauerstoffübertragenden Ferments der Atmung — auf dem die katalytische Oxydation in der lebenden Substanz beruht."

References p. 403

In the same Nobel Lecture (1931), Warburg[3] considers the possibility that MacMunn's three histohematins are inserted in the respiratory chain between the oxygen-transferring ferment and the substrate, and he concludes that it is still too early to decide

"ob die Atmung nicht eine einfache, sondern eine vierfache Eisenkatalyse ist."

In 1933, Warburg, Negelein and Haas[4] presented a scheme for a respiratory chain composed of four iron compounds undergoing changes in iron valency: the respiratory enzyme and the three cytochrome components, valency changes in the iron of these compounds being involved.

The biological activity of cytochrome oxidase thus involved valency change of iron and the respiratory chain came to be conceived as an electron transfer chain based on the iron-porphyrin system.

3. The discovery of the flavoprotein system as an initial oxidation-reduction system other than the iron-porphyrin one

Warburg has told how he was led to the discoveries related in his monumental paper co-authored with Christian[5] and published in 1936. In it he observes, as stated by Colowick, van Eys and Park[6] that

"many fundamental problems were solved and the groundwork was laid for much of modern biochemistry."

The background of this story goes back to Warburg's visit to Baltimore where he delivered the Herter lecture at Johns Hopkins Hospital in the fall of 1929 and visited Clark's laboratory. To quote Warburg[7], in his 1949 book

"1929 sah ich in Baltimore bei Barron und Harrop, wie man in roten Blutzellen durch Zusatz von Methylenblau eine langhaltende regelmässige Atmung erzeugen kann, eine Verbrennung von Glucose unter Entwicklung von Kohlensäure. Die Erscheinung schien mir so interessant, dass ich nach Dahlem zurückgekehrt, eine chemische Untersuchung über den Mechanismus der Methylenblauwirkung begann. Die Ergebnis war die Entwicklung der Wasserstoffübertragenden Fermente."

Until then, Warburg had expressed scorn for experiments involving methylene blue. Indeed, one of his reasons for dismissing the concept of hydrogen activation was that in Thunberg's experiments non-natural reagents such as methylene blue or nitrophenols were used. In Baltimore, however, he clearly saw, for the first time in his experience, a dehydrogenase at work even though methylene blue was involved.

Back in Berlin, Warburg proceeded to enquire into the methylene blue effect. In line with his theories involving iron catalysis, he suggested that the oxidation of sugar by the red cells in the presence of methylene blue was an oxidation by the ferric iron of the methemoglobin resulting from the oxidation of the hemoglobin by the methylene blue (Warburg, Kubowitz and Christian[8]). According to the views he had adopted in the twenties, Warburg regarded the oxidation of hemoglobin to methemoglobin by methylene blue as a surface reaction due to methylene blue adsorption on the surface of the red blood cells (Warburg, Kubowitz and Christian[9]). But shortly afterwards experiments convinced him that such surface effects were non-existent. He concluded this from his observation that, although the ability to oxidize glucose in the presence of methylene blue was lost when the erythrocytes were hemolysed in such cell-free suspensions, glucose-6-P (the Robison ester) was oxidized in the presence of methylene blue. The supernatant of the centrifugation of the hemolysed erythrocyte was fractionated by dialysis and it was shown that the hydrogen-transferring mechanism involved an enzyme (a high-molecular component) and a coenzyme (a heat-stable, low-molecular component) (Warburg and Christian[10]).

In 1932, Warburg and Christian[11–13] isolated from Lebedev's juice of yeast a yellow substance they called "yellow ferment". It was found to be attached to a macromolecular carrier. The authors observed that it was reduced (decolorized to a leuco form) by a system composed of hexosemonophosphate plus a protein fraction obtained from yeast (which they called Zwischenferment) plus a coenzyme from red blood cells (later identified as NADP). This system was reoxidized by methylene blue and by oxygen[11–13], showing that, in this system, hydrogen was geared towards oxygen by the "yellow enzyme", an observation which subsequently gave rise to flavine enzyme studies. Warburg's use of the term Zwischenferment instead of "dehydrogenase" is noteworthy. In fact he was referring to the protein fraction ("apoenzyme" fraction) of pyridine-linked dehydrogenase. As Negelein and Haas[14] wrote in a 1935 paper

"Das Zwischenferment ist aber so wenig eine Dehydrogenase wie das Globin des Hämoglobins ein Sauerstoffübertrager ist."

The importance of these remarkable experiments was further increased when it was recognized that the "yellow enzyme" was related to vitamin B_2. Splitting the yellow ferment with acidic methanol and irradiating it

References p. 403

in alkaline solution, Warburg and Christian[12] obtained a derivative which they crystallized and found to fit the empirical formula $C_{13}H_{12}N_4O_2$; they named it "luminoflavin". Warburg and Christian[11–13] suggested the following respiratory chain:

$$O_2 \rightarrow \text{hemin } Fe^{2+} \rightarrow \text{hemin } Fe^{3+} \rightarrow \begin{array}{c}\text{Leucoform}\\ \text{of}\\ \text{pigment}\end{array} \rightarrow \text{pigment} \rightarrow \text{reducing system}$$

"Lumiflavin" was identified as a methylated alloxazine derivative by Stern and Holiday[15] in 1934. Simultaneously, in 1933, Kuhn, Gyorgi and Wagner-Jauregg[16] on the one hand, and Ellinger and Koshara[17] on the other, isolated from egg white and from whey a yellow compound they called lactoflavin. Its composition corresponded to $C_{17}H_{20}N_4O_6$ and, on illumination, it formed the same "luminoflavin" as that obtained by Warburg.

The background of this work is found in studies in the field of vitamin B_2, to be considered in a subsequent chapter. In brief, dietary lack of vitamin B_2 was associated with a disease called pellagra. Recognized in 1926 as belonging to the vitamin B complex[18], vitamin B_2 became a subject of study in several laboratories interested in the organic chemistry of natural compounds. The "lactoflavin" was active in promoting the growth of rats but not in curing pellagra. Therefore the antipellagra factor was recognized by Goldberger as distinct from vitamin B_2, and the denomination of vitamin B_2 remained attached to the flavin compound.

The recognition by Stern and Holiday[15] of the alloxazine nature of the "luminoflavin" obtained by Warburg was of great help in the studies of organic chemists who elucidated its structure and confirmed it by synthesis[19 20].

Riboflavin, shown to correspond to vitamin B_2, is converted into luminoflavin by the elimination of a carbohydrate side chain during the irradiation in alkaline solution. It was shown to be identical with 6,7-dimethyl-9-D-ribityl isoalloxazine and the site of the oxidation-reduction reaction was located.

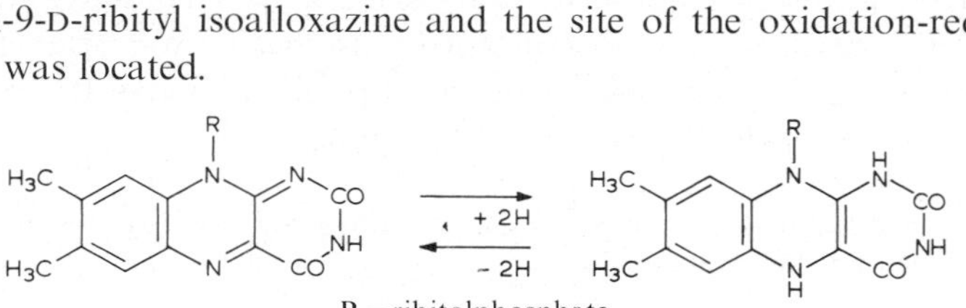

R = ribitolphosphate

Theorell[21] took up the purification of the "yellow ferment" and showed that the high molecular carrier is a protein, a fact which was not obvious at a time when the protein nature of enzymes had not yet been generally accepted. The "yellow ferment" (now known as the "old yellow enzyme") was crystallized by Theorell and Åkeson[22] in 1956. Until 1934, it was considered that riboflavin was the active part of the "old yellow enzyme". But Theorell observed that its splitting with methanol always produces phosphate[23].

Theorell[24] was able in 1935 to isolate a riboflavin-phosphate from the yellow enzyme. In order to purify it, he submitted the yellow enzyme to electrophoresis in a two-compartment apparatus. At the pH of the experiment the yellow enzyme migrated cathodically into the other compartment. A strongly yellow preparation nearly homogeneous in electrophoresis was obtained after repeated fractional precipitations with ammonium sulfate.

To quote Theorell[25]

> "The yellow color could be separated from the high molecular fraction after dialysis at slightly acid reaction and was found to be a colorless protein. The enzyme activity disappeared when the parts separated, but returned when they were recombined at neutral reaction. The yellow dye could not be riboflavin, which was much less active in restoring the activity than the natural dye, even though the absorption band and fluorescence were identical. But there was one difference: the natural dye migrated as an acid in electrophoresis, whereas riboflavin was neutral. Could the "active group" of the old yellow enzyme be a phosphoric acid ester of riboflavin? I had no fresh material for an analysis except an old, heavily infected small quantity of solution that certainly no longer contained flavin — but I knew how much there had been. Yet, it could still be used for an elementary analysis of phosphorus.
>
> I calculated how much there could be, assuming one P per mole of flavin. The result was 14 μg. I endured nervous hours until the result came from the analytical laboratory: 14 μg. There are golden moments in a scientist's life — but they are few.
>
> The fact that enzymatic activity could be restored in a protein by the addition of stoichiometric quantities of a low molecular compound proved beyond doubt that at least this enzyme is a protein. We know now that this holds true for all other enzymes as well."

Finally, in 1936, Kuhn, Rudy and Weygand[26] announced the synthesis of riboflavin 5′-phosphate, which was recognized to be identical with the riboflavinphosphate isolated by Theorell.

The "old yellow enzyme" has so far been isolated only from brewer's bottom yeast after completion of the brewing process. Nevertheless it has great significance in the history of biological oxidations as well as in that of biochemistry as a whole. As stated by Åkeson, Ehrenberg and Theorell[27]:

> "In addition to being the first isolated member of a new class of enzymes, the flavoproteins, the studies made possible for the first time representation of an enzyme action by simple chemical formulation. Also, the necessity of a vitamin for the catalytic activity gave rise to a better understanding of vitamin function."

References p. 403

Plate 109. Hugo Theorell.

The discovery of the "old yellow enzyme", though it never found a "signified" (see Chapter 1 of Vol. 29A) in intermediary metabolism, opened the way for the discovery of a number of flavin-linked dehydrogenases either containing, as a prosthetic group, flavin mononucleotide (FMN) (NADH dehydrogenase) or flavin adenine dinucleotide (FAD) (succinate dehydrogenase, D-amino acid oxidase, xanthine oxidase, *etc.*). Flavin mononucleotide corresponds to the riboflavin phosphate present in what Warburg, referring to the "old yellow enzyme", had called "alloxazine nucleotide". Flavin adenine dinucleotide (FAD) was also discovered by Warburg and Christian[28] when they purified D-amino acid oxidase.

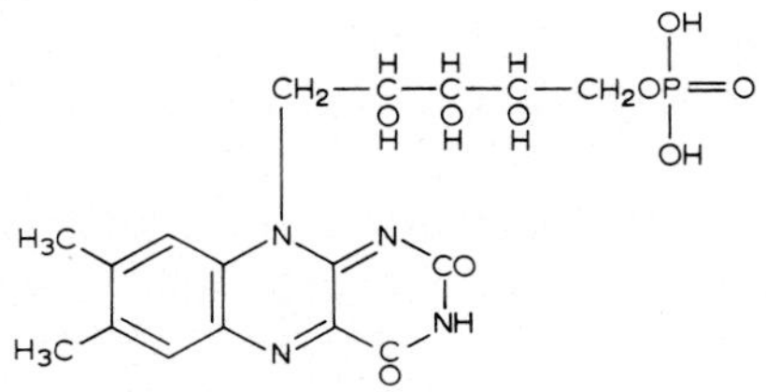

Riboflavin phosphate (flavin mononucleotide, FMN)

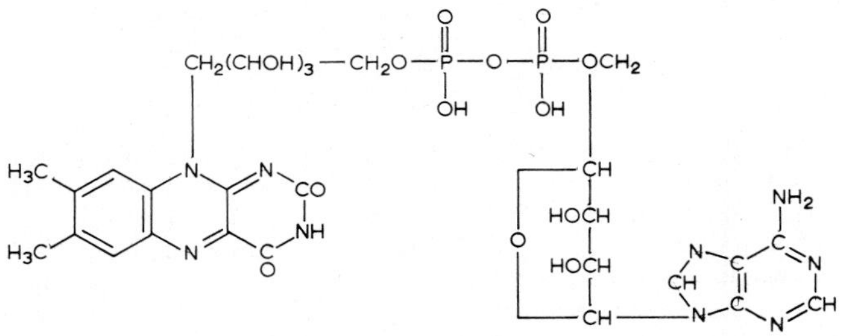

Flavin adenine dinucleotide (FAD)

In this chapter, the essential aspect is the recognition that the enzymes linking the oxidation of NADH or of NADPH to cytochrome *c* or to dyes are the flavoproteins. Its history is retraced by Singer in volume 14 of this Treatise. Succinate dehydrogenase was first isolated from beef heart by Singer and Kearney[29] in 1954. NADH dehydrogenase has been isolated from mitochondrial fragments by Ringer, Minnekami and Singer[30] in 1963. This was no mean achievement as it called for special methods to free the enzyme from other mitochondrial constituents.

References p. 403

The partner of the flavoprotein system, *i.e.* the system which accepts electrons transferred from NADH or from succinate, remained undetermined at the time.

4. Discovery of pyridine-linked dehydrogenases

While Warburg recognized that in some systems, such as those of glucose oxidase, xanthine oxidase, *etc.* the flavoprotein (FAD) accepts hydrogen atoms directly from the organic substrate, he knew also that in the case of the red cell system a coenzyme was required. Warburg and Christian[31] observed that this coenzyme released adenine on hydrolysis. This was also the case with the "cozymase" of von Euler (see Chapter 23). After removal of the adenine, Warburg and Christian observed that other basic compounds remained in the hydrolysate, one of which, called Base I by them, appeared to be related to the coenzyme activity.

A few months later, this compound was the subject of a concise note by Warburg and Christian[32] which reads as follows:

"Co-Fermentproblem

"Base I aus Co-Fermentpräparat* schmilzt bei 125° und hat die Zusammensetzung:

4,052 mg: 8,771 mg CO_2, 1,799 mg H_2O
4,176 mg: 0,811 ccm N (19,5°, 755 mm)
$C_6H_6N_2O$ Ber.: C 59,0%, H 4,92%, N 22,95%
gef.: C 59,04%, H 4,97%, N 22,52%

"Herr Walter Schoeller machte uns darauf aufmerksam, dass Base I in Bezug auf Zusammensetzung und Schmelzpunkt übereinstimmt mit Nicotinsäure-amid**. Ein Vergleich beider Substanzen ergab, dass sie identisch sind.

"Der Misch-Schmelzpunkt von Base I und Nicotinsäure-amid ist derselbe, wie der Schmelzpunkt der beiden reinen Substanzen. Pikrolonate und Platinate kristallisieren in denselben Formen. Der Zersetzungspunkt des Pikrolonats des Nicotinsäure-amids war 218°, des Pikrolonate der Base I 219°, wenn beide Substanzen gleichzeitig erhitzt wurden.

"Nicotinsäure-amid spaltet 1 Mol Ammoniak ab, wenn man eine Lösung in 33%iger Kalilauge 1 Stunde auf 100° erhitzt.

5,072 mg Base I wurden mit 10 ccm 33%iger Kalilauge 1 Stunde auf 100° erhitzt, das gebildete Ammoniak wurde in n/100 HCl aufgefangen. Gefunden 4,0 ccm n/100 NH_3, berechnet für 1 Mol 4,15 ccm NH_3."

* O. Warburg und W. Christian, *diese Zeitschr.* 274 (1934) 112.
** C. Engler, *Ber.*, 27 (1894) 1784.

The preparation of the picrolonate was a decisive step in the identification of Base I with nicotinamide. At the time, Theorell was in Warburg's

Institute on a postdoctoral fellowship. He had joined the laboratory in the fall of 1933 and, as was the custom with the pair of postdoctoral fellows accepted each year, he introduced his own technical knowledge. As he had had experience with electrophoresis in Sweden, an electrophoresis apparatus was built for him in Dahlem. Theorell showed that Warburg's coferment was a phosphoric acid ester.

To quote Theorell[25]

"Warburg and Christian crystallized the active part as picrolonate in December, 1933. Because Warburg suspected that von Euler and Myrbäck were on the same track with their cozymase from yeast, he did not like my idea of going home to Stockholm for Christmas. He finally agreed, but advised me, "I am going to kill you if you mention the word 'picrolonic acid' in Stockholm". Warburg very soon had data on molecular weight, melting point, and elementary composition, but the structure remained difficult to determine because of the small amounts of material available. However, Warburg's friend, Professor Walter Schoeller, found the formula of nicotinic acid amide in a textbook; it had been synthesized some 50 years earlier, long before it was recognized as the antipellagra vitamin."

Warburg, Christian and Griese[33] showed that the pyridine nucleus was the active group of the coferment of red cells. At this stage, they figured out the system for the oxidation of the hexosemonophosphates by oxygen in their model, as follows

$$P + R{-}COH + H_2O = P{<}^{H}_{H} + R{-}COOH \quad \text{-----} \quad (1)$$

$$P{<}^{H}_{H} + A = P + A{<}^{H}_{H} \quad \text{-----} \quad (2)$$

$$A{<}^{H}_{H} + O_2 = A + H_2O_2 \quad \text{-----} \quad (3)$$

$$RCOH + H_2O + O_2 = RCOOH + H_2O_2$$

They noted that a protein is required in reaction (1) and in reaction (2). In reaction (1) this protein is the Zwischenferment (corresponding to the apoenzyme of the pyridine-linked dehydrogenase). In reaction (2), the protein (isolated by Theorell[34] in Warburg's laboratory) is the "Träger-protein" of the flavin enzyme.

Warburg, Christian and Griese[33] concluded that:

References p. 403

"Der Pyridinkörper des Co-Ferments ist seine Wirkungsgruppe, weil die katalytische Wirkung des Co-Ferments auf einem Wechsel des Oxydationszustandes seines Pyridinteils beruht. Der als Nicotinsäureamid aus dem Verband des Co-Ferments gelöste Pyridinkörper ist katalytisch unwirksam."

In a second, much more elaborate paper by Warburg, Christian and Griese[35], also published in 1935, the Warburg coferment ("Wasserstoffübertragende Co-Ferment") isolated from red cells is characterized as follows:

"Nimmt man an, dass in dem Co-Ferment 1 Molekül Adenin, 1 Molekül Nicotinsäureamid, 3 Molekül Phosphorsäure und 2 Moleküle Pentose unter Austritt von 6 Molekülen Wasser vereinigt sind, so erhält man die Formel $C_{21}H_{28}N_7P_3O_{17}$. Dazu stimmt die Elementaranalyse, sowie der Prozentgehalt an Adenin, Nicotinsäureamid und Phosphorsäure. Das Molekulargewicht wäre 743, während wir durch Messung der Gefrierpunktserniedrigung 870 finden."

In a paper which appeared in 1936 Warburg and Christian[5] note that Warburg's coferment always appears to coexist with von Euler's cozymase* and with the phosphorylation coenzyme now recognized as corresponding to ATP. As was pointed out in Chapter 23, von Euler had considered cozymase to correspond to adenylic acid. In 1935, after the publication by Warburg's laboratory on Warburg's coferment, von Euler, Albers and Schlenk[36] recorded a series of data on the basis of which von Euler[37] concluded that nicotinamide was a constituent of his cozymase. This conclusion became the subject of one of those scathing remarks which often characterized Warburg's scientific arguments. In a footnote to the paper of Warburg and Christian[5] we read

"Auf die Mitteilung dieses Analysenergebnisses gründet von Euler einen Beteiligungsanspruch auf das Nikotinsäureamid der Co-Zymase den kein Chemiker anerkennen wird."

The nicotinamide component of cozymase was isolated for the first time by Warburg and Christian[38] in 1936, and during the same year these authors[39] discovered the dihydropyridine derivative of cozymase and its dehydrogenation by acetaldehyde. It is worth noting that in their 1936 paper on the composition of cozymase, Warburg and Christian[5] still call the phosphorylation coenzyme *Adeninnucleotid*, leaving open the question of its corresponding to ATP (as Lohmann believed) or to diadenosinepentaphosphoric acid (Ostern's[40] view).

Warburg and Christian[5] showed that the composition of cozymase (they

* In this presentation, "coenzyme" designates the coenzyme of Harden and Young, "cozymase" the coenzyme of oxidoreduction (later called NAD); Warburg's "coferment" refers to what was to become NADP.

called it "Fermentation coferment") differed from that of their "*Wasserstoff-übertragende Co-Ferment*" by having two phosphate groups instead of three.

To quote Warburg and Christian[5]

"Nach ihrer Wirkungsgruppen nennen wir die beiden wasserstoffübertragenden Co-Fermente "Pyridinnucleotide" und unterscheiden sie, nach ihren Gehalt an Phosphorsäure, als "*Triphospho-Pyridinnucleotid*" und "*Diphospho-Pyridinnucleotid*". Triphospho-Pyridinnucleotid ist das frühere "Wasserstoffübertragende Co-Ferment"; Diphospho-Pyridinnucleotide ist die frühere "Co-Zymase"."

The undesirable English translations "diphosphopyridine nucleotide" (DPN) and "triphosphopyridine nucleotide" (TPN), which give the impression that the two compounds are different categories of nucleotides (an impression not given by Warburg's original terms) led the Enzyme Commission of the International Union of Biochemistry[41] to support Malcolm Dixon's suggestion that the terms be changed into "nicotinamide adenine dinucleotide" (NAD) and "nicotinamide adenine dinucleotide phosphate" (NADP)*, respectively.

The enzymatic reduction and oxidation of both nucleotides were studied by Warburg and Christian[39,42] who discovered the dihydro derivative of NAD and its dehydrogenation by acetaldehyde.

Solutions of pyridine dinucleotide are reduced by dithionite but they are not reduced by biological reductants such as hexosephosphates. But Warburg and Christian brought about an enzymatic reversible hydrogenation of the pyridine dinucleotides. To quote Warburg and Christian[42]

"... if specific proteins present in yeast or other cells are added to the solutions of pyridine nucleotides, the nucleotides and the proteins will combine into a compound in which the pyridine ring can be reduced by carbohydrate compounds. These pyridine nucleotide protein compounds are the hydrogen transmitting enzymes, the active group of which is the pyridine ring."

"Hence, in addition to the enzyme whose active group is hemin bound iron, and in addition to the yellow enzyme whose active group is alloxazine, there now emerges a third class of enzymes, the action of which can be accounted for by a simple chemical reaction" (from the English translation in Kalckar[43]).

The concept of the hydrogenation of NAD was derived from the consideration of another simpler pyridine derivative, trigonelline, the methylbetain of nicotinic acid. Another simple pyridine compound which was reduced reversibly and which is even more closely related to nicotinamide is the

* As stated in Chapter 23, the designations "codehydrogenase I" as well as "cozymase I" and "old cozymase" were also applied to NAD as were the terms "codehydrogenase II", "cozymase II" and "new cozymase" to NADP.

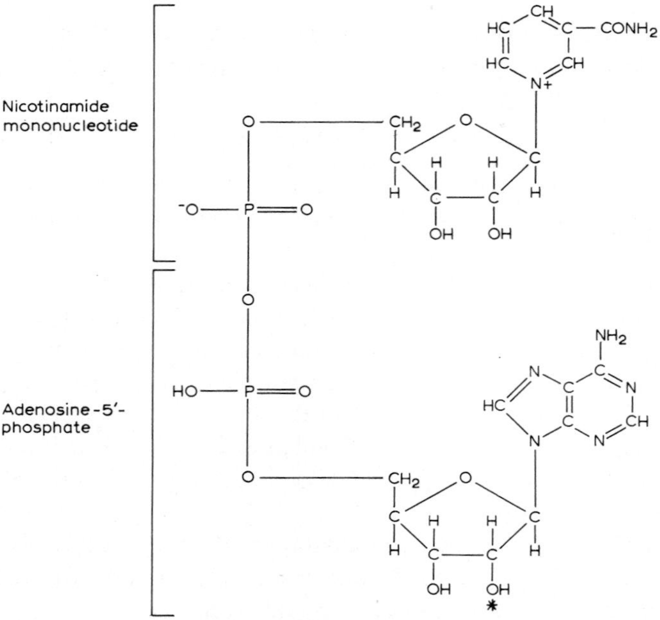

Nicotinamide mononucleotide

Adenosine-5′-phosphate

Nicotinamide adenine dinucleotide (NAD)

The asterisk designates the location of the third phosphate in nicotinamide adenine dinucleotide phosphate (NADP). On the chemical structure and properties of NAD and NADP, see Chapter 1 of Volume 14 of this Treatise, by Colowick, van Eys and Park.

methyliodide derivative of nicotinamide, as suggested to Warburg by Karrer. Both trigonellin and Karrer's compound, when hydrogenated, showed a characteristic absorption band in the long-wavelength ultraviolet region.

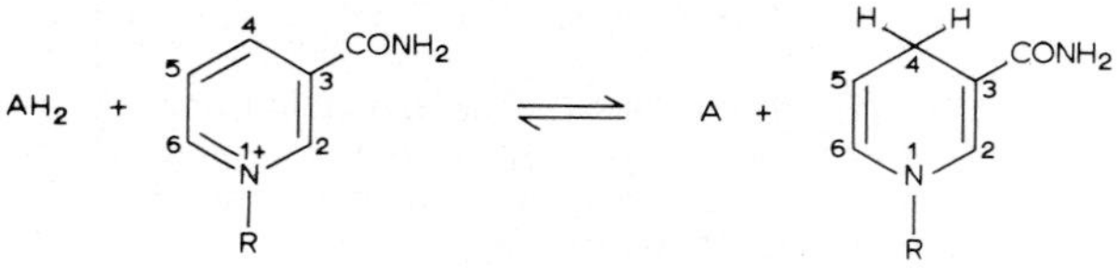

$$AH_2 + NAD^+ \rightleftharpoons NADH + A + H^+$$

The white fluorescence which was elicited by the dihydro compound when irradiated in ultraviolet light made the hydrogenation and dehydrogenation of the pyridine compound visible to the naked eye under the quartz lamp.

The use of this method greatly added to the body of knowledge concerning the compounds requiring a pyridine nucleotide for their dehydrogenation, such as malate, glutamate, β-hydroxybutyrate, *etc.* Recent developments relating to dehydrogenation processes are reported by Colowick, van Eys and Park in Chapter 1, Volume 14 of this Treatise.

We recall again the quotation from Warburg's 1949 book (see the beginning of this section)

"Das Ergebnis war die Entwicklung der wasserstoffübertragenden Fermente."

This was protested by Thunberg[44] who claimed that this discovery had been made a decade earlier and mainly by Wieland.

Who could have realized at that time that Warburg had foreseen as early as 1936 that it took so much work to finally grasp that a unique system exists in mitochondria through which flow the electrons assuring the main amount of ATP synthesis?

To quote Krebs[45] in his biography of Warburg

"The discovery of nicotinamide and the elucidation of its mode of action was a monumental achievement. Many had tried before Warburg to isolate and identify the active principle in the coenzyme fraction of oxido-reduction, especially of fermentations. The existence of coenzyme had been known since the pioneer work of Harden in 1906 but its function in yeast fermentation remained entirely obscure. The presence of adenine, pentose and phosphate had been established by von Euler in impure fractions of the coenzymes of alcoholic fermentations, however, as long as the substances were impure, the significance of the components isolated from it remained uncertain. Warburg made full use of the facilities for large-scale operations which he had installed in his new institute in 1931. To illustrate the scale of his operation: the starting material for the isolation of the coenzyme was 100 litres of washed horse erythrocytes[35]. These were lysed with 200 litres of water and at once treated with 500 litres of acetone. This yielded 4.8 g of "coenzyme". Micro-test monitored the coenzyme content at each stage of the purification procedure."

5. The sequence of electron transfer

The analysis of the orderly sequence of the electron transfer chain in mitochondria owed most of its constituent concepts to the application of highly selective inhibitors of the process. The terminal reaction with oxygen is inhibited by cyanide and H_2S at low concentrations[46].

Rotenone, a plant product used as insecticide, acts at the flavoprotein level[47]; consequently, in the scheme adopted nowadays (Fig. 41), it blocks the electron transfer from flavoprotein dehydrogenases to coenzyme Q. Amytal, a barbituric acid, acts at the same site[48]. Antimycin A, a mixture of closely related antibiotics, blocks the respiratory chain between cytochrome *b* and cytochrome c_1 (refs. 49 and 50). As stated above, Keilin had

Plate 110. Eric Glendenning Ball.

reconstituted, by means of an association of the cytochrome oxidase of heart muscle preparation and of cytochrome *c* of baker's yeast, a powerful catalytic system which oxidized cysteine to cystine.

Haas[51] attempted to measure the role of cytochrome in baker's yeast respiration, using a spectroscopic technique. He measured the concentration of cytochrome in the yeast, determined for each molecule of cytochrome the number of reductions and oxidations per unit interval, and calculated the percentage of the total respiration proceeding through cytochrome. In 1940, Keilin and Hartree[52] produced a cell-free tissue preparation from heart muscle which subsequently was often used for the study of electron transfer chains. This preparation catalyses the oxidation of succinate by molecular oxygen through the cytochrome system.

It was Ball[53], working in Warburg's laboratory, who suggested in 1938 that the order in which the haems of the cytochromes reacted in the oxidation-reduction chain coincided with that of their respective oxidation-reduction potentials. He relied on an acute approach as he observed the change in intensity of the absorption bands of the different cytochromes in media of different potentials and noted that the three cytochromes were reduced by one half, *a* at 0.29 V, *b* at −0.04 V and *c* at 0.27 V. The sequences in the chain would therefore be:

$$b\text{–}c\text{–}a$$

from the hydrogen end.

Also in 1938, Keilin and Hartree[54] proposed the concept of cytochrome a_3 (as the carbon monoxide-combining part of cytochrome *a*; see Chapter 29). Later, the same authors adopted the designation of "cytochrome oxidase" proposed by Dixon[13] in 1929. They[52] concluded in 1940 that the system of cytochromes involved in the oxidation of succinate operated as follows:

$$b\text{–}c\text{–}a\text{–}a_3$$

By 1939, the extensive knowledge of oxidation-reduction potentials of the compounds involved in biological oxidations permitted the construction, by Ball[56], of the diagram shown in Fig. 41. This diagram was published in a review article on the role of flavoproteins and consequently the emphasis is placed on these compounds. In 1939, the involvement of a flavoprotein in the oxidation of reduced NAD had not yet been established.

Warburg's demonstration of the role played by the yellow enzyme in

References p. 403

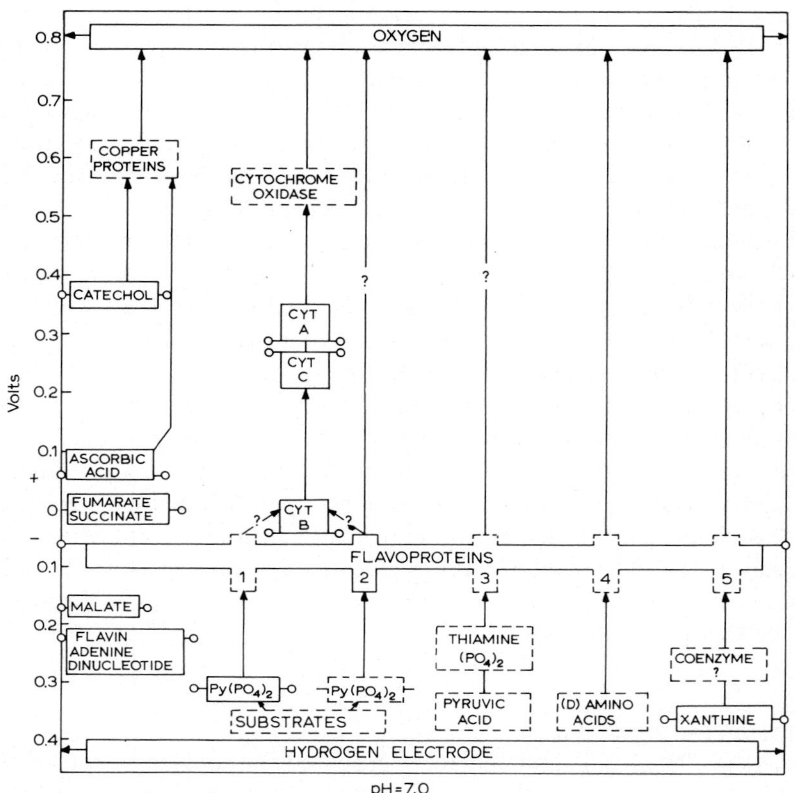

Fig. 41. (Ball[56]). A schematic representation of the role of flavoprotein in biological oxidations $Py(PO_4)_2$ = NAD; $Py(PO_4)_3$ = NADP. (Literature in Ball's paper.)

gearing NADPH to methylene blue was at the basis of the suggestion that a flavoprotein might be acting in a similar way between NADH and a compound not yet identified. The assumption was made in Fig. 41 that electrons are transferred from a reduced flavoprotein to cytochrome oxidase. By showing that the components could be listed in the order of their respective potentials, Ball left scope for several conclusions of great importance.

The indication that cytochrome *b* formed an oxidation-reduction system the potential of which was close to the potential of the methylene blue system, made it possible to interpret the large amount of experimental work accomplished (by Thunberg and others) with methylene blue in the test tube. But even more important, by way of background for the studies on oxidative phosphorylation (see Chapter 37), was the knowledge derived from the oxidation-reduction potentials, concerning the amount of energy that could be converted by interaction between any two components of the chain.

6. Discovery of cytochrome c_1

In 1940, Yakushiji and Okunuki[57] observed in suspensions prepared from beef heart muscle, washed with buffers at pH 4.6 and 7.0 and with succinate and cyanide, as well as the absorption bands a and b, a new band (552 nm) which they thought belonged to a still unknown cytochrome, to which they gave the name "c_1". Slater[58] confirmed these observations but since he could not detect the presence of an unknown cytochrome he proposed that a displacement, due to protein denaturation, of the α-band of cytochrome *c* might explain the above observation. When Keilin and Hartree[59] developed the technique of low-temperature spectroscopy, they clearly recognized the presence of a new sharp and narrow band at 552 nm, in yeast, thoracic muscles of bees, heart muscle preparation and spermatozoa, and they attributed it to a new cytochrome which they called cytochrome *e*. Widmer, Clark, Neufeld and Stotz[60] in the course of observations on heart muscle obtained solutions in which, in the presence of sodium dithionite ($Na_2S_2O_4$), a band was recognized at 554 nm besides the familiar band of cytochrome *b* at 563 nm. They attributed the band at 554 nm to cytochrome c_1. In liver mitochondria deprived of cytochrome *c* by saline extraction, a similar band was observed by Estabrook[61]. The same year, Keilin and Hartree[50] recognized that what they had called cytochrome *e* was in reality identical to the component previously designated by Yakushiji and Okunuki[57] as cytochrome c_1, a name which has since prevailed. In 1941, Okunuki and Yakushiji[62] had placed cytochrome c_1 between cytochromes *b* and *c* in the respiratory chain:

$$-b-c_1-c-a-a_3-$$

The concept was based on the observed acceleration, due to addition of cytochrome *c*, of the reduction of cytochrome oxidase by ferrocytochrome c_1. The position of cytochrome c_1 in the chain was confirmed by Keilin and Hartree[50] who observed that the reduction, by NADH or succinate, of cytochrome c_1, but not hat of cytochrome *b* was inhibited by antimycin A.

References p. 403

Plate 111. Kazuo Okunuki.

7. The gearing from substrate activated to the cytochrome system

On the basis of a consideration of oxido-reduction potentials, Ball[63] proposed (in 1942, *i.e.* at a time when the participation of cytochrome c_1, of coenzyme Q, of non-heme iron and of copper was not yet recognized) the following orderly sequence of electron transport including NAD and the flavoprotein NADH dehydrogenase:

AH_2 / A; NAD^+ / NADH + H^+; FH_2 / F; 2 cyt *b* Fe^{3+} / 2 cyt. *b* Fe^{2+}; 2 cyt *c* Fe^{2+} / 2 cyt *c* Fe^{3+}; 2 cyt *a* Fe^{3+} / 2 cyt *a* Fe^{2+}; H_2O / $\frac{1}{2}O_2$

This scheme was based on the consideration of the value of oxido-reduction potentials (37°, pH 7): NAD(P)−0.3; flavoprotein −0.1; cytochrome *b*−0.04; cytochrome *c*+0.26; cytochrome *a*+0.29. Ball[63] refers to energy release in biological oxidations, as follows

> "The energy liberated when substrates undergo air oxidation is not liberated in one large burst as was thought, but is released in stepwise fashion. At least six separate steps appear to be involved. The process is not unlike that of locks in a canal. As each lock is passed in the ascent from a lower to a higher level a certain amount of energy is expended. Similarly, the total energy resulting from the oxidation of foodstuffs is released in small units of parcels, step by step. The amount of free energy released at each step is proportional to the difference in potential of the system comprising the several steps".

By 1950, it was generally recognized that the system of succinate dehydrogenase and the systems of the hydrogen donors of the tricarboxylic acid cycle requiring NAD (or NADP in the case of isocitrate) also required the system of the cytochromes to reach oxygen. What remained unclear was whether the sequence of carriers was identical.

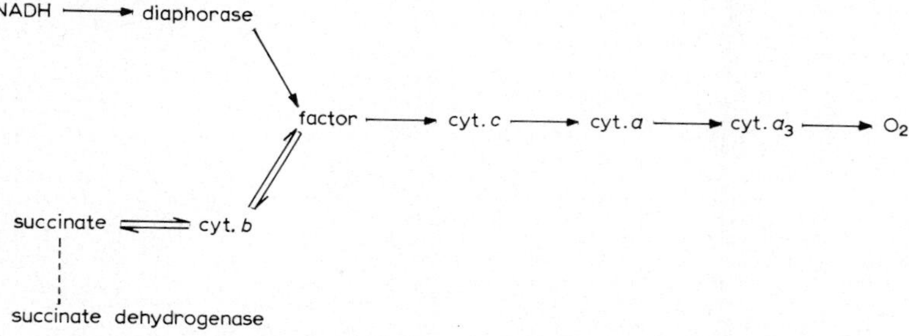

Fig. 42. (Slater[64]). The scheme of electron transport as described in 1950.

References p. 403

Plate 112. E. C. Slater

Examining the oxidation of NADH, Slater[64] observed that the preparation of Keilin and Hartree catalyses both the oxidation of NADH by O_2 (NADH-oxidase activity) and that of NADH by cytochrome *c* (NADH-cytochrome *c* reductase activity). In this study, Slater used inhibitors to determine the nature of the electron transfer chain. Cyanide, CO, azide and fluoride inhibit cytochrome oxidase. BAL (British Anti-Lewisite; 2,3-dimercapto-1-propanol), antimycin A, amytal interfere with electron transfer between cytochrome *c* and succinate or Q and NADH. Slater concluded that the electron transfer chain may be represented as in Fig. 42, showing a convergence of two distinct pathways at the level of the "BAL sensitive factor", which implies that both systems compete for the same final sequence. The concept of this competition was confirmed, in Shanghai, by Wu and Tsou[65] in 1955. These authors demonstrated that succinate and NADH each inhibit the other's oxidation, a fact which was confirmed by Davis, Blair and Mahoney[66] who showed that when both substrates were present, the rate of oxidation by beef mitochondria never exceeded the rate which prevailed with NADH alone. Wu and Tsou[65] observed that, at concentrations less than $1 \cdot 10^{-6}\,M$, NADH inhibits the oxidation of succinate by about 20 per cent. By contrast, succinate at a concentration of $1 \cdot 10^{-2}\,M$ inhibits by approximately 5 per cent the oxidation of glutamate by NAD^+.

In the 1950 scheme (Fig. 42) two possibilities were left open. One of these was the direct reduction of cytochrome *b* by succinate (cytochrome *b* identical with succinate dehydrogenase). The other possibility was that a separate enzyme (succinate dehydrogenase) was at work. The suggestion that succinate dehydrogenase was identical with cytochrome *b* had been made by Bach, Dixon and Zerfas[67] and by Ball, Anfinsen and Cooper[68]. It was refuted by Tsou[69] who showed that in the Keilin–Hartree heart-muscle preparation, cytochrome *b* could be oxidized and reduced even after the dehydrogenase had been completely inactivated. Like NADH dehydrogenase, succinate dehydrogenase is a flavoprotein whose flavin-prosthetic group transfers hydrogen atoms. It was simultaneously isolated from heart muscle Keilin-Hartree preparation by Wang, Tsou and Wang[70] and from heart mitochondria by Singer, Kearney and Bornath[71]. The post-1950 schemes thus altered the fate of succinate electrons as shown in Fig. 42, the new sequence being:

$$\text{succinate} \rightarrow \begin{array}{c}\text{succinate}\\ \text{dehydrogenase}\end{array} \rightarrow \text{cytochrome } b$$

8. Revival of Wieland's theory of hydrogen removal

Wieland had thought that hydrogen can be transferred directly from a reductant to an oxidant. By 1950, however, biochemists pointed out that the donated hydrogen would have to dissociate and so produce protons in the medium; thereafter the dissociation donor would be able to donate electrons to the acceptor (see Clark[72] or Kalckar[73]). This misconception was discarded as a result of observations by Westheimer, Fisher, Conn and Vennesland[74 75] who showed in experiments with alcohol dehydrogenase, 1,1-dideuteroethanol (CH_3–CD_2OH) and NAD, that the reduced NAD contained 1 atom of deuterium per molecule. This proved that a hydrogen atom had been directly transferred from the α-carbon of ethanol to the nicotinamide ring. Other experiments by the same authors also confirmed the hypothesis of direct hydrogen transfer.

In analysing the causes of this misconception and its prevalence for some time, Vennesland and Westheimer[76] regarded it an example of the tendency to interpret theories pertaining to thermodynamics as though they were applicable to mechanisms as well. To quote Vennesland and Westheimer[76]

"A unified theoretical treatment of the *equilibria* of oxidation-reduction reactions was achieved by defining them as electron transfer, with or without the participation of oxygen and hydrogen atoms. This theoretical treatment probably enhanced the tendency among biochemists to think of the *mechanism* of the biological oxidations sequence also in terms of electron transfer; any loss or gain of hydrogen, required by the stoichiometry, was obtained by expelling ions into, or taking them up from, the solvent. The application of deuterium as a tracer, however, has made it possible to show that some of the pyridine nucleotide dehydrogenase reactions require an actual transfer of hydrogen (or deuterium) atoms between substrate and coenzyme. For these reactions, at least, simple electron transfer is not a correct interpretation of the facts."

Reduced NAD is a dihydropyridine but the abbreviation "NADH" might suggest that it is the product of a 1-electron reduction when in fact it involves the uptake of two electrons and one proton. This ambiguity is avoided by writing the oxidized form as NAD^+, indicating a quaternary pyridinium structure

$$NAD^+ + RH_2 \rightleftharpoons NADH + R + H^+$$

This means that the pyridine-linked dehydrogenase directly transfers two reduction equivalents, one appearing as a hydrogen atom in the reduced pyridine nucleotide and the other as an electron in this reduced pyridine nucleotide; the remaining hydrogen atom appears as a proton in the medium.

The *Report of the Commission on Enzymes*[41] of the International Union of Biochemistry recommended the use of the designation "$NADH_2$"; Clark[77], however, rightly criticized this recommendation on the ground that no such stable compound exists.

9. Cytochrome *b*

In a 1958 review paper Slater[78] suggested that a compound in order to merit inclusion in the chain should satisfy three criteria:

(*1*) Amounts present in the enzyme preparation recognized as being commensurate with enzyme activity;

(*2*) Inactivation after removal from the enzyme preparation (either by extraction or by combination with an inhibitor), restoration of activity by introduction of the compound;

(*3*) Oxidation and reduction of the compound during operation of the enzymatic system, at a rate in agreement with overall enzyme activity.

Applying these criteria in a study of the Keilin–Hartree system, Slater concluded that cytochromes a_3, *a*, *c* and *b* as well as flavoprotein passed the test while the evidence remained inconclusive for other candidates. Slater[78] proposed the scheme shown in Fig. 43.

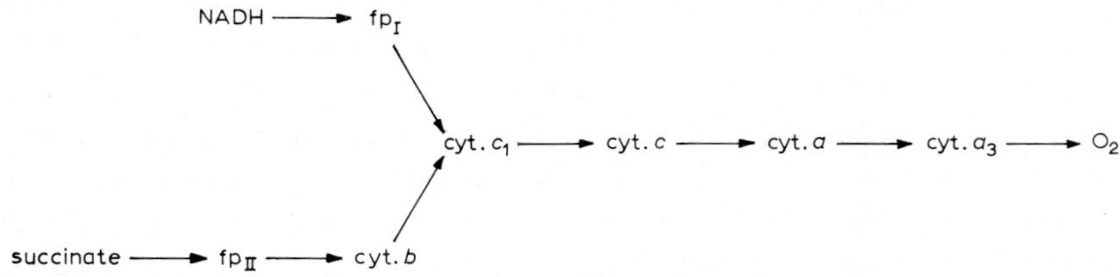

Fig. 43. (Slater[78]). The respiratory chain according to Slater. fp_I represents the flavoprotein NADH dehydrogenase (diaphorase) and fp_{II} succinate dehydrogenase.

This scheme is based in part on the now abandoned notion that the respiratory chain of heart muscle preparations involved in NADH oxidation does not involve cytochrome *b*. Shortly thereafter Chance[79], who used rapid spectrophotometric methods, likewise concluded that the rate of reduction of cytochrome *b* is too slow for that cytochrome to participate in the chain of succinate oxidation by a heart muscle Keilin–Hartree

References p. 403

preparation. As Figs. 42 and 43 show it was granted for some time that the NADH oxidase system bypasses cytochrome *b*. But cytochrome *b* was later shown to be involved in the oxidation of NADH by phosphorylating systems. In his studies on the intracellular NADH dehydrogenase system of yeast, Chance[80] concluded that a pigment spectroscopically similar to a cytochrome of type *b* is involved. On the basis of their study of four different types of transition of the oxidative phosphorylation system, Chance and Williams[81] in 1955 found cytochrome *b*

> "to function rapidly enough to claim a position in the respiratory chain with as much justification as the other components."

Using a special spectrophotometric procedure involving double-beam and split-beam recording spectrophotometers, Chance and Williams were able to observe directly the respiratory chain in intact mitochondria, the electron transfer chain and oxidative phosphorylation acting simultaneously. They alternately exposed the mitochondria to an excess of NADH and O_2, and to anaerobic and aerobic conditions. In contrast to some views stemming from a study of the Keilin–Hartree preparation, they observed that cytochrome *b* is functional in NADH oxidation. Chance and Williams[81] also obtained direct information on the sequence of carriers in the respiratory chain. For this purpose they used several methods, one of which was to make O_2 available to mitochondria in which the carriers were reduced, and then to observe the order in which they became reoxidized. They found that cytochrome a_3 was the first to undergo oxidation, followed by cytochrome *a*, cytochrome *c*, cytochrome *b*, flavoprotein and NADH.

They concluded that cytochromes *c*, *a* and a_3 participate in both types, phosphorylating or not, while cytochrome *b* functions in the phosphorylating chain but not in the Keilin–Hartree preparation. This conclusion (which implies that cytochrome *b* is one of the carriers in the chain of NADH oxidation) was confirmed in studies by Chance[80] on yeast and ascites tumour cells[82].

As stated above, Chance and Williams[81] held that, when oxidative phosphorylation occurs, cytochrome *b* is situated in the main respiratory chain, starting either with succinate or with NADH. The same authors considered that the non-phosphorylating systems on which much had been based in the past are artefacts in which cytochrome *b* had lost its capacity to participate. Discussions on the issue are recalled in Slater's 1958 review paper[78] (and in Fig. 43), written before coenzyme Q was recognized as a participant in the respiratory chain. To quote Slater's[78] conclusion

"The reviewer concludes, therefore, that there is not yet convincing evidence that cytochrome *b* should be removed from the direct pathway for succinate oxidation even in the non-phosphorylating preparation. This raises the question whether it should still be placed on a side-path for the oxidation of NADH. The earlier conclusion was based on the slow rate of reduction by cytochrome *b* by the addition of relatively low concentrations of NADH (but sufficient to saturate the NADH oxidase system) to heart-muscle preparation in the absence of air, or in the presence of air and cyanide[83]. It has now been found that cytochrome *b* is reduced much more rapidly if a larger concentration of NADH ($10^{-3}\,M$) is allowed to consume all the oxygen from the suspension (Colpa-Boonstra and Slater, unpubl.). In fact, under these conditions, cytochrome *b* is as rapidly reduced by NADH as by succinate. This difference between the rates of reduction of cytochrome *b* by NADH, in these two different types of experiments, is not found when succinate is used instead of NADH. An explanation of this discrepancy must await further work."

Meanwhile, Slater maintains his earlier conclusion which goes back to his 1958 scheme. This is one of the uncertainties characterizing the whole situation concerning cytochrome *b* even today.

10. Discovery of coenzyme Q (ubiquinone)

Fortuitous discoveries concerning biochemical systems have often revealed unsuspected participants belonging to other systems.

This was the case with coenzyme Q. The investigations leading to the discovery of ubiquinone or coenzyme Q were led by two groups, one from the Biochemistry Department of the University of Liverpool, the other from the Institute of Enzyme Research at the University of Wisconsin. In brief, the work of the Liverpool group (which was much concerned with vitamin A studies),

"developed from studies on the lipids found in intestinal mucosae and from observations of abnormally high concentrations of certain rat liver lipid constituents brought about by vitamin A deficiency. Two substances provisionally designated SA and SC were separated by chromatography of liver unsaponifiable matter on alumina. Each had a distinctive ultra-violet absorption spectrum by the aid of which the separations could be followed. SA was found to be very widely distributed in animal tissues and in yeast; it was readily reduced and the product reoxidized." (Morton[84]).

For background literature on these studies the reader is referred to Morton[84,85].

The discovery began with the observation by Lovern, Edisbury and Morton[86] in 1937 that large amounts of vitamin A are present in the intestinal mucosa of fish. This observation was made in the context of

References p. 403

an inquiry into the processes that take place in the intestinal tissues. It was found that in this tissue retinene$_1$ is reduced to, and carotene converted to vitamin A; it was later established that cholesterol is dehydrogenated in the guinea pig mucosa to provitamin D_3 (7-dehydrocholesterol) (literature on these metabolic properties in the intestinal mucosa in Morton[85]).

This study of intestinal mucosa led to the isolation, by Festenstein[87], of ubiquinone from the lining of horse's gut, Festenstein then being in the midst of preparing a Ph.D. thesis in Morton's laboratory. Festenstein found small amounts of such intermediates between carotene and vitamin A. Studying by chromatography on alumina the lipid or the unsaponifiable matter of the tissue, Festenstein isolated a compound labile to alkali. It showed an absorption peak at 272 nm, an inflexion at 330 nm and a weak band at 400 nm. He took it to be a new, unknown compound. In fact, as early as 1940 Moore and Rajapogal[88] had shown the presence, in extracts of rat liver and gut by lipoid solvents, of a substance showing an absorption maximum at 275 nm.

To define the new compound, Cain and Morton[89] began the fractionation of liver unsaponifiable fraction. This was not the best source, but it must be remembered that these studies were started in a laboratory much concerned with vitamin A and that the 272 nm compound was investigated as a possible intermediate between carotene and vitamin A. Nevertheless, the presence of the new substance in liver was established, though on prolonged processing artefacts were derived from carotenoids. One of these, cholest-3 : 5-dien-7-one, has an absorption peak around 270 nm and is very similar to the spectrum of the new substance, a circumstance that caused many difficulties.

Another inquiry, again in the context of vitamin A studies, was started in Morton's laboratory. It concerned the problem of the systemic mode of action of vitamin A and the comparison of the unsaponifiable fractions of liver in vitamin A-deficient and in normal rats. It led to the discovery by Morton's group[90–93] of larger amounts of SA (the 272 nm substance), and of SC, as defined above in vitamin A-deficient rats. SA was obtained from rat liver in sufficient amounts to permit an attempt at characterization.

As noted by Morton[85], this preparation was probably a mixture of coenzyme Q_9 and coenzyme Q_{10}. An erroneous determination of the molecular weight by the Rast method (about half the true value) for some time steered the investigation towards the possibility of a polycyclic substance of a steroid nature (literature in Morton[84]). Late in 1956, Morton sought the help of Dr. O. Isler of Hoffmann-La Roche at Basel, not only on account of the competence of his group in the synthesis of polyisoprenoid systems, but also to take advantage of the facilities of industrial laboratories for the handling and provision of large amounts of costly material. For example (cited by Morton[85]), the Basel laboratory obtained 120 g of pure ubiquinone-50 and 40 g of less pure material from 3200 kg of pig heart after using 1200 kg of potash, 40 kg of pyrogallol, 150 kg of alumina and 3000 l of solvents.

The contribution made by the Wisconsin group started from their interest in the electron transport chain of mitochondria and especially in the role of their lipid constituents.

Crane, Hatefi, Lester and Widmer[94] in 1957; Crane, Fechner and Ambe[95] in 1959, and Crane, Widmer, Lester and Hatefi[96] again in 1959, isolated from ox-heart mitochondria a water-insoluble quinone with an absorption peak at 275 nm in ethanol, which was crystallized and exhibited the reversible oxidation-reduction properties of a quinone. Hatefi, Lester and Ramasarma[97] identified the compound in various types of electron transport particles (see below) in amounts which justified its recognition as a member of the electron transport system. It received the name of "coenzyme Q".

As related by Morton[84], in July 1957 Lester sent to him, together with a reprint of the 1957 paper[94], a letter stating

> "After this note was submitted we became aware of the interesting work of your group concerning the characterization of an unsaponifiable lipid material with λ_{max} 272 mμ (*Biochem. J.*, 59 (1955) 558). The ultraviolet and visible absorption of the material we have isolated corresponds very closely to your material. Since on reduction with potassium borohydrate we find considerable absorption remaining in the 290 mμ region we thought that this might indicate a quinone-hydroquinone reaction. We would be happy to have your comments in this regard."

In reply, on July 12, 1957 Morton sent a very full account of the work performed in Liverpool, mentioning that a steroid structure and many naphthoquinone structures had been ruled out and giving data on molecular weight, oxidation-reduction potential, carbon and hydrogen analysis, *etc.*

Lester's reply, acknowledging Morton's letter on behalf of Crane and Hatefi, was sent on August 29, 1957. Lester wrote

> "The generous information which was supplied leaves us in no doubt that SA, if not identical with Q_{275} is indeed a very close brother."

References p. 403

By this time (1957) the studies on SA in the context of work on vitamin A, and those on Q_{275} as a carrier in the respiratory chain had come together. The name "ubiquinone", expressing the compound's widespread occurrence, was suggested by Morton, Wilson, Lowe and Leat[98] in 1957 in a paper that referred to Crane, Hatefi, Lister and Widmer[94] in which these authors had described the quinone nature of the compound isolated from mitochondria (see Chapter 9, Volume 9 of this Treatise, by Morton, on quinones).

The Wisconsin group came to prefer the name "coenzyme Q_{275}", and, later, "coenzyme Q".

As stated above, the Liverpool group had the support of the organic chemists of Isler's group at Basel. The Wisconsin group in its turn was helped by Folkers and his colleagues of Merck, Sharp and Dohme Research Laboratories at Rahway, N.J. The analysis of the structure of ubiquinones and their properties has been described in detail in several review articles; specifically the reader is referred to Schindler[99], Hatefi[100] and Crane[101].

The generic formula (tetrasubstituted benzoquinone) was established independently by the Liverpool group and by the Wisconsin group, as follows

$$\text{2,3-}(H_3CO)_2\text{-5-}CH_3\text{-1,4-benzoquinone-6-}(CH_2{-}CH{=}C(CH_3){-}CH_2)_n H$$

coenzyme Q_{10}	n = 10
coenzyme Q_9	n = 9
coenzyme Q_8	n = 8
coenzyme Q_7	n = 7
coenzyme Q_6	n = 6

The natural members of the series are known as coenzymes Q_{10}, Q_9, Q_8, Q_7 and Q_6 as established by Lester, Crane and Hatefi[102]. As these compounds act as electron carriers in the organisms in which they are found they are designated as the coenzyme Q group.

In animal tissues CoQ_{10} and CoQ_9 are found while in microorganisms the range is from Q_6 to Q_{10}. In plants the chloroplasts contain plastoquinone (trisubstituted quinone), which plays the role of CoQ in the chain (see Volume 14).

In a review article on the coenzyme Q group (ubiquinones) Crane[101] records three different methods which have led to locating coenzyme Q in the respiratory chain: the use of specific inhibitors; the extraction of coenzyme Q and the determination of the enzyme action stopped and restored after the addition of coenzyme Q; the fragmentation of the mitochondria and the identification of the fragments containing coenzyme Q. From these studies (literature in Crane[101]), it was concluded that coenzyme Q_{10} functions between succinic dehydrogenase or NADH dehydrogenase and cytochrome c_1, providing a primary partner for the flavoprotein in the electron transfer.

Nevertheless, at the time of this writing the primary partner for flavoprotein still has not been unambiguously recognized.

If the weight of evidence favours CoQ, nevertheless cytochrome *b* and cytochrome *c* have also been suggested as primary reaction partners for the electron transfer from the oxidation of NADH, NADPH or succinate oxidase by the action of flavin enzyme.

11. Coenzyme Q as a shuttle between the flavoprotein and the cytochrome complexes

This concept was first formulated by Green[103]. Coenzyme Q is accessible, not only to NADH and succinate but also to fatty acyl CoA (Kröger and Klingenberg[104]) and to α-glycerophosphate (Klingenberg[105]). Choline (Drabikowska and Szarkowska[106]) and proline (Ericinska and Szarkowska[107]) have been shown to reduce mitochondrial coenzyme Q.

Kröger and Klingenberg[104] recognize coenzyme Q as a common hydrogen acceptor for flavin-linked substrates. That Q was the merger of the succinate and NADH sources was supported by the experiments conducted by Ernster, Lee, Norling and Persson[108]. They showed that Q-depleted preparations showed little or no succinate-oxygen or NADH-oxygen oxidoreductase activity. When Q was reincorporated in amounts corresponding to those present in the original samples, both activities were restored. Since they showed that in their preparations the flavoproteins as well as the span from cytochrome *b* to oxygen was active, they concluded that Q was the shuttle between flavoproteins and cytochrome *b*.

12. Functional complexes of the respiratory chain

While in the study of many metabolic pathways a reassembly *in vitro* of the purified constituents has confirmed the activity of the chain, no such assembly has been possible in the case of the respiratory chain. So far, for example, no rapid reaction has been observed *in vitro* between flavoprotein

References p. 403

Plate 113. Youssef Hatefi.

and coenzyme Q or any of the cytochromes. Several causes may account for this failure. It may be due to modifications of the flavoproteins or of the cytochromes in the course of their isolation, or to the necessary addition of another substance or other substances. Then again, the cause may lie in the requirement of an integrated structure which does not reconstitute when the participant compounds are mixed *in vitro*. It is of interest to recall that before he attempted to deal with purified enzymes and so insured the success of the biochemistry of the thirties, Warburg had long been a promotor of the necessity of structure (see for instance his paper entitled *Über Beziehungen zwischen Zellstruktur und biochemischen Reaktionen I.*[109]). But while efforts to reassemble purified catalysts and carriers in a working respiratory chain have failed so far, reconstitution from particulate fragments of the respiratory chain turned out to be more successful. The concept of reconstitution is akin, in this context, to the reconstitution of an enzyme from the apoenzyme and the coenzyme, or to the reconstitution of tobacco mosaic virus from its nucleic acid core and protein envelope, as accomplished by Fraenkel-Conrat and Williams[110]. We have retraced in Chapter 29 the evolution of ideas which resulted in the very existence of cytochrome a_3 being questioned, and in the consideration of cytochrome oxidase as a complex of cytochrome *c*, cytochrome *a* and oxygen.

Since 1959, Hatefi and his group have (literature in Chapter 4, by Hatefi, Volume 14 of this Treatise) succeeded in dividing up the mitochondrial electron transfer system into four enzyme complexes respectively involved in four parts of the respiratory chain, and such that their association reconstitutes the chain:

$$\mathrm{NADH} + \mathrm{H}^+ + \mathrm{Q} \rightarrow \mathrm{NAD}^+ + \mathrm{QH}_2 \quad (1)$$

$$\text{Succinate} + \mathrm{Q} \rightarrow \text{fumarate} + \mathrm{QH}_2 \quad (2)$$

$$\mathrm{QH}_2 + \text{ferricytochrome } c \rightarrow \mathrm{Q} + 2 \text{ ferrocytochrome } c + 2\,\mathrm{H}^+ \quad (3)$$

$$2 \text{ ferrocytochrome } c + 2\,\mathrm{H}^+ + \tfrac{1}{2}\,\mathrm{O}_2 \rightarrow 2 \text{ ferricytochrome } c + \mathrm{H_2O} \quad (4)$$

These four complexes have been studied by Hatefi and his school and referred to as:

Complex I—Reduced NAD-coenzyme Q reductase
Complex II—Succinic-coenzyme Q reductase
Complex III—Reduced coenzyme Q—cytochrome *c* reductase
Complex IV—Cytochrome oxidase

References p. 403

Complexes I and II contain flavoprotein, Fe as a transition metal, and lipid. Complexes III and IV contain hemoprotein and lipid, the transition metal being Fe in the case of III and copper in the case of complex IV. The complexes I, II and III are functionally linked by way of coenzyme Q, while III and IV are linked by way of cytochrome *c*. While the flavoproteins, the three non-heme irons and cytochromes *a*, *b* and c_1 are fixed components, coenzyme Q and cytochrome *c* are easily extracted and restored. Reconstitution experiments and other studies have led Hatefi *et al.*[111,112] to propose the scheme shown in Fig. 44. This scheme shows the now prevailing concept of a respiratory chain of function complexes which can be separated and reunited into a functional whole.

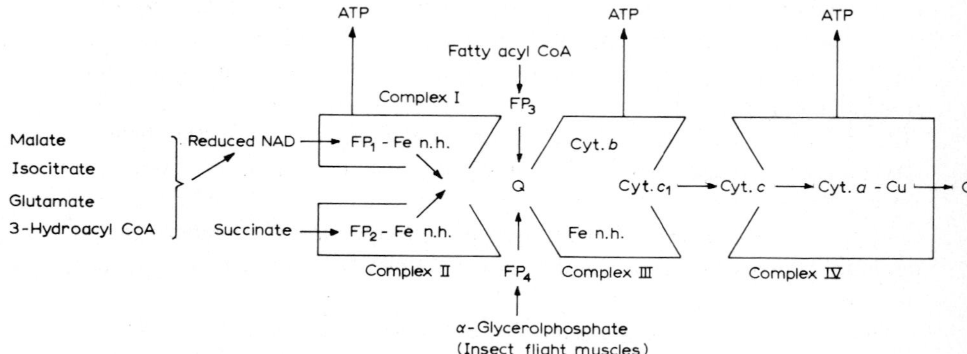

Fig. 44. A schematic representation of the functional complexes and their arrangement in the electron-transfer system. FP, flavoprotein; Fe n.h., non-heme iron. The figure also illustrates the concept of the synthesis of one molecule of ATP in each complex as a pair of electrons traverses the chain of that complex.

The shift in focus is from biomolecules aligned in a single orderly series according to their oxido-reduction potentials to a chain of enzyme complexes which can be separated and which reassemble spontaneously to reconstitute the elementary particle of the mitochondria.

13. Transition metal ions

Besides quinoid structures (flavins, coenzyme Q) and iron porphyrin chelates-heme groups of cytochromes *a*, *b*, c_1 and *c*, the mitochondrial electron transfer system has been shown to contain transition metal ions not chelated with porphyrin (non-heme iron, copper). Non-heme iron (appearing as a polymeric array of iron atoms) has been found at three sites in the respiratory chain: between NADH dehydrogenase and coenzyme Q, between succinate dehydrogenase and coenzyme Q and between cytochromes b and c_1. The concept of non-heme iron as part of the respiratory chain related to its high level (eight times more than heme iron) of non-heme iron present in submitochondrial particles (see Green[113]). Kearney and Singer[114] found that succinate dehydrogenase contains iron and suggested that it may shuttle its electrons from succinate to acceptor through iron as well as through flavin.

Beinert and Sands[115] first introduced a technique for the determination of electron paramagnetic resonance (EPR), at $-100°$, in the study of mitochondrial preparations. As stated by Wainio[116], referring to Beinert and his collaborators,

"Although they were able to cause a signal to be altered with NADH, they did not fully appreciate the wide distribution of the component responsible for the major $g=1.94$ signal and the minor $g=2.01$ signal until a few years later. The asymmetric $g=1.94$ signal was seen for the first time[118] when they examined the EPR spectra of different succinate dehydrogenase preparations after reduction with succinate or dithionite...

...In 1962, Beinert, Heinen and Palmer[117] stressed the fact that many of the materials which showed the $g=1.94$ signal contained only non-heme iron. There was a complete absence of heme iron, and the flavin-iron complex was ruled out as giving rise to the signal because similar signals were found with proteins that did not contain flavin, as with reduced hemerythrin and the cytochrome b-b_1 complex from beef-heart mitochrondria."[119]

Complex I contains as many as 17 atoms of non-heme iron per molecule of flavin (Hatefi, Haavik and Griffiths[120]) and Complex II as many as 8 (Zeigler and Docq[121]). Complex III (in which EPR signals were for some time undetected) has been questioned with respect to the functional significance of the iron.

Rieske, Hansen and Zaugg[122] have reexamined Complex III by EPR spectroscopy and found a signal at 1.90 similar in type to the signal at 1.94 and showed that the component involved is closely associated electronically with cytochrome c_1. In subsequent research, Rieske, Zaugg and Hansen[123] have shown that two indigenous atoms of non-heme iron are found per molecule of cytochrome c_1.

Another transition metal ion, copper, is associated with cytochrome a in a cluster of six (Takomori, Sekezu and Okunuki[124]). The controversies on the copper content of cytochrome oxidase has been retraced by Wainio[116]. Copper is introduced in the scheme of Fig. 44 as a transition metal in electron transfer towards oxygen, on the basis of the copper content of purified cytochrome oxidase preparations and on the deter-

References p. 403

mination of the reduction of copper (determined by EPR spectroscopy and chemical procedures) which parallels the reduction of heme *a* under various conditions (literature in Hatefi's chapter in Volume 14 of this Treatise).

14. Remaining questions

The scheme in Fig. 44 illustrates the present concept of the respiratory chain as an orderly arrangement of specific hemochromogens, held together with protein and phospholipid constituents, in which quinoid structures (flavin, coenzyme Q) and transition metals play their role, and in which the hydrogen atoms and later the electrons alone pass from an active group to the next, ever higher in oxidation-reduction potential.

The sequence of electron transfer reactions is consistent with the concept of the potentials becoming more positive in the successive steps. It is also consistent with the experimental evidence afforded by the experimentation *in vitro* with isolated carriers. For instance, NADH cannot reduce cytochrome *b* directly, and the same is true of cytochrome *c* and cytochrome *a*. NADH dehydrogenase cannot reduce cytochrome *a* without the presence of cytochromes *b* and *c*. Nevertheless some questions remain, *e.g.* with respect to cytochrome *b*. Then again, while it is generally assumed that between NADH and coenzyme Q the electron transfer occurs in two-electron steps, and between cytochrome *b* and oxygen in one-electron steps, the matter cannot be considered settled. A total of four electrons is required to reduce O_2 to two OH^- ions, but it is as yet uncertain how these electrons are provided by the respiratory chain.

In the present recurrent history (see the *Introduction*) our knowledge of the general picture of the respiratory chain appears as well established in its main features. No mention has so far been made of unlucky candidates which have from time to time been suggested as constituents of the respiratory chain, without however gaining recognition. One of the most ancient is glutathione which was shown by Ogston and Green[125] to have little effect on the oxygen uptake. No direct experimental ground has so far been discovered in favour of the inclusion of vitamin K in the chain[126]. α-Tocopherol has also been mentioned[127] but so far without convincing demonstration.

REFERENCES

1 O. Warburg, *Naturwissenschaften*, 15 (1927) 546.
2 D. Keilin, *Nature*, 119 (1927) 670.
3 O. Warburg, *Nobel Address*, 1931.
4 O. Warburg, E. Negelein and E. Haas, *Biochem. Z.*, 266 (1933) 1.
5 O. Warburg and W. Christian, *Biochem. Z.*, 287 (1936) 291.
6 S. P. Colowick, J. van Eys and J. H. Park, in M. Florkin and E. Stotz (Eds.), *Comprehensive Biochemistry*, vol. 14. Amsterdam, 1966.
7 O. Warburg, *Wasserstoffübertragende Fermente*, Freiburg i. Br., 1949.
8 O. Warburg, F. Kubowitz and W. Christian, *Biochem. Z.*, 221 (1930) 494.
9 O. Warburg, F. Kubowitz and W. Christian, *Biochem. Z.*, 227 (1930) 245.
10 O. Warburg and W. Christian, *Biochem. Z.*, 242 (1931) 206.
11 O. Warburg and W. Christian, *Naturwissenschaften*, 20 (1932) 688, 980.
12 O. Warburg and W. Christian, *Biochem. Z.*, 254 (1932) 438.
13 O. Warburg and W. Christian, *Biochem. Z.*, 257 (1933) 492.
14 E. Negelein and E. Haas, *Biochem. Z.*, 282 (1935) 206.
15 K. G. Stern and E. R. Holiday, *Ber. deut. chem. Ges.*, 67 (1934) 1104.
16 R. Kuhn, P. György and T. Wagner-Jauregg, *Ber. deut. chem. Ges.*, 66 (1933) 315.
17 Ph. Ellinger and W. Koshara, *Ber. deut. chem. Ges.*, 66 (1933) 315.
18 J. Goldberger, G. A. Wheeler, R. D. Lillie and L. M. Rogers, *Public Health Rept. (U.S.)*, 41 (1962) 297.
19 R. Kuhn, R. Reinemund, H. Kaltschmitt, R. Ströbele and H. Trishmann, *Naturwissenschaften*, 23 (1935) 260.
20 P. Karrer, K. Schöpp and F. Benz, *Helv. Chim. Acta*, 18 (1935) 426.
21 H. Theorell, *Naturwissenschaften*, 22 (1934) 289.
22 H. Theorell and Å. Åkeson, *Arch. Biochem. Biophys.*, 65 (1956) 439.
23 H. Theorell, *Biochem. Z.*, 275 (1934) 37.
24 H. Theorell, *Biochem. Z.*, 275 (1935) 344.
25 H. Theorell, *Vitamins and Hormones*, 28 (1970) 151.
26 R. Kuhn, H. Rudy and F. Weygand, *Ber. deut. chem. Ges.*, 69 (1936).
27 Å. Åkeson, A. Ehrenberg and H. Theorell, in P. Boyer, H. Lardy and K. Myrbäck (Eds.), *The Enzymes*, 2nd ed., vol. 7, New York and London, 1963.
28 O. Warburg and W. Christian, *Biochem. Z.*, 298 (1938) 368.
29 T. P. Singer and E. B. Kearney, *Biochim. Biophys. Acta*, 15 (1954) 151.
30 R. L. Ringler, S. Minekami and T. P. Singer, *J. Biol. Chem.*, 238 (1963) 801.
31 O. Warburg and W. Christian, *Biochem. Z.*, 274 (1934) 112.
32 O. Warburg and W. Christian, *Biochem. Z.*, 275 (1935) 464.
33 O. Warburg, W. Christian and A. Griese, *Biochem. Z.*, 279 (1935) 143.
34 H. Theorell, *Biochem. Z.*, 278 (1935) 263.
35 O. Warburg, W. Christian and A. Griese, *Biochem. Z.*, 282 (1935) 157.
36 H. von Euler, H. Albers and F. Schlenk, *Z. physiol. Chem.*, 237 (180) I, 240 (1936) 113.
37 H. von Euler, *Biochem. Z.*, 286 (1936) 140.
38 O. Warburg and W. Christian, *Biochem. Z.*, 285 (1936) 156.
39 O. Warburg and W. Christian, *Biochem. Z.*, 286 (1936) 81.
40 P. O. Ostern, *Biochem. Z.*, 270 (1934) 1.
41 *Report of the Commission in Enzymes of the International Union of Biochemistry*, Oxford, 1961.
42 O. Warburg and W. Christian, *Helv. Chim. Acta*, 19 (1936) E 79.
43 H. Kalckar, *Biological Phosphorylations. Development of Concepts*, Englewood Cliffs, N.J., 1969.

44 T. Thunberg, *Kungl. Fysiografiska Sällskapets, I Lund. Förh.*, 20 (1950) 1.
45 H. A. Krebs, *Biogr. Mem. of Fellows of the Royal Soc.*, 18 (1972) 629.
46 D. Keilin, *Erg. Enzymf.*, 2 (1933) 239.
47 P. E. Lindahl and I. C. E. Öberg, *Exptl. Cell Res.*, 23 (1961) 228.
48 D. J. Horgan and T. P. Singer, *J. Biol. Chem.*, 243 (1968) 834.
49 V. R. Potter and A. E. Reif, *J. Biol. Chem.*, 194 (1952) 287.
50 D. Keilin and E. F. Hartree, *Nature*, 176 (1955) 200.
51 E. Haas, *Naturwissenschaften*, 22 (1934) 207.
52 D. Keilin and E. F. Hartree, *Proc. Roy. Soc. Lond., Ser. B.*, 129 (1940) 277.
53 E. G. Ball, *Biochem. Z.*, 295 (1938) 262.
54 D. Keilin and E. F. Hartree, *Nature*, 141 (1938) 870.
55 M. Dixon, *Biol. Rev. Cambr. Philos. Soc.*, 4 (1929) 352.
56 E. Ball, *Cold Spring Harbor Symp.*, 7 (1939) 100.
57 E. Yakushiji and K. Okunuki, *Proc. Imp. Acad. Jap.*, 16 (1940) 299.
58 E. C. Slater, *Nature*, 163 (1949) 532.
59 D. Keilin and E. F. Hartree, *Biochem. J.*, 44 (1949) 205.
60 C. Widmer, H. W. Clark, H. A. Neufeld and E. Stotz, *J. Biol. Chem.*, 210 (1954) 861.
61 R. W. Estabrook, *Federation Proc.*, 14 (1955) 45.
62 K. Okunuki and E. Yakushiji, *Proc. Imp. Acad. Jap.*, 17 (1941) 263.
63 E. G. Ball, in *A Symposium on Respiratory Enzymes*, Madison, Wis., 1942.
64 E. C. Slater, *Biochem. J.*, 46 (1950) 484.
65 C. Y. Wu and C. L. Tsou, *Sci. Sinica*, 4 (1955) 137.
66 E. J. Davis, P. V. Blair and A. J. Mahoney, *Biochim. Biophys. Acta*, 172 (1969) 574.
67 S. J. Bach, M. Dixon and L. G. Zerfas, *Biochem. J.*, 40 (1946) 229.
68 E. G. Ball, C. B. Anfinsen and O. Cooper, *J. Biol. Chem.*, 168 (1947) 257.
69 C. L. Tsou, *Biochem. J.*, 49 (1951) 512.
70 Y. L. Wang, C. L. Tsou and T. Y. Wang, *Abstr. 3rd Intern. Congr. Biochem., 1955, Supplement* (1955).
71 T. P. Singer, E. B. Kearney and P. Bernath, *J. Biol. Chem.*, 223 (1956) 599.
72 W. M. Clark, *Oxidation-Reduction Potentials of Organic Systems*, Baltimore, 1960.
73 H. M. Kalckar, *Chem. Rev.*, 28 (1941) 71.
74 H. F. Fischer, E. E. Conn, B. Vennesland and F. H. Westheimer, *J. Biol. Chem.*, 202 (1953) 687.
75 F. H. Westheimer, H. F. Fischer, E. E. Conn and B. Vennesland, *J. Am. Chem. Soc.*, 73 (1951) 2403.
76 B. Vennesland and F. H. Westheimer, in W. D. McElroy and H. B. Glass (Eds.), *The Mechanism of Enzyme Action*, Baltimore, Md., 1954.
77 W. M. Clark, *Science*, 141 (1963) 995.
78 E. C. Slater, *Advan. Enzymol.*, 20 (1958) 147.
79 B. Chance, *Nature*, 169 (1952) 215.
80 B. Chance, in D. M. McElroy and B. Glass (Eds.), *The Mechanism of Enzyme Action*, Baltimore, Md., 1954.
81 B. Chance and G. R. Williams, *J. Biol. Chem.*, 217 (1955) 429.
82 B. Chance, *Trans. N.Y. Acad. Sci.*, 16 (1953) 74.
83 E. C. Slater, *Biochem. J.*, 46 (1950) 484.
84 R. A. Morton, *Nature*, 182 (1958) 1764.
85 R. A. Morton, *Vitamins and Hormones*, 19 (1961) 1.
86 J. A. Lovern, J. R. Edisbury and R. A. Morton, *Nature*, 140 (1937) 776.
87 G. N. Festenstein, *Ph.D. Thesis, Liverpool, 1950* (cited by Morton[65,66]).
88 T. Moore and K. R. Rajapogal, *Biochem. J.*, 34 (1940) 335.
89 J. C. Cain and R. A. Morton, *Biochem. J.*, 60 (1955) 274.

90 J. S. Lowe, R. A. Morton and R. G. Harrison, *Nature*, 172 (1953) 716.
91 G. N. Festenstein, F. W. Heaton, J. S. Lowe and R. A. Morton, *Biochem. J.*, 59 (1955) 558.
92 F. W. Heaton, J. S. Lowe and R. A. Morton, *Biochem. J.*, 60 (1955) XVIII.
93 F. W. Heaton, J. S. Lowe and R. A. Morton, *Biochem. J.*, 67 (1957) 208.
94 F. L. Crane, Y. Hatefi, R. L. Lester and C. Widmer, *Biochim. Biophys. Acta*, 25 (1957) 220.
95 F. L. Crane, W. Fechner and K. Ambe, *Arch. Biochem. Biophys.*, 81 (1959) 277.
96 F. L. Crane, C. Widmer, R. L. Lester and Y. Hatefi, *Biochim. Biophys. Acta*, 31 (1959) 476.
97 Y. Hatefi, R. L. Lester and T. Ramasarma, *Federation Proc.*, 17 (1958) 238.
98 R. A. Morton, G. M. Wilson, J. S. Lowe and W. M. F. Leat, *Chem. and Ind.*, (1957), 1649.
99 O. Schindler, *Fortschr. Chem. Org. Naturw.*, 20 (1962) 73.
100 Y. Hatefi, *Advan. Enzymol.*, 25 (1963) 275.
101 F. L. Crane, *Progr. Chem. Fats and other Lipids*, 7/11 (1964) 267.
102 R. L. Lester, F. L. Crane and Y. Hatefi, *J. Am. Chem. Soc.*, 80 (1958) 4751.
103 D. E. Green, *Comp. Biochem. Physiol.*, 4 (1962) 81.
104 A. Kröger and M. Klingenberg, *Current T. Bioeng.*, 2 (1967) 151.
105 M. Klingenberg, in Goodman and Lindberg (Eds.), *Biological Structure and Function*, Vol. 2, London, New York, 1961.
106 A. K. Drabikowska and L. Szarkowska, *Acta Biochem. Polon.*, 12 (1965) 387.
107 M. Ericinska and L. Szarkowska, *3rd FEBS Meeting*, Warsaw, 1966, Abstr. No. M_7.
108 L. Ernster, I. Y. Lee, B. Nordling and B. Persson, *European J. Biochem.*, 9 (1969) 299.
109 O. Warburg, *Arch. Ges. Physiol.*, 145 (1912) 277.
110 H. Fraenkel-Conrat and R. C. Williams, *Proc. Natl. Acad. Sci. (U.S.)*, 41 (1955) 690.
111 Y. Hatefi, A. G. Haavik and D. E. Griffiths, *Biochem. Biophys. Res. Commun.*, 4 (1961) 441, 447.
112 Y. Hatefi, A. G. Haavik, R. L. Fowler and D. E. Griffiths, *J. Biol. Chem.*, 237 (1962) 2661.
113 D. E. Green, in O. H. Gaebler (Ed.), *The Enzymes: Units of Biological Structure and Function*, New York, 1956.
114 E. B. Kearney and T. P. Singer, *Biochim. Biophys. Acta*, 14 (1954) 572.
115 H. Beinert and R. H. Sands, *Biochem. Biophys. Res. Commun.*, 1 (1959) 171.
116 W. W. Wainio, *The Mammalian Mitochondrial Respiratory Chain*, New York and London, 1970.
117 H. Beinert, W. Heinen and G. Palmer, *Brookhaven Symp. Biol.*, 15 (1962) 22, 229
118 H. Beinert and R. H. Sands, *Biochem. Biophys. Res. Commun.*, 3 (1960) 41.
119 H. Beinert, in A. San Pietro (Ed.), *Non-heme Iron Proteins*, Yellow Springs, Ohio, 1965.
120 Y. Hatefi, A. G. Haavik and D. E. Griffiths, *J. Biol. Chem.*, 237 (1962) 1676.
121 D. M. Ziegler and K. A. Doeg, *Arch. Biochem. Biophys.*, 97 (1962) 41.
122 J. S. Rieske, R. E. Hansen and W. S. Zaugg, *J. Biol. Chem.*, 239 (1964) 3017.
123 J. S. Rieske, W. S. Zaugg and R. E. Hansen, *J. Biol. Chem.*, 239 (1964) 3023.
124 S. Takemori, I. Sckezu and K. Okunuki, *Nature*, 188 (1960) 593.
125 F. J. Ogston and D. E. Green, *Biochem. J.*, 29 (1935) 1983.
126 C. Martius, in *Quinones*, Ciba Symp., 1961.
127 A. Nason and I. R. Lehman, *J. Biol. Chem.*, 222 (1956) 511.

Chapter 37

Oxidative Phosphorylation

As the history of oxidative phosphorylation has been retraced in detail in this Treatise by a master of the art (E.C. Slater, in Chapter VII of Volume 14) we will limit ourselves here to general historical aspects of the concept.

1. The discovery of aerobic phosphorylation

At the time when the "Meyerhof cycle" was widely accepted, it was believed that the anaerobic breakdown of metabolites, as described in studies on glycolysis for instance, occurred during respiration.

If the products of glycolysis (alcohol, lactic acid) were not detectable in aerobic conditions it was because they were, in these conditions, converted back to carbohydrate at the expense of a partial combustion of alcohol or of lactic acid. In such a perspective, respiration was a method for getting rid of the products of glycolysis (the energy-yielding process) by burning part of them.

Meyerhof[1] considered this a general metabolic feature providing an explanation for the Pasteur effect* (known for some time as the "Pasteur–Meyerhof effect"), which consists in the inhibition of glycolysis by respiration.

In 1930, Engelhardt published a study on mammalian erythrocytes in which, as in the experiment conducted by Barron and Harrop which was to be of such importance to Warburg's fundamental studies (Chapter 36), respiration was observed after the addition of methylene blue. He observed that when glycolysis was inhibited by the addition of fluoride, the methylene-blue respiration was accompanied by an esterification of phosphate. When he turned to nucleated erythrocytes of the pigeon as experimental material, Engelhardt observed that when the respiration of the bird

---

* The Pasteur effect has recently been the subject of an historical study by Krebs[41], to which the reader is referred.

References p. 420

Plate 114. Vladimir Alexandrovich Engelhardt.

erythrocytes was inhibited by cyanide, a rapid increase of inorganic phosphate took place as the result of ATP splitting. To explain these observations, two hypotheses were possible. The simpler was to suppose that respiration inhibited ATP hydrolysis. No evidence could be found, however, in favour of this hypothesis. The second was akin to the Meyerhof "cycle". It supposed that ATP breakdown continued unchecked under all conditions but that a resynthesis of ATP occurred under aerobic conditions.

To quote Engelhardt,

"Es wird angenommen, dass die Spaltung von Phosphorsäureverbindungen (darunter auch von Pyrophosphat) ständig in der Zelle vor sich geht, auch unter aeroben Bedingungen. Nur wird diese Spaltung durch die Atmung rückgängig gemacht."

A second paper, on the respiration of pigeon nucleated erythrocytes, was published by Engelhardt[3] in 1932. In it, Engelhardt described experiments in which he maintained the nucleated red cells in nitrogen and in oxygen respectively for consecutive periods lasting approximately one hour, and observed alternating ATP splitting and ATP synthesis. Engelhardt considered that this confirmed and extended the "principle of resynthesis" formulated by Meyerhof in his theory on the resynthesis of glycogen in muscle.

"Durch die gegenseitige Koppelung der Abbauprozesse mit den als treibende Kraft für den Wiederafbau dienenden Reaktionen, dadurch, dass der Abbau seine eigene Negation herbeiführt, ist einerseits der Mechanismus der selbständigen Regulation des ganzen Vorganges gegeben, andererseits die Voraussetzung zum ständigen Gang des Prozesses, die Quelle der ununterbrochenen chemischen Umsetzungen, die im Grund der Lebenserscheinungen der Zelle liegen."

Though Engelhardt's experiments and his interpretation were based on the errouneous concept of the Meyerhof cycle, his experimental data and their graphical representation were of striking interest to at least a few people who subsequently were the first to draw attention to the aerobic phosphorylation process. Further studies on this topic were published in 1935 by Runnström and Michaelis[4] and by Lennerstrand and Runnström[5]. The latter authors experimented with yeast maceration juice poisoned by fluoride which inhibited the transformation of phosphoglycerate, and consequently the specific acceptor in glycolysis. Oxygen thus became the only hydrogen acceptor and the following reaction took place:

$$\text{2 phosphoglyceraldehyde} + O_2 \ (\textit{via}\ \text{flavoprotein}) + 2\ P_i + \text{hexose}$$
$$\rightleftharpoons \text{2 phosphoglycerate} + 2\ H_2O + \text{F-1,6-PP}$$

References p. 420

Plate 115. Herman Moritz Kalckar.

The amount of oxygen utilized and that of phosphoglycerate formed were found to be equivalent. The conclusion drawn from these experiments was that the coupling mechanism is independent of the nature of the hydrogen acceptor.

2. The concept of transferable phosphate yielded aerobically by reactions other than those of glycolysis

As noted by Slater, the active interest in aerobic phosphorylations is due to Kalckar's work. In 1937, Kalckar[6,7] found in experiments similar to those of Runnström (but carried out with kidney-cortex extracts) that a vigorous aerobic phosphorylation of glucose took place but with no detectable trace of anaerobic phosphorylation. With the kidney extracts, no production of phosphoglycerate was observed, in contrast to Runnström's experiments with yeast juice. It was therefore concluded that no fermentation-like coupling with phosphoglyceraldehyde oxidation occurred. With glucose as substrate, F-1,6-P_2 was the main product of phosphorylation. Other compounds such as glycerol were also phosphorylated in aerobic conditions. These experiments led to the suggestion that aerobically transferable phosphate was obtained by reactions other than those of glycolysis, described in Chapter 24.

In his monograph on biological phosphorylations, Kalckar[7] points out that Lundsgaard's[8a] second article on alactacid muscle contraction (1930) had already shown that respiration in iodoacetate-poisoned sartorii can be coupled to the rephosphorylation of creatine and hence can maintain a longer series of contractions that can anaerobic-poisoned sartorii. Oxidation through other pathways than those of glycolysis must be coupled with phosphorylation.

Kalckar observed, for each atom of oxygen used, the esterification of about one mole of phosphate ($P:O=1$).

Kalckar[7,8] has retraced the circumstances which led him to conduct his experiments with kidney-extracts, a most significant contribution to biological oxidation studies. What gave rise to these experiments was an interest in the active transport of glucose in intestine and kidney. Lundsgaard had, by 1933, shifted his interest to this field of research and one of the tools he used was phlorhizin, a drug which in mammals produced glycosuria. This excretion of glucose into the urine is the result of an inhibition of the active glucose reabsorption taking place in the renal

References p. 420

tubes. In 1937 Kalckar, who was a student of Lundsgaard in Copenhagen, became interested in the molecular mechanism of glucose reabsorption and the action of phlorhizin.

Lundsgaard[9] considered that a cycle of phosphorylation-dephosphorylation was involved in the active transport of glucose and that phlorhizin was active through its effects on phosphatase. Kalckar was not satisfied with the then prevalent notion that phosphatases might catalyze both phosphorylations and dephosphorylations. He was therefore compelled to search for an active phosphorylation system in kidney cortex, the site for glucose and amino acid recapture, which had been shown by P. György *et al.*[10] to exhibit the highest Q_{O_2} and the lowest glycolysis of mammalian tissues.

Kalckar[6] used minced kidney cortex, and as he considered respiration to be a critical factor, he performed his experiments with the Warburg apparatus and flasks.

In his first experiments Kalckar[6] compared respiration and phosphorylation in briefly dialysed kidney extracts, in the presence of fluoride, with or without added glucose, and he found respiration as well as phosphorylation proceeding rapidly.

The type of phosphorylation identified by Kalckar[11,12] was new, differing from the aerobic rephosphorylation in muscle and from the aerobic phosphorylation described by Engelhardt in the presence of redox dyes with glucose as metabolite, and resulting in a liberation of glucose oxidation products (Warburg, Kubowitz and Christian[13]). In the preparation he used Kalckar showed that aerobic phosphorylation has a number of specific characteristics. It has, for example an absolute need for oxygen and is not sustained by glycolysis. Then again, respiration and aerobic phosphorylation in the kidney cortex dispersions are stimulated by the addition of dicarboxylic acids and of tricarboxylic acids. Addition of glucose, glycerol or 5′-adenylic acid to kidney cortex dispersions in the presence of phosphate and fluoride produces an accumulation of hexosephosphate, of α-glycerophosphate, of ATP and ADP, respectively. In offering evidence for the production of phosphorylated hexose in parallel with respiration in a tissue in which glycolysis was inhibited, Kalckar had formulated the concept of aerobic phosphorylation as distinct from anaerobic phosphorylation.

3. Aerobic phosphorylation linked to electron transfer

As stated above, Kalckar had described a "respiratory" synthesis of phosphorylated hexoses under conditions of inhibited glycolysis and he had observed, for each atom of oxygen used, the esterification of about one mole of phosphate ($P:O = 1$). This quantitative aspect received attention in a 1939 paper by Belitser and Tsybakova[14]. These authors experimented with chopped or ground muscle (pigeon-breast or rabbit-heart), carefully washed with a low ionic-strength saline solution, with NAD, creatine and Mg^{2+} added. In this mixture they introduced one of the substrates pyruvate, citrate, ketoglutarate, succinate, fumarate or lactate. With each of these they observed vigorous respiration and a synthesis of creatine phosphate unhampered by either iodoacetic acid or by fluoride.

To clarify their theoretical position the authors recall the oxido-reduction steps of glycolysis and state that

"On the basis of these findings, one is led to presume that in respiration, as in glycolysis, phosphorylation is connected with a "glycolytic" oxidation-reduction. From a number of indirect findings, however, it is apparent that some oxidizing process may be linked with phosphorylation without having any direct connection with glycolysis. Braunstein[15] and Severin[16] showed that oxidation of pyruvic acid, ketobutyric acid and glutamic acid, as well as of alanine, brings about a stabilization of adenosine triphosphate in nucleated erythrocytes. Grimlund[17] found that oxidation of lactic acid, pyruvic acid, and succinic acid increases the working capacity of a muscle in which glycolysis is obliterated. This was previously found to be true in the case of lactic acid by Meyerhof and his coworkers[18,19] who also stated that lactic acid oxidation brings about stabilization of phosphagen in muscles poisoned by iodoacetate. It can be presumed that stabilization of phosphorylated esters is the result of their resynthesis; then, on the basis of the investigations quoted above, a "strictly respiratory" synthesis of these esters might well be inferred. This interpretation has received additional and more direct support. Kalckar[6] has described a "respiratory" synthesis of phosphorylated hexoses in kidney cortex preparations, under conditions in which glycolysis is arrested. About the same time one of us[20,21] found that minced muscle tissue under aerobic conditions can sustain a synthesis of phosphagen from creatine and inorganic phosphate even after poisoning with monobromo-acetic acid." (translation in Kalckar[7])

Colowick, Welch and Cori[22] confirmed Kalckar's findings on kidney cortex suspensions, concerning aerobic phosphorylation. They used as substrates a number of dicarboxylic acids and observed that the step succinate to fumarate was associated with phosphate esterification, as had been noted by Belitser and Tsybakova[14].

4. The concept of respiratory chain phosphorylation (oxidative phosphorylation)

As Kalckar had done in the case of kidney-cortex suspensions, Belitser and

Plate 116. Vladimir Alexandrovich Belitser.

Tsybakova[14] measured P:O ratios and found values greater than unity (between 1.9 and 2.1). They therefore concluded that such a high efficiency shows

"that phosphorylation is not only coupled with the primary steps of dehydrogenations of substrate molecules... but also with certain oxidation–reductions which participate in further hydrogen transport."

Ochoa[23,24], experimenting in Sir Rudolph Peters' laboratory in Oxford with extracts of brain or of heart, observed (in 1941 and 1942) for α-ketoglutarate oxidation to succinate a P:O value of three as against only one "energy-rich" bond in the case of the oxidation of succinate to fumarate as defined by Lipmann in his 1941 review paper. In heart extracts Ochoa did not find phosphorylation associated with the oxidation of large quantities of NADH.

The high values obtained by Ochoa were first questioned by Ogston and Smithies[26] who suggested that P:O values as high as 2.0 were unlikely in light of a number of (unjustified) assumptions one of which was that only the oxidation-reduction potential below the level of cytochrome *c* can be used for generating energy-rich bonds. That Ochoa's figures were in the correct range was confirmed by Loomis and Lipmann[27], Cross, Taggart, Cova and Green[28] and Hunter and Hixon[29].

Commenting on the discovery by Belitser and Tsybakova and by Ochoa of high P:O ratios in what has since been called "Oxidative phosphorylation", Kalckar[7] writes:

"Both of these studies revealed coupling between phosphorylation (P) and oxygen (O) of a magnitude which prompted a further remodeling of our thinking."

To quote Belitser and Tsybakova[14]:

"The high efficiency of respiratory synthesis (in oxidation of a molecule of hexose, no less than 24 molecules of phosphagen are synthesized) shows that phosphorylation is not only coupled with the primary steps of dehydrogenations of substrate molecules (like dehydrogenation of phosphotriose or of pyruvic acid, the process of complete oxidation of the hexose molecule has to include 12 primary dehydrogenations), but also with certain oxidation reductions which participate in further hydrogen transport." (translation in Kalckar[7]).

As Ochoa[30] noted:

"It would be difficult to understand how the removal of 2 H would give rise to an uptake higher than 1 atom phosphorus until their further catalytic transfer is also linked with phosphorylation."

References p. 420

As stated by Slater (Chapter VII of Volume 14), these quotations illustrate the conclusion reached by both groups that

> "phosphorylation must occur not only when the substrate is dehydrogenated—as in the oxidation reaction of glycolysis—but also during the further passage of the hydrogen atoms (or electrons) along the respiratory chain to oxygen. Thus these workers were the first to postulate what is now known as respiratory-chain phosphorylation."

There was no immediate recognition of the insight (formulated by Belitser and by Ochoa) that the form of phosphorylation discovered by Kalckar is essentially different (redox chain) from that involved in glycolysis (dehydrogenation). The theory developed by Szent-Györgyi (Chapter 30) was still widely accepted at the time. Krebs' important distinction between the carbon and the hydrogen pathway had not yet been clarified; on the contrary, and it was still widely believed that the succinate-fumarate and the malate-oxaloacetate couples were involved in the respiratory chain.

Around 1940, the energy aspects of biological oxidations came to be seen in a new light as a result of new insights into oxidation–reduction as described in reviews of the respiratory chain of electron transfer (see Chapter 37). As Ball[31] stated in 1942, "the energy resulting from the oxidation of foodstuffs" was believed to be "released in small units or parcels, step by step", the amount of energy released at each step being "proportional to the difference in potential of the system comprising the several steps". This thermodynamical treatment and the quantitative data it yielded played an essential role in the development of the concept of oxidative phosphorylation as linked with the respiratory chain, and though they are not of the nature of mechanisms explaining theories, they remain the firm foundation of the concept.

The recognition of the existence of an electron pathway (see Chapter 36) oxidizing NADH resulting from the dehydrogenation of most of the Krebs cycle intermediates and composed of flavoproteins and the cytochrome system, was essential, in combination with the energy relations defined by Ball, to the development of the concept of a coupling of the respiratory chain with the phosphorylation processes. From then on, "oxidative phosphorylation" became synonymous with phosphorylation in the redox chain, though the coupling of phosphorylations with individual steps of that redox chain remained to be unravelled.

5. Experimental demonstration that oxidative phosphorylation is coupled to the electron transfer in the respiratory chain

That phosphorylation sites exist above the primary NAD level in the respiratory chain was directly demonstrated by Friedkin and Lehninger[32,33]. These authors showed that aerobic incubation of NADH with a properly complemented suspension of particulate elements of rat liver resulted in an extensive incorporation of inorganic phosphate labelled with ^{32}P into ATP. But no such incorporation was detectable in anaerobic conditions (under nitrogen) nor when NADH was replaced by NAD^+, or in the absence of Mg^{2+} or ATP. The authors concluded that oxidative phosphorylation had occurred in conjunction with the passage of electrons from NADH to oxygen *via* the cytochrome system. In their aerobic experiment Kennedy and Lehninger[33a] were able to locate in mitochondria the incorporation, into ATP, of inorganic phosphate labelled with ^{32}P.

As tracer techniques give no information on the P:O ratio a possible interpretation could have been that the ^{32}P incorporation observed corresponded to relatively minor phosphate exchanges.

To put aside this possible interpretation Lehninger[34], in non-isotopic experiments using highly purified NADH and also by measurement of the phosphate uptake showed in a direct way that phosphorylation sites exist above the primary level of NADH in the respiratory chain. The ratio P:NADH was measured (NADH disappearance spectrophotometrically and the phosphate exchange colorimetrically). From the value of P:NADH, Lehninger[34] obtained a P:O value as high as 1.89.

"Comparative experiments showed that electron transport from NADH to oxygen very nearly accounts for all the phosphorylation observed when β-hydroxybutyrate undergoes aerobic oxidation to acetoacetate *via* the NAD-linked β-hydroxybutyric dehydrogenase. The data presented provide direct support for predictions based on thermodynamic considerations that a large part of phosphate uptake during oxidative phosphorylation takes place during electron transport between primary coenzyme and oxygen".

6. Identification of phosphorylation steps in the respiratory chain

That dinitrophenol may act as un uncoupler of oxidative phosphorylation was suggested in 1945 by Lardy and Elvehjem[35] and demonstrated by Loomis and Lipmann[27], who concluded:

"These results indicate that DNP reversibly uncouples phosphorylation from oxidation, an effect that can also be obtained with atebrin (mepacrine) in 10^{-3} *M* concentration. Although

References p. 420

sodium azide can lower the P:O ratio, it cannot replace phosphate in the system and is, in slightly higher concentration, a powerful inhibitor of respiration as well. DNP does not inhibit respiration except in high concentration."

Though the use of inhibitors helped in the study of phosphorylation steps, the location of their effects at the level of definite reactions was sometimes difficult.

Another problem in experimentation came from the activity of the ATPase of mitochondria, which made the recognition of ATP synthesis difficult. On the other hand, in the complex media used in these studies barriers sometimes hindered the penetration of a reagent into a definite site.

In 1951, Judah[36] found a P:O ratio of approximately four in the case of the oxidation of α-ketoglutarate and of approximately 3 in the case of β-hydroxybutyrate. Slater, in Chapter VII of Volume 14 of this Treatise, has recalled the efforts undertaken in many laboratories resulting in definitions of the following P:O ratios and of the number of phosphorylation steps (ATP formation)

ketoglutarate $+\frac{1}{2}\,O_2 \rightarrow$ succinate $+ CO_2$ — 3 to 4

hydroxybutyrate $+\frac{1}{2}\,O_2 \rightarrow$ acetoacetate $+ H_2O$ — 2 to 3

succinate $+\frac{1}{2}\,O_2 \rightarrow$ fumarate $+ H_2O$ — 1 to 2

The nature of the relationship between respiratory chain and phosphorylation as indicated in Fig. 44 (p. 400) has been confirmed, by means of entirely different methods, by Chance and Williams[37].

7. Mechanisms of oxidative phosphorylation

The reader is referred to Slater's presentation of the history of oxidative phosphorylation, in Chapter VII of Volume 14 of this Treatise, which comprises an extensive review of the different aspects of the problem.

In essence, the chemical theory postulates the participation of high-energy compounds ($A \sim C$) as intermediates in the chain of reactions leading from oxidation to ATP synthesis. The chemiosmotic (Mitchell) theory conjectures the formation of a membrane potential for the synthesis of ATP. Historical aspects of this theory are retraced by Mitchell himself in Chapter IV of Volume 22 of this Treatise. A supposition common to both theories is the accumulation of energy (energy-linked reaction), in the absence of P_i and ADP, as $A \sim C$ or as membrane potential, *i.e.* in a form different from the pyrophosphate bond of ATP.

An experimental demonstration of the possible existence of such an "energy-linked reaction" was provided by the so-called "reversal of the respiratory chain". In 1956, Chance[38] showed that addition of succinate to mitochondria in the absence of ADP results in a reduction of NAD^+. Chance and Hollunger[39,40] suggested that this is the result of a reversal of the respiratory chain, the energy coming from the oxidation of succinate (see Slater's Chapter for a discussion and for evidence showing that energy can accumulate in mitochondria without being converted to ATP). Isolation of postulated A$\sim$C compounds has failed so far.

8. Energy aspects

The complete oxidation of pyruvate is accompanied by a free energy change of -270 kcal per mole. For the biological oxidation of pyruvate 5 moles of oxygen are required, leading to the formation of approximately 15 moles of ATP. As the mechanism of oxidative phosphorylation remains unknown it is unsafe to go beyond the conclusion that oxidative phosphorylation results in a production of energy-rich phosphate bonds of ATP (approximately 15 per mole of pyruvate), yielding free energy for the accomplishment of work.

REFERENCES

1 O. Meyerhof, in H. Geiger and K. Scheel (Eds.), *Handbuch der Physik*. vol. II, Berlin. 1926.
2 V. A. Engelhardt, *Biochem. Z.*, 227 (1930) 16.
3 V. A. Engelhardt, *Biochem. Z.*, 251 (1932) 343.
4 J. Runnström and L. Michaelis. *J. Gen. Physiol.*, 18 (1935) 717.
5 A. Lennerstrand and J. Runnström, *Biochem. Z.*, 283 (1935) 12.
6 H. M. Kalckar, *Enzymologia*, 2 (1937) 47.
7 H. M. Kalckar, *Biological Phosphorylations. Development of Concepts*, Englewood Cliffs, N.J. 1969.
8 H. M. Kalckar, personal communication to the author.
8a E. Lundsgaard, *Biochem. Z.*, 227 (1930) 51.
9 E. Lundsgaard, *Biochem. Z.*, 264 (1933) 269.
10 P. György, W. Keller and T. Brehme, *Biochem. Z.*, 200 (1928) 356.
11 H. M. Kalckar, *Enzymologia*. 5 (1938) 365.
12 H. M. Kalckar. *Fosforyleringer i Dyrisk Vaev*, Copenhagen, 1938) excerpt from English abstract in Kalckar[7]).
13 O. Warburg, F. Kubowitz and W. Christian, *Biochem. Z.*, 227 (1930) 245.
14 V. A. Belitser and E. T. Tsybakova, *Biokhimiya*, 4 (1939) 516 (English translation in Kalckar[7]).
15 A. E. Braunstein, *Report to the XVth Intern. Physiol. Cong. in Leningrad (1935)*.
16 V. A. Severin, *Biokhimiya*, 2 (1937) 60.
17 K. Grimlund, *Skand. Arch. Physiol.*, 73 (1936) 109.
18 O. Meyerhof and E. Boyland, *Biochem. Z.*, 237 (1931) 406.
19 O. Meyerhof, Ch. Gemmill and G. Benato, *Biochem. Z.*, 258 (1933) 371.
20 V. A. Belitser, *Biokhimiya*, 2 (1937) 334.
21 V. A. Belitser, *Biokhimiya*, 3 (1938) 80.
22 S. P. Colowick, M. S. Welch and C. Cori, *J. Biol. Chem.*, 133 (1940) 359.
23 S. Ochoa, *J. Biol. Chem.*, 151 (1943) 492.
24 S. Ochoa, *J. Biol. Chem.*, 138 (1941) 751.
25 F. Lipmann, *Advan. Enzymol.*, 1 (1941) 99.
26 A. G. Ogston and O. Smithies, *Physiol. Rev.*, 28 (1948) 283.
27 W. F. Loomis and F. Lipmann, *J. Biol. Chem.*, 173 (1948) 807 (reprinted in Kalckar[7]).
28 R. J. Cross, J. V. Taggart, G. A. Cova and D. E. Green, *J. Biol. Chem.*, 174 (1949) 655.
29 F. E. Hunter Jr. and W. S. Hixon, *J. Biol. Chem.*, 181 (1949) 73.
30 S. Ochoa, *Nature*, 146 (1940) 267 (reprinted in Kalckar[7]).
31 E. Ball, in *A Symposium on Respiratory Enzymes*, Madison, 1942.
32 M. Friedkin and A. L. Lehninger, *J. Biol. Chem.*, 174 (1948) 757.
33 M. Friedkin and A. L. Lehninger, *J. Biol. Chem.*, 178 (1949) 611.
33a E. P. Kennedy and A. L. Lehninger, *J. Biol. Chem.*, 179 (1949) 957, 964, 969.
34 A. Lehninger, *J. Biol. Chem.*, 190 (1951) 345 (reprinted in Kalckar[7]).
35 H. A. Lardy and C. A. Elvehjem, *Ann. Rev. Biochem.*, 14 (1945) 16 (excerpt reprinted in Kalckar[7]).
36 J. D. Judah, *Biochem. J.*, 49 (1951) 271.
37 B. Chance and G. R. Williams, *Advan. Enzymol.*, 17 (1956) 65.
38 B. Chance, in O. H. Gaebler (Ed.), *Units in Biological Structure and Function*, New York, 1956.
39 B. Chance and G. Hollunger, *Federation Proc.*, 16 (1957) 163.
40 B. Chance and G. Hollunger, *Nature*, 185 (1960) 666.
41 H. A. Krebs, *Essays Biochem.*, 8 (1972) 1.

Section III

Autotrophic Phosphorylations

Chapter 38

Photosynthetic Phosphorylation

1. Introduction

Ample space will be devoted to photosynthesis in Part IV (*Biosynthesis*) of this *History*. It is nevertheless fitting to deal, in Part III, with its primary photochemical aspect: the absorption of light quanta by chlorophyll, producing oxidizing and reducing equivalents. Photosynthetic phosphorylation converts light energy into energy-rich phosphate bonds. Since this process takes place in chloroplasts it requires, like oxidative phosphorylation in mitochondria, an electron transport system. However, instead of the participation of oxygen it requires that of the oxidants and reductants present in quantasomes.

The history of photosynthetic phosphorylation goes back to 1954. It is reviewed in Vol. 14 of this Treatise (Chapter VIII, by Frenkel and Cost), by one of its pioneers and we shall therefore limit the present section to general historical viewpoints along the lines adopted in previous chapters of Part III of the present *History*.

2. Energy transfer

It appears obvious that since light was recognized as the source of energy in photosynthesis, a transfer of light energy to another form should have been recognized as necessary for its participation in biosynthetic reactions. Nevertheless, energy transfer was for a time ignored. Willstätter and Stoll[1], for instance, held that a molecule of CO_2 is reduced to carbohydrate by its combination with chlorophyll.

The concept of energy transfer was hinted at by Gaffron and Wohl[2] in 1936. They calculated that to account for the observed rates of CO_2 reduction in the process considered by Willstätter and Stoll, the participation of 2000 molecules of chlorophyll would be required for the reduction of one molecule of CO_2. The arguments put forward by Gaffron

References p. 430

Plate 117. Albert W. Frenkel.

and Wohl were weakened when it was understood that some sort of reducing power is generated which does not require more energy than one quantum at a time (see Wassink[3]).

3. Participation of ATP in photosynthesis

The publication by Lipmann of his epoch-making review paper in 1941, repeatedly quoted in previous chapters, prompted those specialized in photosynthesis to include the concept of energy-rich phosphates in their conceptual constructions. In tracer experiments with ^{14}C and ^{32}P (see Buchanan *et al.*[4]) it became clear that, instead of the free organic compounds, their phosphate esters are involved in the CO_2 reduction of photosynthesis. It was also clear that in the Calvin cycle, phosphoglyceric acid has to be phosphorylated to diphosphoglyceric acid (a high-energy ester with one P attached to the carboxyl). ATP is needed for this phosphorylation. Accordingly, when a cell begins to photosynthesize the concentrations of high-energy phosphate bonds and P_i should vary. An effect of photosynthesis on the phosphate species was indicated experimentally (literature in Wassink[3]).

The phosphate level of *Chlorella* in light and in darkness has also been studied by Kandler[5] and he concluded that ATP participates in photosynthesis.

4. Source of free energy for ATP formation in photosynthesis: photosynthetic phosphorylation

That the source of the free energy for ATP formation was to be found in the partial oxidation of an intermediate (for example NADPH) was suggested by several authors (Emerson, Stauffer and Umbreit[6]; Ruben[7]).

We recognize, in these theories of partial oxidation providing the energy for biosynthesis, the sequels of the theory of Meyerhof's cycle.

The first direct demonstration of a formation of ATP from ADP and P_i in light we owe to Frenkel[8], who experimented with sonically disintegrated *Rhodospirillum rubrum*.

Arnon, Whatley and Allen[9], in 1954, demonstrated with chloroplasts the same phenomenon, to which they gave the name "photosynthetic phosphorylation"; subsequently, the same authors[10,11] showed that phosphorylation accompanied the reduction of exogenous NADP or ferricyanide.

References p. 430

Plate 118. Daniel Israel Arnon.

5. The electron transport system of chloroplasts

Between photosynthetic pigments, the existence of an energy transfer was experimentally recognized by the method of sensitized fluorescence (literature in Wassink[3]), but the transfer to other kinds of molecules could not be explained (as in the case of pigments) by the concept of inductive energy resonance involving overlapping of mutual absorption and fluorescence bands. Two hypotheses were put forward. One asserted the formation of a complex between the ultimate pigment in their sequence and the "energy adaptor". According to the other, an electron is carried through a protein (semi-conductors).

The discussion still continues, the assumption being generally accepted that an electron transport system can accept electrons from the photochemical reductant and transfer them to the photochemical oxidant, energy-rich phosphate bonds being generated in the course of this electron transport.

The definition of the essential mechanism of the primary process as an electron flow mechanism, similar to the respiratory chain of mitochondria, was already suggested by Szent-Györgyi[12] in 1941, and an electron-flow scheme was proposed by Arnon[13] in 1961.

In the studies undertaken by Arnon's school, electron flow was believed to be hastened by the presence of phosphate and ADP, and a stoichiometric relation was established between the number of moles of ATP formed and the number of electrons transferred. An electron-flow scheme for photo-induced electron transport is given in Fig. 45.

In Volume 14 of this Treatise, Frenkel and Cost have reviewed the history of our knowledge concerning the components of the electron-transport apparatus of photosynthetic phosphorylating systems (chlorophyll, accessory pigments, quinones and naphthoquinones, flavins, flavones, ferredoxin, pyridine nucleotide transhydrogenase, NADPH, NADH-cytochrome *c* reductase, NADPH diaphorase, succinate dehydrogenase, haem proteins, hydrogenase, plastocyanin, allagochrome).

The scheme of Fig. 45 shows photosynthesis as dependent upon two different primary light reactions. This concept was introduced by Duysens. The history of this development has been reviewed by Duysens and Amesz in Volume 27 of this Treatise.

Duysens observed in the red alga *Porphyridium cruentum* a more efficient cytochrome oxidation at 680 nm than at 560 nm. He suggested that the

References p. 430

Plate 119. L. M. N. Duysens.

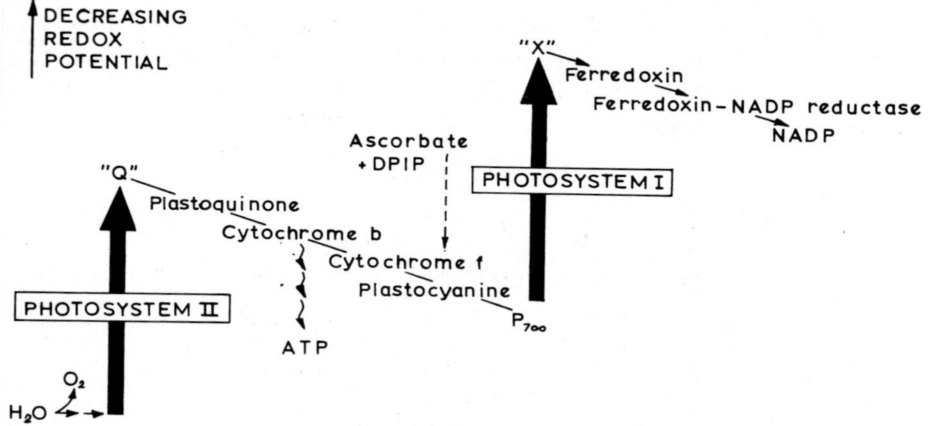

Fig. 45. An electron flow scheme for photo-induced electron transport in isolated chloroplasts. (Avron[14]).

two systems are of different pigment composition, system II containing fluorescent chlorophyll *a* and a large amount of phycobilin, and active in cytochrome reduction, and system I containing non-fluorescent or weakly fluorescent chlorophyll *a* and less phycobilin, and active in cytochrome oxidation. That NADP (and not NAD) is specific in photosynthetic phosphorylation was demonstrated by Arnon *et al.*[10].

The scheme of Fig. 45 (which is speculative) considers that a flow of electrons results from the absorption of a quantum of light by photosystem II, leading to the loss of an electron by water and the reduction of Q, the result being the formation of the strong reductant "reduced Q" ($E'_0 = \pm 0.1$ V at pH 7.4) from the poor reductant water.

From reduced Q the electrons flow through an electron transport chain (analogous to the respiratory chain of mitochondria) to reduced P_{700} (and not to oxygen as in the respiratory chain). In the course of this kind of respiratory chain, the drop in potential is utilized to form ATP. Another quantum is absorbed by photosystem I and leads to the oxidation of P_{700} and to the reduction of NADP.

"Overall, the absorption of four light quanta, two in each photosystem, will transfer two electrons through the chain, produce one atom of oxygen, one molecule of ATP, and one molecule of NADP" (Avron[14]).

References p. 430

Although the source of free energy in photosynthetic phosphorylation is defined by the above studies, the subject is not yet amenable to any form of recurrent history as the mechanism involved remains controversial.

REFERENCES

1 R. Willstätter and A. Stoll, *Untersuchungen über Chlorophyll*, Berlin, 1913.
2 H. Gaffron and K. Wohl, *Naturwissenschaften*, 24 (1936) 81, 103.
3 E. C. Wassink, in M. Florkin and H. S. Mason (Eds.), *Comparative Biochemistry*, Vol. V, New York and London, 1963.
4 J. G. Buchanan, J. A. Bassham, A. A. Benson, D. F. Bradley, M. Calvin, L. L. Daus, M. Goodman, P. M. Hayes, V. H. Lynch, L. T. Norris and A. T. Wilson, in W. D. McElroy and B. Glass (Eds.), *Phosphate Metabolism*, Vol. II, Baltimore, 1952.
5 O. Kandler, *Z. Naturforsch.*, 5b (1950) 423.
6 R. L. Emerson, J. F. Stauffer and W. W. Umbreit, *Am. J. Botany*, 31 (1944) 107.
7 S. Ruben, *J. Am. Chem. Soc.*, 65 (1943) 279.
8 A. Frenkel, *J. Am. Chem. Soc.*, 76 (1954) 5568.
9 D. I. Arnon, F. R. Whatley and M. B. Allen, *J. Am. Chem. Soc.*, 76 (1954) 6324.
10 D. I. Arnon, F. R. Whatley and M. B. Allen, *Nature*, 180 (1957) 182.
11 D. I. Arnon, F. R. Whatley and M. B. Allen, *Science*, 127 (1958) 1026.
12 A. Szent-Györgyi, *Science*, 93 (1941) 609.
13 D. I. Arnon, in W. D. McElroy and B. Glass (Eds.), *Light and Life*, Baltimore, 1961.
14 M. Avron, *Curr. T. Bioeng.*, 2 (1967) 1.

Chapter 39

Sources of Free Energy in Chemoautotrophs

In Chapter 37, Section 7, we referred to the mechanism of oxidative phosphorylation and mentioned the participation in the electron-transfer chains of so-called "energy-linked reactions" in the chemical theory of these phosphorylations. Whereas, in glycolysis, the formation of ATP takes place in a coupled reaction resulting from the oxidation of 3-phosphoglyceraldehyde by phosphoglyceraldehyde dehydrogenase and NAD, in an electron-transfer chain such as that of oxidative phosphorylation or of chemoautotrophic phosphorylation, our present conception is that of the cleavage of intermediates of oxidative phosphorylation ($A \sim C$) or possibly of ATP, the hydrolysis of which is the energy-donating reaction (energy-linked reactions).

Chemoautotrophs transfer the energy of their inorganic substrates into the energy of ATP energy-rich bonds, establishing by the production of ATP and the reductant NADH the conditions necessary for biosynthesis and for carbon assimilation in particular.

Chemoautotrophic bacteria were discovered in 1888 by Winogradsky. They are autotrophs, *i.e.*, they synthesize their organic constituents *de novo* as do the photoautotrophic bacteria, without making use of preformed organic material. In contrast with photoautotrophic bacteria, in which the free energy is derived from quanta of light, the free energy converted into ATP by chemoautotrophs has its source in the oxidation of simple inorganic compounds available in the environment and introduced into the cells. While in the biological oxidations of heterotrophs organic compounds are oxidized, chemoautotrophic respiration oxidizes inorganic substrates. The nature of the inorganic substrate varies. It is ammonium or nitrite ions in the nitrifiers Nitrosomonas and Nitrobacter, sulfide or thiosulphate ions in the non-photosynthetic sulfur bacteria, hydrogen in Hydrogenomonas or ferrous ions in Ferrobacillus. The history of the concepts pertaining to the assimilation of carbon by chemoautotrophs will be considered in a later chapter (in Part IV of this *History*). The present historical review deals only with the conversions of the energy

References p. 438

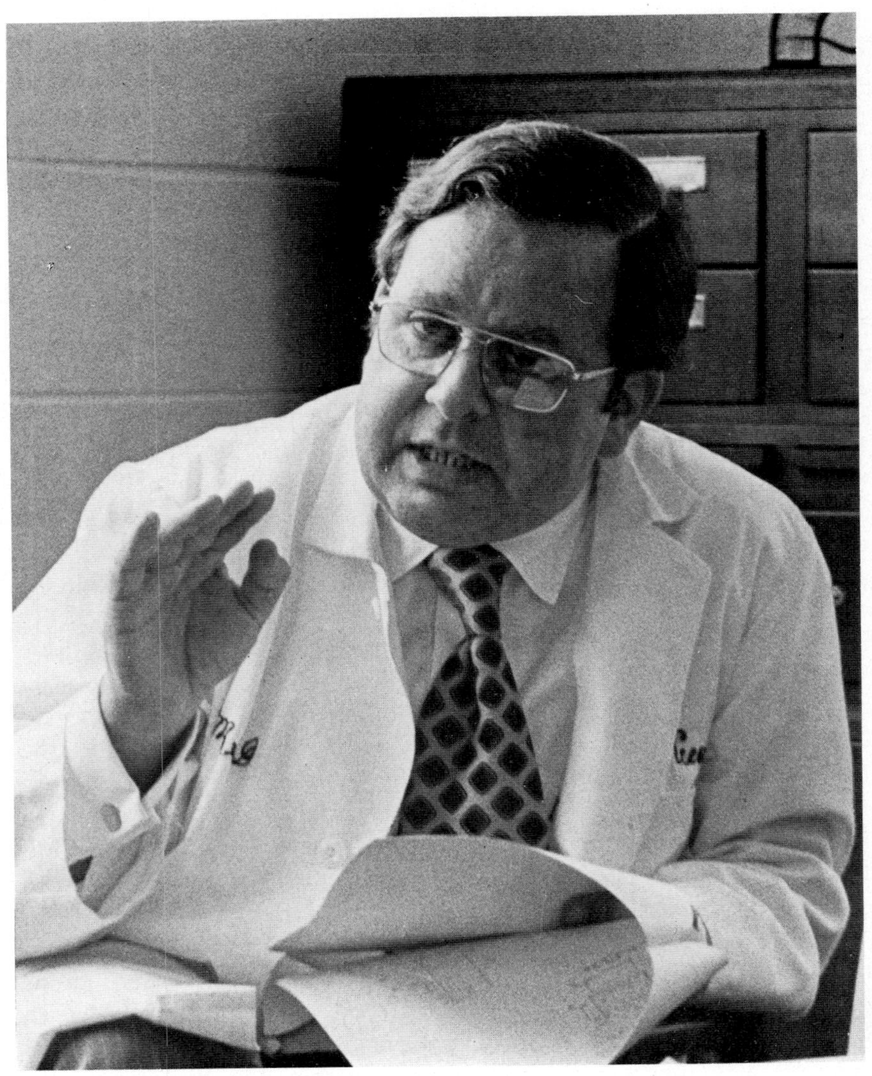

Plate 120. Lutz Kiesow.

obtained by the oxidation of the exogenous inorganic substrate into the energy of ATP with a concomitant formation of the reductant NADH.

The first theory was formulated by Meyerhof[2–4], who introduced what has been designated as the respiratory theory of chemoautotrophy. He considered it a process similar to cellular respiration, consisting of a transfer of electrons from the nitrite introduced in the *Nitrobacter* cell, for instance, to oxygen. Such theories were favoured in the light of a number of subsequent observations such as the discovery of cytochrome in chemoautotrophic bacteria by Lees and Simpson[5], by Aleem and Alexander[6] and by Aleem and Nason[7]. In addition, a synthesis of ATP was coupled to the oxidation of the inorganic substrate, as shown by Aleem and Nason[8].

This apparently satisfying picture was disturbed by Kiesow[9] in 1962. Kiesow's observations on *Nitrobacter winogradskyi* led him to conclude that "the oxidation of nitrite requires respiratory energy in the form of ATP" and "that a simple electron transport from nitrite *via* cytochrome towards oxygen is unlikely". These observations indicated the participation, in the oxidation of the inorganic substrates of the chemoautotrophs, of ATP-consuming steps (energy-linked reactions). The next step forward was the demonstration by Kiesow[10] and by Aleem *et al.*[11] of the "energy-linked reduction of pyridine nucleotides". The values of the redox potential of the ferrous ions, for instance, serving as substrates to *Ferrobacillus*, or of the nitrite ions serving as substrates to *Nitrobacter* are much too positive to permit the reduction of NAD (this does not apply to hydrogen gas, the substrate of *Hydrogenomonas*, which reduces NAD directly as shown by Packer and Vischniac[12]).

Kiesow[10] showed that the reduction of NAD in *Nitrobacter* was not only coupled to an oxidation of nitrite but also required oxygen.

The formulation of the reaction reads as follows:

$$NO_2^- + H_2O = NO_3^- + 2\,H^+ + 2^e \qquad (1)$$

$$NAD^+ + 2\,H^+ + 2^e = NADH + H^+ \qquad (2)$$

$$NAD^+ + NO_2^- + H_2O = NADH + NO_3^- + H^+ \qquad (3)$$

This formulation did not account, however, for the consumption of oxygen nor for the fact that the reaction did not take place in anaerobic conditions. Furthermore, if one adds up the redox potentials of equations (1) and (2) it appears that the formulation is thermodynamically impossible. This led Kiesow to suggest that the oxygen-consuming process takes place in parallel thereby providing the energy necessary to make the former process possible.

References p. 438

This was cleared up by Aleem *et al.*[11] who by direct spectroscopic examination of cell-free extracts of *Nitrobacter* in conditions precluding aerobic generation, showed that the reduction of NAD by reduced mammalian cytochrome *c*, added as an electron donor, takes place only in the presence of ATP.

In 1964, Kiesow[13] was able to show directly with Nitrobacter the reduction of NAD upon addition of nitrite and the prevention of this reaction by the uncouplers of oxidative phosphorylation. Later on, the energy-linked NAD reduction by the inorganic substrate was demonstrated with *Nitrosomonas europaea*[14] and with *Thiobacillus novellus*[15].

A new trail of research started when Kiesow[13] observed a very unexpected result. He found that in anaerobic conditions *Nitrobacter* reduced nitrate to nitrite. In cell-free systems the reduction of nitrate could also be obtained with NADH (not NADPH) as the reductant. In this pathway Kiesow[13] demonstrated the participation of the two cytochromes which had been shown by Lees and Simpson[5] to participate in the aerobic nitrite oxidation by intact cells of *Nitrobacter*. A flavoprotein was identified among the participants of the system (Kiesow[13]) which has since been purified by Straat and Nason[16]. Furthermore, the system is located in the same particles obtained from the double membranes of the cells, where the oxidation of nitrite takes place. This corroborated the view that the reduction of nitrate consists of a complete or partial reversion of the oxidation of nitrite. The nitrate reduction by NADH was furthered by magnesium ions, by inorganic phosphate and by ADP; also, a synthesis of ATP was demonstrated (Kiesow[13]). Since the formation of ATP was uncoupled by 2,4-dinitrophenol without the electron transfer being inhibited, the analogy with the oxidative phosphorylation linked to the mitochondrial respiratory chain was clear except that in nitrate reduction, oxygen is not required. There, the final electron acceptor is nitrate and not oxygen.

Kiesow[1], in a review to which the present chapter is highly indebted, formulates as follows the process of the reduction of nitrate by NADH and the coupled phosphorylation of ADP:

$$\mathrm{NADH + H^+ = NAD^+ + 2\,H^+ + 2^e} \quad (4)$$

$$\mathrm{NO_3^- + 2\,H^+ + 2^e = NO_2^- + H_2O} \quad (5)$$

$$n\,\mathrm{ADP^{3-}} + n\,\mathrm{HPO_4^{2-}} + n\,\mathrm{H^+} = n\,\mathrm{ATP^{4-}} + n\,\mathrm{H_2O} \quad (6)$$

$$\text{Sum: } \mathrm{NADH + NO_3^-} + n\,\mathrm{ADP^{3-}} + n\,\mathrm{HPO_4^{2-}} + (n+1)\mathrm{H^+} = \mathrm{NAD^+ + NO_2^-} + n\,\mathrm{ATP^{4-}} + (n+1)\mathrm{H_2O} \quad (7)$$

Equations (4) and (5) represent equations (1) and (2) above, in reverse.

To quote Kiesow[1],

> "One can now visualize quite easily that lack of oxygen causes a reversion of that part of the electron transfer chain of Nitrobacter which is ordinarily responsible for the energy-linked reduction of pyridine nucleotide as observed by Kiesow (1963). The function of ATP in the coupled system would then be to transfer electrons against the thermodynamic gradient. These electrons originate from nitrite and are trapped by NAD; simultaneously ATP hydrolysis provides the missing, but necessary, free energy... It would be the sole role of oxygen to provide the ATP necessary to overcome the thermodynamic barriers of electron transfer."

But what aerobic pathway does provide ATP in autotrophic organisms?

Ever since oxidative phosphorylation linked to respiratory chains had been discovered, the formation of energy-rich phosphate bonds had been looked for in chemoautotrophs and the formation of such bonds described as Vogler and Umbreit[17] did (in *Thiobacillus*) just after the publication of Lipmann's epoch-making review.

But progress required exploration of the field by means of studies on cell-free systems. This necessary step was taken by Aleem and Alexander[6], and Aleem and Nason[7,8], who demonstrated the formation, by cell-free preparations of *Nitrobacter*, of ATP coupled to nitrite oxidation.

Another line of research concerned what was known as "residual respiration". As far back as 1939, Bömeke[18] had demonstrated in nitrifying bacteria deprived of nitrite the existence of a type of respiration which had been considered an alternate mechanism distinct from the oxidation of nitrite and destined to ensure, by the oxidation of its own substance, the survival of bacteria in periods of nitrite need, a logical consequence of the application of common sense (and of finalism).

A reinvestigation of "residual respiration" in Nitrobacter demonstrated that the seat of this respiration was located in the same particles of the double membranes in which took place the oxidation of nitrite as well as the reversed reduction of nitrate. The residual respiration, although it produced ATP, did not permit the fixation of CO_2 by the chemoautotrophic cells (Kiesow[10]). This indicated that the chemoautotrophic reaction must produce another substance besides ATP.

Kiesow[13] showed that addition of small amounts of NADH to cell-free particle preparations in the presence of oxygen did prime the oxidation of nitrite. On the other hand, the presence of an ATP trap (glucose-hexokinase) inhibits the priming effect of NADH. The conclusion is that the NADH priming of the oxidation of nitrite was mediated by ATP.

References p. 438

Kiesow[13] inferred from these experiments the existence of a pathway permitting the oxidation of NADH by oxygen resulting in a coupled reaction, ATP priming the nitrite oxidation. The enzymatic oxidation of NADH by O_2 and the simultaneous formation of ATP was understood to be mediated by the same cytochrome that mediates the aerobic nitrite oxidation and the anaerobic nitrate reduction. The "residual respiration" appeared in this perspective as resulting from the presence of any NAD-dependent oxidation of endogenous substrate following the course of the aerobic oxidation of NADH as it runs in nitrite oxidation.

The isolation of the particular electron transfer chain of *Nitrobacter*, the main chemoautotroph so far studied, has been reviewed in Kiesow's article[1] to which the reader is referred. The main steps as outlined by Kiesow are as follows.

At least two cytochromes have been identified in *Nitrobacter* (Lees and Simpson[5]). One is a cytochrome of the *c* type (Butt and Lees[19]) which upon reduction presents an α-absorption band at 552 nm (cytochrome 552). A second cytochrome shows, when reduced, an α-band at 590 nm (cytochrome 590). As both cytochromes display the characteristics of their reduced forms after addition of nitrite they are considered to take part in the respiratory chain from nitrite to oxygen. The following steps

$$NO_2^- \rightarrow \text{cytochrome 590} \rightarrow \text{cytochrome 552} \rightarrow \text{cytochrome oxidase} \rightarrow \text{oxygen}$$

can be demonstrated in the chain, which runs its full course provided that either oxygen or ATP are present. Cytochrome 590 is the primary site of interaction between the inorganic substrate and the sequence of electron carriers is from cytochrome 590 (Kiesow[20]), the first reaction not being energy-linked. However, the reaction from cytochrome 590 to cytochrome 552 involves an energy-linked reduction, ATP being required. Once the reduction of cytochrome 552 has been achieved, the electron flow has two possible pathways. Along the thermodynamic gradient it may proceed towards oxygen *via* cytochrome oxidase and produce ATP.

This electron flow is also responsible for the aerobic oxidation of nitrite and, since cytochrome oxidase is involved, this pathway is inhibited by KCN and CO at low concentrations. Since the energy-linked reduction of cytochrome 552 is involved in the pathway, uncouplers of oxidative phosphorylation inhibit the oxidation of nitrite. This pathway provides the ATP required for energy-linked reactions and the ATP active in CO_2 fixation. The other pathway involves the oxidation, by NAD, of reduced cytochrome 552, leading to the energy-linked production of NADH, and is sensitive to uncouplers of oxidative phosphorylation.

Another electron flow (not inhibited by uncoupling) accounts for the endogenous (residual) respiration, the electrons deriving from endogenous substrates (not from nitrite oxidation) or from NADH. This flow also accounts for the anaerobic reduction of nitrate in the reverse reaction and for the aerobic oxidation of NADH by molecular oxygen. When oxygen is the terminal acceptor, the system is sensitive to KCN and CO. The participants in this electron transfer system are

$$\text{NAD linked dehydrogenases} \rightarrow \text{NADH} \rightarrow \text{flavoproteins} \rightarrow$$
$$\text{cytochrome 552} \rightarrow \text{cytochrome oxidase} \rightarrow \text{oxygen} \rightarrow$$

ATP is generated by oxidative phosphorylation coupled with this electron transfer chain (of which cytochrome 590 is a side-track).

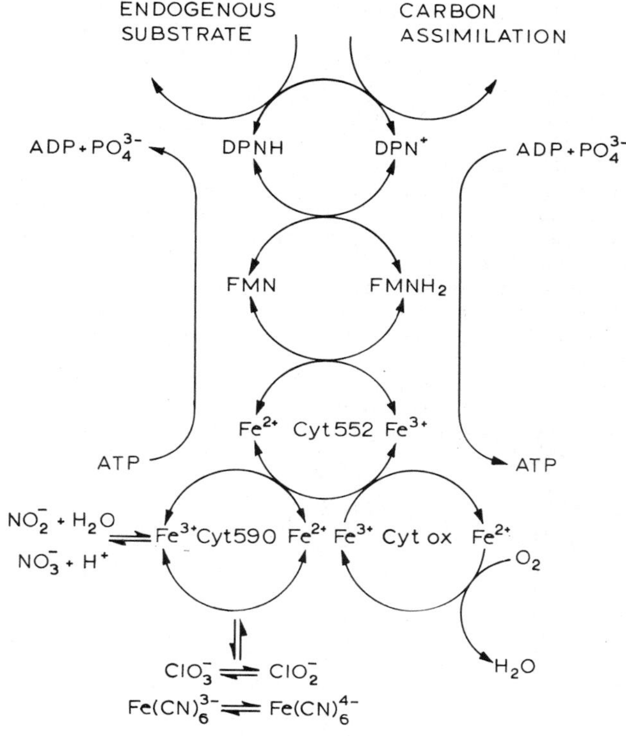

Fig. 46. The electron transport system of *Nitrobacter*. (Kiesow[1]).

References p. 438

If we consider the reverse reaction in which nitrate (or chlorate, or ferricyanide) is the electron acceptor, the electrons bypass the cytochrome oxidase and flow along the chain

NAD-linked dehydrogenases → NADH → flavoprotein →

cytochrome 552 dehydrogenases → cytochrome 590 → NO_3^-

Thus an understanding of the sources of energy in chemoautotrophic organisms was provided by the detection of the participating energy-linked reactions.

The series of chemical interactions starts out with the inorganic substrate.

"By harvesting and channeling the energy of the inorganic molecule into the energy of NADH and ATP, the prerequisites of carbon assimilation are provided." (Kiesow[1]).

REFERENCES

1 L. A. Kiesow, *Cur. T. Bioeng.*, 2 (1967) 195.
2 O. Meyerhof, *Arch. Ges. Physiol.*, 164 (1916) 353.
3 O. Meyerhof, *Arch. Ges. Physiol.*, 165 (1917) 229.
4 O. Meyerhof, *Arch. Ges. Physiol.*, 166 (1917) 240.
5 H. Lees and J. R. Simpson, *Biochem. J.*, 65 (1957) 297.
6 M. I. H. Aleem and M. Alexander, *J. Bacteriol.*, 76 (1958) 510.
7 M. I. H. Aleem and A. Nason, *Biochem. Biophys. Res. Commun.*, 1 (1959) 323.
8 M. I. H. Aleem and A. Nason, *Proc. Natl. Acad. Sci. (U.S.)*, 46 (1960) 763.
9 L. Kiesow, *Z. Naturforsch.*, 17b (1962) 455.
10 L. Kiesow, *Biochem. Z.*, 338 (1963) 400.
11 M. I. H. Aleem, H. Lees and D. J. D. Nicholas, *Nature*, 200 (1963) 759.
12 L. Packer and W. Vischniac, *Biochim. Biophys. Acta*, 17 (1955) 153.
13 L. Kiesow, *Proc. Natl. Acad. Sci. (U.S.)*, 52 (1964) 980.
14 M. I. H. Aleem, *Biochim. Biophys. Acta*, 113 (1966) 216.
15 M. I. H. Aleem, *J. Bacteriol.*, 91 (1966) 729.
16 P. A. Straat and A. Nason, *J. Biol. Chem.*, 240 (1965) 412.
17 K. G. Vogler and W. W. Umbreit, *J. Gen. Physiol.*, 26 (1942) 157.
18 H. Bömeke, *Arch. Mikrobiol.*, 10 (1939) 385.
19 W. D. Butt and H. Lees, *Nature*, 182 (1958) 732.
20 L. Kiesow, unpubl. results, mentioned in ref. 1.

Appendix

On "Active Hexosephosphates" (see Chapter 21)

In a paper by Meyerhof and Kiessling[1] published in 1936 (submitted in October 1935) we find a statement concerning both active hexosephosphates and their stabilization products, and the role of ATP as energy transmitter:

"Es könnte nämlich die Spaltungsenergie des Adenylpyrophosphats ähnlich wie bei der Synthese der Kreatinphosphorsäure auch in unserem Fall bei der Umesterung nicht verloren gehen, sondern zur Hauptsache auf das synthetisierte Produkt übertragen werden. Dann würde zunächst ein Zuckerester von einem grösseren Energieinhalt entstehen als die stabilen Hexosephosphorsäuren, der einen radikalartigen Character haben dürfte. Dies würde das entscheidende Moment für die Reaktionsfähigkeit des neu phosphorylierten Produkts sein. Wenn allerdings während der Lebensdauer des Radikals keine solche Weiterreaktion, die in unserem Falle eine durch Brenztraubensäure beziehungsweise Acetaldehyde induzierte Oxydation wäre, erfolgen kann, so ginge das Primärprodukt in die relativ stabilen Hexosephosphorsäuren über (wieweit dabei die Veresterung des anorganischen Phosphats mit der Umesterungsreaktion gekoppelt ist, ist noch vorzeitig, zu diskutieren)".

The point is that the energy of ATP splitting is preserved in the synthesized product by the formation of a radical of higher energy content than the stable hexosephosphates. If during the lifetime of the radical no reaction takes place, the primary activated product is converted into the stabilized product. By that time the phosphate transfer by ATP had been established in a number of cases, for instance by Lohmann[2] between ATP and creatine, by Parnas, Ostern and Mann[3] between phosphoenolpyruvate and ADP; by von Euler and Adler[4] between ATP and glucose and by Dische[5] between ATP and hexosemonophosphate. It is therefore astonishing to note the persistence of the concept of "active hexosephosphate". The reason lies in the fact that Meyerhof did not satisfy himself about the existence of strict stoichiometry of the reactions just mentioned, and particularly in the case of the transfer of phosphate from ATP to glycogen as the phosphate acceptor.

The concept of "active hexosephosphates" and their stabilization products was abandoned when Cori and Cori[6], in 1936, described the enzymatic steps of the formation of G-6-P from G-1-P derived from

References p. 440

glycogen plus inorganic phosphate, as the formation of G-6-P from glucose and ATP was recognized, not as "stabilization product" of an "active" intermediate, but as an intermediate in the pathway and as the starting of the phosphate cycle in muscle.

REFERENCES

1 O. Meyerhof and W. Kiessling, *Biochem. Z.*, 283 (1935) 83.
2 K. Lohmann, *Biochem. Z.*, 271 (1934) 264.
3 J. K. Parnas, P. Ostern and T. Mann, *Biochem. Z.*, 272 (1934) 64; 275 (1934) 74.
4 H. von Euler and E. I. Adler, *Z. physiol. Chem.*, 235 (1935) 122.
5 Z. Dische, *Biochem. Z.*, 280 (1935) 248.
6 C. F. Cori and G. T. Cori, *Proc. Soc. Exptl. Biol. Med.*, 34 (1936) 702.

Name Index

Subject Index

COMPREHENSIVE BIOCHEMISTRY

Section I — Physico-Chemical and Organic Aspects of Biochemistry
Volume 1. Atomic and molecular structure
Volume 2. Organic and physical chemistry
Volume 3. Methods for the study of molecules
Volume 4. Separation methods

Section II — Chemistry of Biological Compounds
Volume 5. Carbohydrates
Volume 6. Lipids — Amino acids and related compounds
Volume 7. Proteins (Part 1)
Volume 8. Proteins (Part 2) and nucleic acids
Volume 9. Pyrrole pigments, isoprenoid compounds, phenolic plant constituents
Volume 10. Sterols, bile acids and steroids
Volume 11. Water-soluble vitamins, hormones, antibiotics

Section III — Biochemical Reaction Mechanisms
Volume 12. Enzymes — general considerations
Volume 13. (third revised edition). Enzyme nomenclature (1972)
Volume 14. Biological oxidations
Volume 15. Group-transfer reactions
Volume 16. Hydrolytic reactions; cobamide and biotin coenzymes

Section IV — Metabolism
Volume 17. Carbohydrate metabolism
Volume 18. Lipid metabolism
Volume 19. Metabolism of amino acids, proteins, purines, and pyrimidines
Volume 20. Metabolism of cyclic compounds
Volume 21. Metabolism of vitamins and trace elements

Section V — Chemical Biology

Volume 22. Bioenergetics
Volume 23. Cytochemistry
Volume 24. Biological information transfer
Volume 25. Regulatory functions
Volume 26. Part A. Extracellular and supporting structures
Volume 26. Part B. Extracellular and supporting structures (continued)
Volume 26. Part C. Extracellular and supporting structures (continued)
Volume 27. Photobiology, ionizing radiations
Volume 28. Morphogenesis, differentiation and development
Volume 29. Part A. Comparative biochemistry, molecular evolution
Volume 29. Part B. Comparative biochemistry, molecular evolution (continued)

Section VI — A History of Biochemistry

Volume 30. *Part I.* Proto-biochemistry
Part II. From proto-biochemistry to biochemistry
Volume 31. *Part III.* History of the identification of the sources of free energy in organisms
Volume 32. *Part IV.* The unravelling of biosynthetic pathways
Volume 33. *Part V.* History of molecular interpretations of physiological and biological concepts, and of the origins of the conception of life as the expression of a molecular order